Biofouling: Detection, Prevention and Control

Biofouling: Detection, Prevention and Control

Tommy Brooks

Callisto Reference
www.callistoreference.com

Callisto Reference,
118-35 Queens Blvd., Suite 400,
Forest Hills, NY 11375, USA

Visit us on the World Wide Web at:
www.callistoreference.com

ISBN: 978-1-64116-791-8 (Hardback)

Cataloging-in-Publication Data

Biofouling : detection, prevention and control / Tommy Brooks.
p. cm.
Includes bibliographical references and index.
ISBN 978-1-64116-791-8
1. Fouling. 2. Fouling organisms. 3. Fouling--Measurement.
4. Fouling organisms--Control. I. Brooks, Tommy.

TD427.F68 B56 2023
628.96--dc23

Table of Contents

	Preface	VII
Chapter 1	An Overview of Biofouling	1
	Problem Analysis	19
Chapter 2	Reverse Osmosis and Nanofiltration: Biofouling and Membrane Performance	20
	Development of a Methodology	43
Chapter 3	Development and Testing of a Membrane Fouling Simulator	44
Chapter 4	Early Biofouling Detection and Pressure Drop Measurements	69
Chapter 5	Biofouling Studies and the Application of MRI	88
Chapter 6	Developing 3D Computational Model	104
	Fundamental Studies	140
Chapter 7	Biofouling, Pressure Drop and the Impact of flux	141
Chapter 8	Biomass and Pressure Drop Development: Role of Feed Spacer	159
Chapter 9	Evaluation of Experimental Data through the 3D Numerical Model	181

	Insights on Control Studies	208
Chapter 10	Feed Water and Linear Flow Velocity: Significance of Substrate Load and Substrate Concentration	209
Chapter 11	Biofilm Accumulation and Morphology: Role of Flow Regime	231
Chapter 12	Reduction of Membrane Biofouling and Phosphate Limitation	252
	Future Perspectives	273
Chapter 13	Biofouling Control: Potential Complimentary Perspectives	274
	Permissions	
	Index	

Preface

Biofouling refers to the buildup of plants, microorganisms, small animals or algae on surfaces like the hulls of submarines and ships, and parts of grates, water inlets and pipework. The degradation caused by biofouling compromises an item's ability to perform its primary function. A variety of ultraviolet equipment can be used that can detect and prevent accumulation of biofouling. The different strategies and methods for preventing and controlling biofouling include antifouling, coatings, ultrasonic antifouling and energy methods. Anti-fouling refers to the capability of especially designed materials, like biocides and non-toxic paints to prevent or eliminate biofouling. Biodispersants can be used in industrial processes to control the buildup of biofouling. Nontoxic strategies which prevent biofouling include usage of zwitterions and nano-engineered superhydrophobic surfaces for creating an ultra-low fouling surface. This book is compiled in such a manner, that it will provide in-depth knowledge about the detection, prevention and control of biofouling. It will serve as a valuable source of reference for those interested in this area of study.

This book is a result of research of several months to collate the most relevant data in the field.

When I was approached with the idea of this book and the proposal to edit it, I was overwhelmed. It gave me an opportunity to reach out to all those who share a common interest with me in this field. I had 3 main parameters for editing this text:

1. Accuracy - The data and information provided in this book should be up-to-date and valuable to the readers.

2. Structure - The data must be presented in a structured format for easy understanding and better grasping of the readers

3. Universal Approach - This book not only targets students but also experts and innovators in the field, thus my aim was to present topics which are of use to all

Thus, it took me a couple of months to finish the editing of this book.

I would like to make a special mention of my publisher who considered me worthy of this opportunity and also supported me throughout the editing process. I would also like to thank the editing team at the back-end who extended their help whenever required.

Tommy Brooks

1

An Overview of Biofouling

INCREASING DEMAND FOR CLEAN FRESHWATER

One of the most pervasive problems afflicting people throughout the world is inadequate access to clean freshwater and sanitation (Montgomery and Elimelech, 2007 and Shannon *et al.*, 2008). More than 1.2 billion people lack reliable access to safe drinking water, 2.6 billion have little or no sanitation, while millions of people die annually from diseases transmitted through unsafe water. Waterborne pathogens have a devastating effect on public health especially in the developing countries. Problems are expected to grow worse in the coming decades. In 2025, the number of people affected by severe water shortages is expected to increase fourfold to sixfold (Cosgrove and Rijsberman, 2000; Miller, 2003 and United Nations Environmental Programme, 2008; Figure 1.1). Among the countries likely to run short of water in the next 25 years are Ethiopia, India, Kenya, Nigeria and Peru. Parts of other large countries like China already face chronic water problems (Hinrichsen *et al.*, 1998 and Tibbetts, 2000). Bahrain, Kuwait, Saudi Arabia and the United Arab Emirates have resorted to the desalination of seawater from the Gulf. Bahrain has virtually no freshwater (Riviere, 1989), while three-quarters of Saudi Arabia's freshwater comes from fossil groundwater, which is reportedly being depleted at an average rate of $5.2\,km^3$ per year (Postel, 1997). Also industrialized nations as Australia and Spain have observed severe water shortages recently (Martín-Rosales *et al.*, 2007 and Ummenhofer *et al.*, 2009). Besides a shortage of fresh water there is also an increasing need for extensive treatment of the existing fresh water resources. In both developing and industrialized nations, a growing number of (micro)pollutants are discharged into water supplies: heavy metals, agricultural chemicals like pesticides, pharmaceutical derivatives, disinfection by-products, endocrine disruptors, and so on. Two key problems are that the amount of suspected harmful contaminants is growing rapidly, and that many of these compounds are toxic in trace quantities (Shannon *et al.*, 2008 and World Health Organization). Regulations on drinking water quality become stricter. The United Nations Millennium Development Declaration (2000) called for the world to halve the proportion of people without access to safe drinking water as well as the proportion of people who do not have access to basic sanitation by 2015. It called upon the international community to develop integrated water resources management and water efficiency plans by 2005; and to support countries in their efforts to monitor and assess the quantity and quality of water resources.

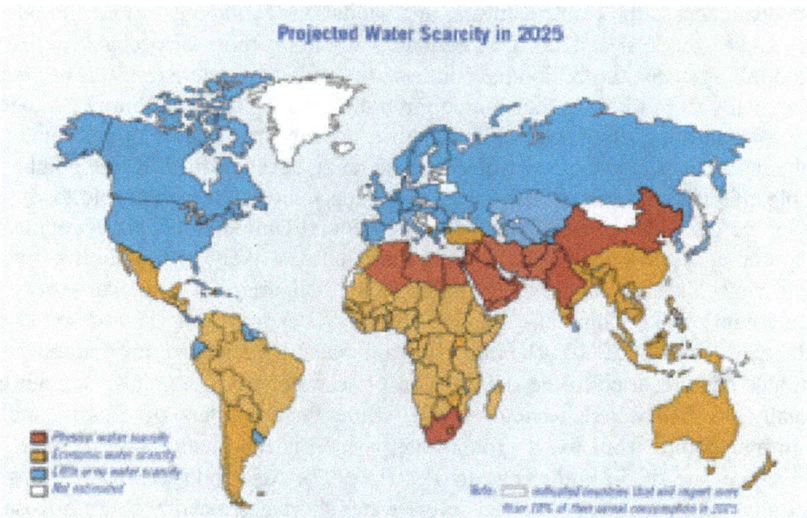

Figure 1.1 Projected global water scarcity in 2025 (International Water Management Institute)

Freshwater is not evenly distributed over the world. The availability varies with geographical region and time (e.g. precipitation over the seasons). Only a small part of the freshwater is a naturally renewable source of freshwater (Miller, 2003). About 97% of the global water resource is salt water (Figure 1.2). Desalination technology can help to quench the world's thirst (Mallevialle *et al.*, 1996 and Stikker, 2002). The global desalination capacity is increasing rapidly in time (Wangnick, 2005; Figure 1.3). The oldest desalination methods are based on evaporating water and collecting the condensate. The newest commercial technology for desalination is based on membrane treatment (Frenkel, 2004).

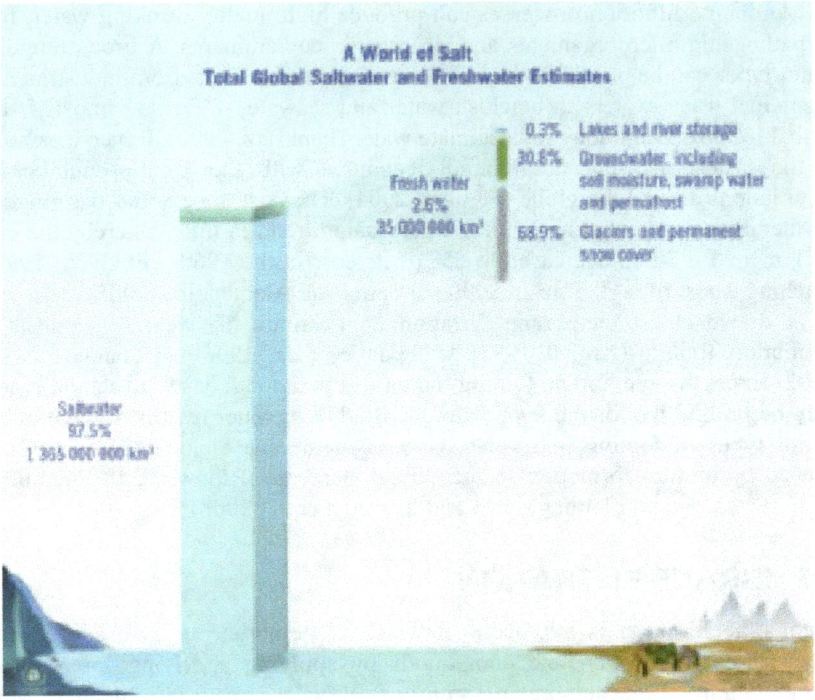

Figure 1.2 Total global saltwater and freshwater estimates (UNESCO, 1999)

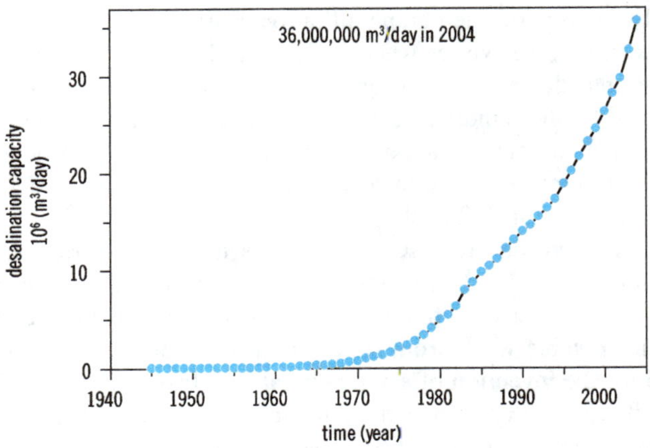

Figure 1.3 Global increase of cumulative installed desalination capacity, 1945–2004 (Wangnick, 2005)

Membrane filtration processes can produce high quality drinking water, free of pathogenic microorganisms and (in)organic contaminants. A broad range of water types can be purified with membrane based treatment: from industrial and municipal waste waters to brackish water and seawater. Reverse osmosis is the global leading technology to desalinate water (Pankratz, 2000). Reverse osmosis is the fastest growing desalination technique with the greatest number of installations around the globe (Frenkel, 2004). The cost of seawater desalination by membrane treatment has shown a decreasing trend in time, whereby the cost of 1 m^3 in 1997 had decreased to 5% of its cost in the 1960s (ESCWA, 1997), reaching a cost of ~€0.5/m^3 in 2000 (El-Fadel and Alameddine, 2005).

A drawback in membrane filtration applications like reverse osmosis is membrane fouling (Amjad, 1993; Mallevialle *et al.*, 1996 and Shannon *et al.*, 2008) Excessive membrane fouling increases operational costs substantially and may be prohibitive for the application of RO/NF in water treatment. One of the major types of fouling in reverse osmosis membrane elements is biofouling, caused by biofilm formation in membrane elements (Ridgway and Flemming, 1996; Patching and Fleming, 2003 and Shannon *et al.*, 2008).

MEMBRANE FILTRATION

Membrane filtration is a process in which a membrane is used as selective physical barrier to separate compounds by applying a driving force across the membrane. In a membrane system a feed water stream is separated in two streams, the product or permeate, containing solutes that passed the membrane and the concentrate containing solutes and particles rejected by the membrane (Figure 1.4A, Amjad, 1993 and Mallevialle *et al.*, 1996).

The early history of membrane filtration started over 250 year ago with the French cleric Abbé Nollet when he observed water transport across a pig bladder covering the mouth of a jar containing wine (Nollet, 1748, 1779 and Lonsdale, 1982). One hundred years later in 1867, Moritz Traube prepared the first artificially membrane (Traube, 1867). In 1950, Hassler introduced the first concept of membrane desalination describing 'salt repelling osmotic membranes' and 'permselective films' (Hassler, 1950 and Glater, 1998). In the late 1950s the basis for modern-day reverse osmosis was laid by research with cellulose acetate membranes by Reid and Breton (1959) and Loeb and Sourirajan (1960, 1963). Reid and Breton were the first to demonstrate that cellulose acetate films could produce potable water from saline solutions. Loeb and Sourirajan (1963) are credited for the invention of asymmetrical cellulose acetate membranes with improved salt rejection and water flux, making membrane desalination practical. The first spiral-wound element was developed by General Atomics in 1963.

The oldest patents for reverse osmosis are dated 1964 (Hassler, 1964 and Loeb and Sourirajan, 1964) and 1965 (University of California, 1965). In the 1970s thin film composite membranes were introduced and in time improvements were made to improve water flux and rejection properties and reduce the feed pressure. The history of membrane science is described in several reviews (Lonsdale, 1982; Glater, 1987; Brandt *et al.*, 1993; Böddeker, 1995 and Baker, 2004).

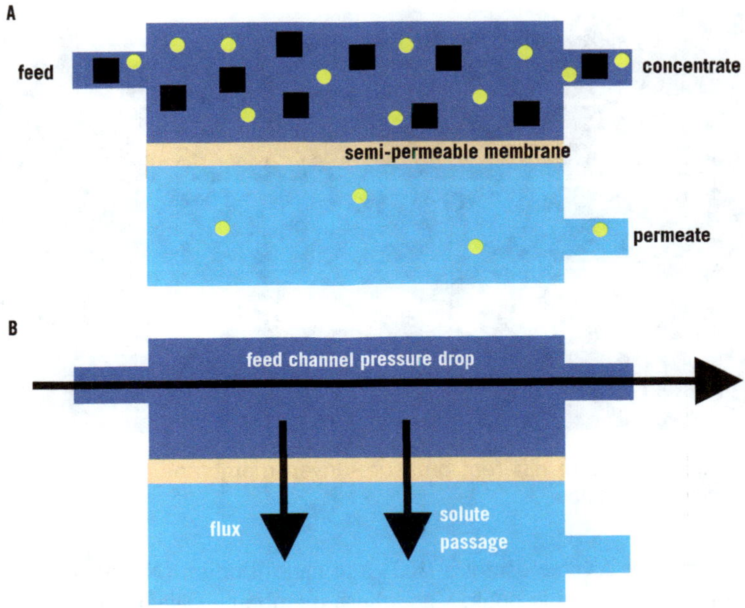

Figure 1.4 Scheme of pressure driven membrane unit (**A**) and membrane performance indicators (**B**): feed channel pressure drop, normalized flux and solute passage

Membrane operations can be classified considering the parameters driving force, separation mechanism and rejection properties. In case of pressure driven membrane processes, the driving force is a pressure difference across the membrane. Four pressure driven membrane filtration processes can be discriminated based on differences in feed pressures and membrane rejection capacities: microfiltration, ultrafiltration, nanofiltration and reverse osmosis, ranked by increasing pressure (Figure 1.5). A classification generally made is low pressure membranes for microfiltration and ultrafiltration and high pressure membranes for nanofiltration and reverse osmosis. Microfiltration screens particles from 0.1 to 0.5 microns and ultrafiltration screens particles from 0.005 to 0.05 microns. Nanofiltration is applied for removal of divalent ions e.g. sulfate and hardness, natural colour (humic acids) and partial removal of monovalent ions e.g. sodium and chloride. Reverse osmosis membranes are able to remove mono- and divalent ions for more than 99%. The pores in NF and RO membranes are smaller than 1 nm.

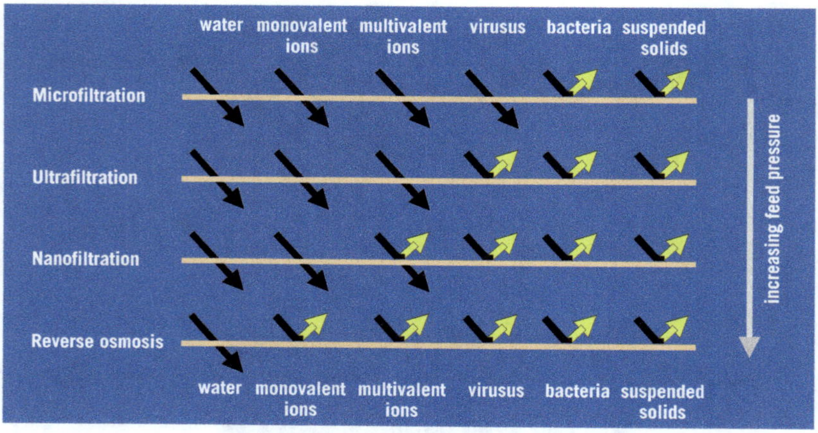

Figure 1.5 Scheme of different pressure-driven membrane filtration processes and rejection capacities

High pressure reverse osmosis (RO) and nanofiltration (NF) membranes are the research focus of this book, since these membrane systems are suitable for rejection of salt (desalination), pathogens and (in)organic micropollutants and numerous reports indicate that biofouling is one of the most serious problems in these membrane systems.

RO and NF are pressure driven membrane separation processes in which a dense membrane allows diffusion of the solvent and solutes. Diffusion of solutes like salts, (low) molecular weight compounds and particles is low compared to water, resulting in a rejection for those substances. RO is the membrane process used for desalination of brackish and seawater. The concept of the process can be described by a system of communicating vessels where a membrane separates high and low salt solutes (Figure 1.6). Water from the low saline solute diffuses through the semi-permeable membrane to the more concentrated saline solute. This diffusion of water through a semi-permeable membrane is called osmosis. The volume of the high saline solute increases and the saline concentration is reduced, while the volume of the low saline solute decreases and the saline concentration increases on the other side of the membrane, until 'osmotic equilibrium' is reached. The difference in height between the concentrated and diluted salt concentration reflects the osmotic pressure difference between both solutions. When pressure is applied to the saline solution, larger than the osmotic pressure, the water flow is reversed and water flows from the concentrated saline solute through the membrane to the diluted solution while dissolved salts and impurities are withheld by the membrane. This process is called reverse osmosis (Table 1.1). NF membranes are more open compared to reverse osmosis membranes, resulting in a poor rejection of monovalent ions and much lower operating pressures. NF and RO membranes can be used for removal of bacteria, viruses, pesticides and multivalent ions such as calcium and magnesium (softening) and organics control (colour), but allow water to pass through the membrane. Membrane performance indicators are the feed channel pressure drop, normalized flux also named permeability, and membrane solute passage (rejection properties, see Figure 1.4B, Amjad, 1993 and Mallevialle *et al.*, 1996).

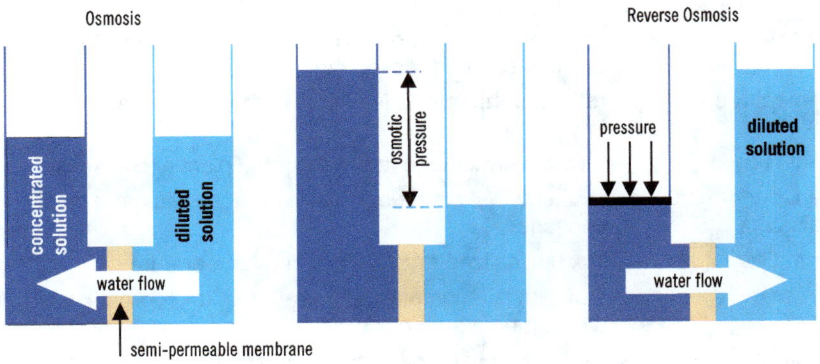

Figure 1.6 Principle of osmosis and reverse osmosis

Two membrane materials make up the bulk of commercial RO membranes, cellulose acetate and aromatic polyamide. Cellulose acetate membranes were broadly used in the past. Cellulose acetate membranes were the first commercialized RO membranes developed in the late 1960s. Some of the reasons that cellulose acetate membranes lost favour to the new polyamide membranes were the lower salt rejection capacities, and the lower thermal, pH, and chemical stability (cellulose acetate tend to hydrolyse in time), susceptibility to biological attack, and higher net drive pressure requirements due to the lower membrane permeability (Mallevialle *et al.*, 1996). The RO and NF membrane presently of choice worldwide is the polyamide thin film composite membrane (TFC). TFC membranes, composed of a strong asymmetric support membrane and a thin dense polyamide toplayer, have a higher permeability and can be used at higher temperatures at a broader pH range. Cellulose acetate membranes are chlorine resistant while TFC membranes have low resistance to chlorine.

Table 1.1 List of definitions for RO membrane filtration

Definitions	Explanation
osmosis	diffusion of water through a semi-permeable membrane into the more concentrated solution
reverse osmosis	solvent flow from an area of high solute concentration, through a membrane, to an area of low solute concentration. Dissolved salts and impurities are withheld by the membrane.
feed	solute fed into membrane installation
permeate/product	solute passing the membrane
concentrate/brine	concentrated solution flowing out of installation. The solution has not passed the membrane.
element recovery	ratio of permeate to feed flow of the element expressed in %
plant recovery	ratio of permeate to feed flow of the plant expressed in %
plant performance indicators	feed channel pressure drop, normalized flux and solute passage
normalized flux	flux normalized for pressure and temperature
biofilm	a biofilm is a complex aggregation of microorganisms growing on a surface
biofouling	biofilm formation causing 'unacceptable' operational problems
concentration polarization	increase of salt concentration near the membrane surface due to permeate flow through the membrane. This phenomenon impacts process performance by increasing the osmotic pressure at the membrane's surface, reducing flux, increasing salt passage and increasing the probability of scale development.

Membrane element

A membrane element is the operational unit containing membranes. Several element configurations have been developed: plate and frame, hollow fibre and spiral wound. Nowadays, the most widely used reverse osmosis and nanofiltration elements in practice have a spiral wound configuration. Spiral wound membrane elements have a surface to volume ratio of 300–1,000 m^2/m^3 enabling small footprint systems and relative low prices per m^2 membrane area (Amjad, 1993 and Mallevialle *et al.*, 1996). Spiral wound membrane elements are produced from membrane sheets which are wound along a central perforated permeate collection tube (Figure 1.7). Two flat-sheet membranes are glued together on the inside of three of its edges, making an envelope. The remaining open edge is connected to the central collection tube. In the envelope, the membranes are separated by a porous mesh named product spacer, facilitating the transport of product water to the central product collection tube. A membrane element contains a number of these envelopes, which are separated from each other on the feed side of the membrane envelopes by a feed spacer. The feed spacer separates the membranes and generates turbulence and mixing, improving mass transport near the membrane surface. Figure 1.8 shows a feed spacer and a membrane element feed side with narrow flow channels containing feed spacers. The wound membranes and spacers with an end cap at each end of the element are cast in a glass fibre casing.

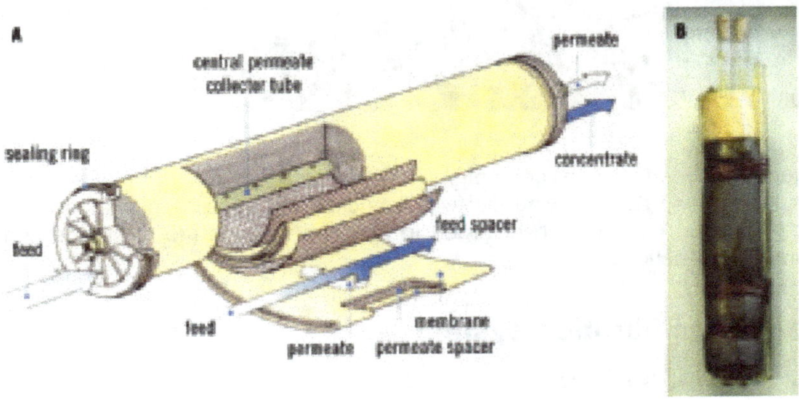

Figure 1.7 Spiral wound membrane element configuration: current (**A**) and one of the 'first' elements (**B**) dated ~1963

The outer dimensions of spiral wound membrane elements are standardized. Nowadays most common is the 8 inch outer diameter element, but, 4 inch outer diameter and more recently 16 inch outer diameter membrane elements are also used in practice. In a spiral wound membrane element, the feed water flows through the feed spacer channels in the membrane element to the concentrate side. The product passes the membrane, flows through the product spacer channel to the perforated product collection tube. The amount of water passing the membrane in an element varies between approximately 7 and 15% of the feed flow, indicating that most water flows along the membrane feed side without passing the membrane.

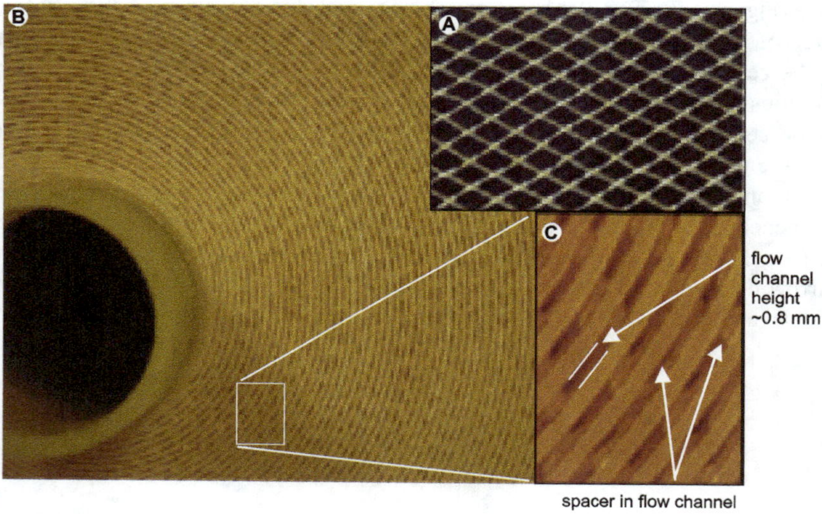

Figure 1.8 Feed spacer geometry (**A**) and spiral wound membrane element without end cap (**B**), showing the feed flow channels containing the feed spacer (**C**). A part of (B) is shown enlarged in (C). The tube on the left is the central permeate collection tube

Membrane filtration system

A membrane system basically consists of a high pressure pump and a large number of pressure vessels, containing up to 8 membrane elements (Figure 1.9). Permeate production by the membrane elements placed in series results in a decline of the feed water flow velocity along the membranes over the pressure vessel. A tapered configuration of pressure vessels is applied to maintain proper flow velocities along the membrane, preventing fouling and minimizing concentration polarization (Mallevialle *et al.*, 1996). The concentrate flow of first stage pressure vessels is fed into a second stage consisting of a lower number of pressure vessels, resulting in proper flow velocities over the second stage. A full-scale seawater RO installation with a permeate production capacity of 54,000 m^3/day is shown in Figure 1.10.

Nowadays, also larger installations are in operation and being build/designed. An example is the RO installation in Ashkelon (Israel) with a permeate production capacity of 330,000 m^3/day achieved by the plant 40,000 membrane elements. The Ashkelon plant produces around 13% of the country's domestic consumer demand.

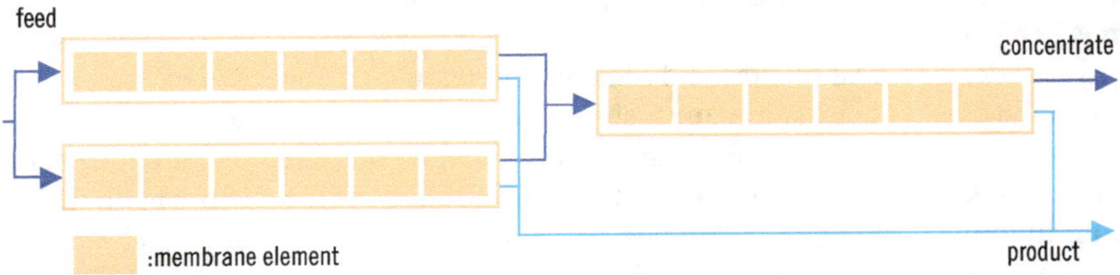

Figure 1.9 Scheme of membrane filtration installation consisting of three pressure vessels, each containing 6 spiral wound membrane elements. The installation has a tapered configuration, the concentrate of the two first stages are fed to one second stage, resulting in a desired velocity profile

Figure 1.10 Seawater reverse osmosis membrane filtration installation in Larnaka, Cyprus. The permeate production capacity is 54,000 m^3/day. The total amount of eight-inch diameter membrane elements is 5,800. The plant size can be estimated from the person shown on the left of the figure

Membrane fouling

A drawback in membrane filtration applications is membrane fouling, resulting in an increase of feed channel pressure drop and/or decline of flux and/or increase of salt passage. The consequences of fouling can be: (i) increase of required feed pressure and consequently higher energy consumption, (ii) frequent chemical cleaning of the membranes, (iii) shortening lifetime of the membranes (Figure 1.11). Membrane autopsy, a destructive membrane element study, is commonly used to study and diagnose membrane fouling (Figure 1.12), which involves among others the analysis of accumulated material. The major fouling mechanisms of NF and RO membrane elements are scaling (inorganic deposits), particulate (colloidal matter) and organic fouling and biofouling. Different types of fouling may occur simultaneously and can influence each other (Flemming, 1993). Scaling by inorganic compounds is usually controlled by an anti-scalant and/or acid. Particulate fouling can be controlled by pretreatment, such as ultrafiltration. Therefore, all types of fouling except biofouling and organic fouling – related types of fouling – are controllable.

Figure 1.11 Spiral wound reverse osmosis elements and cartridge filters used for polishing RO feed water

Numerous authors have described biofouling problems in membrane installations (Kissinger, 1970; McDonough and Hargrove, 1972; Winters and Isquith, 1979; Paul, 1991, 1996; Flemming, 1993; Tasaka *et al.*, 1994; Ridgway and Flemming, 1996; Baker and Dudley, 1998; Huisman and Feng Kong, 2004; Schneider *et al.*, 2005 and Karime *et al.*, 2008). From 70 surveyed U.S. reverse osmosis membrane installations, 58 reported fouling problems, with biofouling as the predominant operational problem (Paul, 1991). Gamal Khedr reported (2000, 2002) that in the Middle East, about 70% of the seawater RO membrane installations suffered from biofouling problems, indicating that biofouling – excessive growth of biomass, i.e. biofilms – is a major type of fouling in spiral wound RO membrane systems.

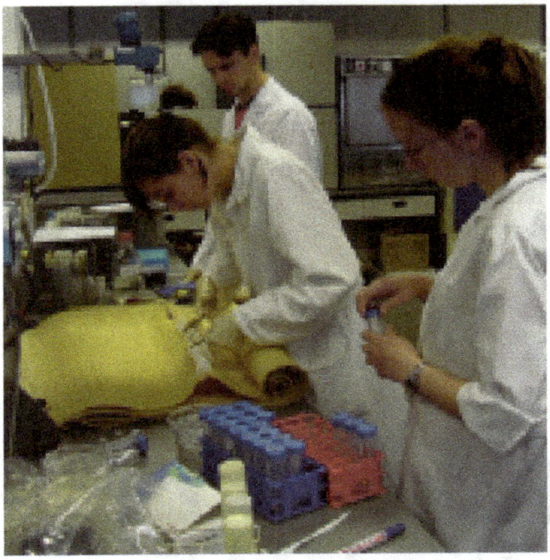

Figure 1.12 Autopsy of spiral wound membrane element for fouling diagnosis

BIOFILMS AND BIOFOULING

Biofilm formation is the accumulation of microorganisms, including extracellular compounds, on a surface due to either deposition or growth or both (Hamilton, 1985; Costerton *et al.*, 1987 and Characklis and Marshall, 1990). Biofouling is the extent of biofilm formation causing unacceptable (operational) problems (Characklis and Marshall, 1990). In this context 'unacceptable' means that operational guidelines are exceeded for e.g. pressure drop increase, flux reduction, salt passage increase. Initially, biomass accumulation on surfaces was called microbial slime and films, bacterial adhesion, attached growth, microfouling and (micro)biological fouling (ZoBell and Anderson, 1936; Lloyd, 1937; Heukelekian and Heller, 1940; Zobell, 1943; Characklis, 1973a,b; DiSalvo and Cobet, 1974; Marszalek *et al.*, 1979 and Winters and Isquith, 1979).

Formation of a biofilm usually involves three subsequent phases: (i) adhesion and attachment of microorganisms to a surface, (ii) followed by growth and (iii) a stationary phase. Especially in the stationary phase in laboratory biofilm systems, biomass detachment is observed by erosion and sloughing. The biofilm is held together by excreted organic polymer matrix of microbial origin called extracellular polymeric substances (EPS, Geesey, 1982; Allison and Sutherland, 1984 and Wingender *et al.*, 1999). Biofilms can contain many different types of microorganisms, e.g. bacteria, protozoa, fungi and algae. Bacteria living in a biofilm can have significantly different properties from free-floating bacteria of the same species. Biofilms are ubiquitous in nature. Biofilms can be found on rocks and pebbles at the bottom of most streams or rivers and on the teeth of most animals as dental plaque. Biofilms grow in hot acidic pools and on glaciers in Antarctica.

Benificial biofilms serve in the water and waste water industry, in bioremediation applications and industrial biotechnology (Bryers, 2000). For example, many sewage treatment plants include a treatment stage in which waste water passes over biofilms grown on filters, which extract and convert organic compounds. In such biofilms, removal occurs of organic matter, suspended solids, pathogens and other microorganisms. Slow sand filters rely on biofilm development in the same way to filter surface water from lake, spring or river sources for drinking purposes.

Biofouling is the undesirable accumulation of microbial biofilm on a surface that significantly degrades equipment performance and/or the useful equipment lifetime (Characklis and Marshall, 1999). Detrimental effects of biofilms can be (i) corrosion of pipelines (Geesey *et al.*, 1994), energy losses by (ii) increased heat transfer resistance (e.g. process heat exchangers) and (iii) increased fluid frictional resistance like in pipelines, on ship hulls, in porous media such as water wells and filters, and in membrane systems (Characklis and Marshall, 1990).

Note that the definition of biofilms is in general term the accumulation of bacteria on a surface. Biofouling is defined as a biofilm leading to problems. Biofouling has therefore always an application context.

Biofouling in membrane systems

Microorganisms are present on all surfaces in contact with water and their presence is not indicative for biofouling (Figure 1.13). Many studies showed that microorganisms are commonly observed on water-exposed surfaces, even in pure water systems (Mittelman, 1991). The presence of microorganisms was already observed on membranes after a short contact time, in the order of minutes (Ridgway and Flemming, 1996 and Schaule, 1992). Ridgway was the first to use the term 'biofouling' in relation to 'membrane' in publications (Scopus database: July 2009) and is one of the pioneers in studying RO membrane biofouling (1983, 1984, 1985). Before 1983, limited information was found because other terminology was used (Winters and Isquith, 1979) or data was reported as conference proceeding (Argo and Ridgway, 1982) or report (Ridgway *et al.*, 1981). The 27 years since the first peer reviewed paper on membrane biofouling show a steadily increasing annual number of papers (Figure 1.14), with 491 papers appearing in 2009.

Controlling biofouling may be achieved by chemical dosage to the feed water. Currently applied thin film composite NF and RO membranes are sensitive for free chlorine. Free chlorine damages the membrane structure causing decrease of membrane rejection. A limited number of plants apply monochloramines in controlling biofouling successfully. A much better membrane resistance to monochloramines compared to chlorine has been reported (DOW, 2009a): 300,000 ppm-hours for chloramine and up to ~1,000 ppm-hours for free chlorine. Since monochloramine is formed by adding ammonia to chlorine, it is possible that free chlorine will be present (for e.g. the ammonia dosing is not correct or fails). Moreover, iron and manganese catalyzes membrane oxidation by monochloramines (Gabelich *et al.*, 2005 and Da Silva *et al.*, 2006). Another reason to avoid chloramination is the formation of N-nitrosodimethylamine (NDMA), a probable human carcinogen. The RO rejection capacities for NMDA are 10 to 50%. Recently, an alternative for chloramines, 2,2-Dibromo-3-Nitrilopropionamide (DBNPA. DOW, 2005, 2009b) is applied successfully in a limited number of plants.

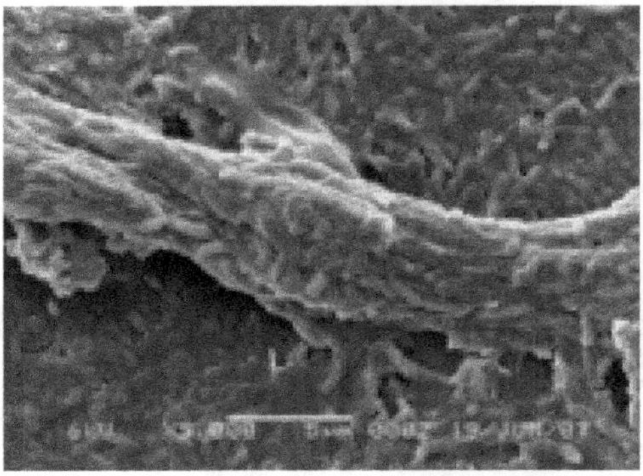

Figure 1.13 Scanning electron microscope image of biofilm on the membrane of a membrane element taken from a full-scale installation

Biofilm accumulation is affecting negatively the performance of NF and RO installations and in several plants, operators struggle with this problem since simple and effective solutions are lacking. Reported studies commonly involved trial and error approaches in (full-scale) installations, of which the obtained data did not allow always good conclusions. Current tools and measurements were not sufficient (Flemming, 1998, 2003 and Greenlee *et al.*, 2009). Most laboratory systems used where not representative in geometry or fluid flow conditions compared to spiral wound membrane systems. This situation in 2006 was the starting point of studies described in this report. One of the objectives was to develop a monitor to simulate biofouling in spiral wound RO elements.

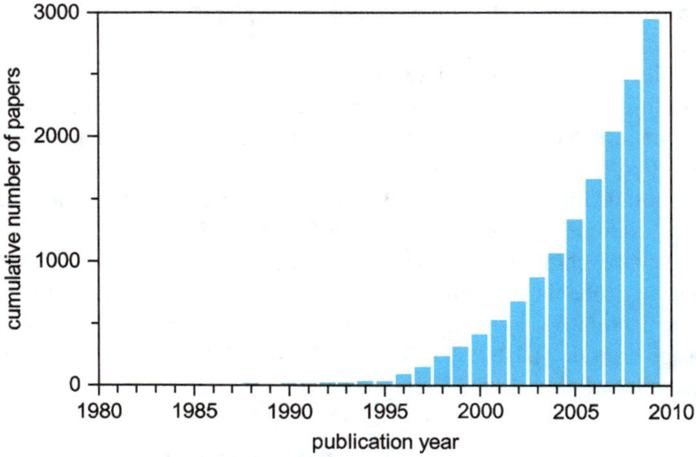

Figure 1.14 Cumulative number of peer reviewed publications on membrane biofouling in the period 1982–2009. The total number of papers is 2941 until 2010 (Scopus database: March 2010)

SCOPE AND OUTLINE

The main goal of this study was to (i) relate biomass accumulation in spiral wound NF and RO membrane elements with hydrodynamics and membrane performance and to (ii) determine key parameters that influence biofouling, aiming at excluding chemical use. The book structure is shown in Figure 1.15.

Quantitative relationships between membrane performance and biomass accumulation were lacking at the time of initiating the studies described in this book. A better understanding of biofouling and membrane performance in practice was needed. Studies at 15 full-scale and pilot plant NF and RO membrane installations were performed to diagnose biofouling. To quantify biofouling, several biomass parameters in membrane elements were investigated during membrane autopsies and compared with the pressure drop increase in membrane installations with different types of feed water. Several biological parameters of the feed water were included in the study to evaluate whether water quality parameters are an appropriate alternative for the laborious destructive membrane studies to quantify biofouling (**Chapter 2**).

A need for a monitor and sensitive pressure drop measurements to study and monitor biofouling became apparent from the fact that data from practice where to scattered and didn't allow a good analysis of the problem. Most laboratory systems used where not representative in geometry or fluid flow conditions compared to spiral wound membrane systems. A small membrane fouling simulator was developed and tested on suitability to study and monitor membrane biofouling. A comparison study of the monitor and spiral wound membrane elements in test rigs and a full-scale installation was performed to determine whether the monitor showed the same development of biofouling (**Chapter 3**). An accurate and sensitive differential pressure drop transmitter was introduced and tested. Optimization of pressure drop measurements for early biofouling detection was part of the study (**Chapter 4**).

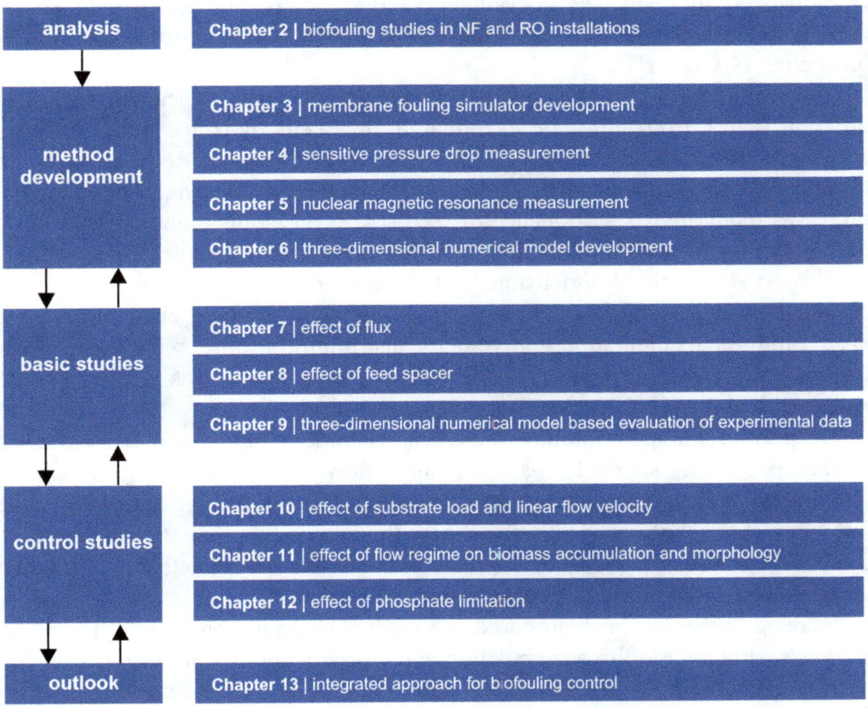

Figure 1.15 Structure of book

For in-situ assessment of biomass localization and hydrodynamics in membrane systems, a Magnetic Resonance Imaging (MRI) method and a flow cell for MRI studies have been developed. To evaluate the applicability of MRI for biofouling studies, the evolution of biofilm development and velocity distribution have been studied in spiral wound membrane elements and a representative plastic small monitor (**Chapter 5**).

To provide mechanistic insight in membrane biofouling and membrane performance a three-dimensional computational model of biofouling and fluid dynamics was developed (**Chapter 6**). The developed methods described in Chapters 3 to 5 were used in the biofouling studies while a close link with practice was maintained by including comparison studies with spiral wound membrane elements.

The relation between biofouling and membrane flux – water volume flowing through the membrane per unit area and time – was studied in spiral wound NF and RO membranes with extensive pretreatment and biofouling monitors. One of the study goals was to evaluate whether the critical flux concept stating that 'below a critical flux no fouling occurs' is a suitable approach to control biofouling (**Chapter 7**).

The effect of biomass accumulation on membrane performance was studied at different scales, from full-scale to miniature flow cells by conventional methods as well as MRI and the monitor sight window. In time, non-destructive in-situ observations on fouling accumulation and velocity profiles were made and pressure drop was monitored. The influence of feed spacer on biomass and pressure drop development was studied in monitors with and without feed spacer (**Chapter 8**).

The developed three-dimensional numerical model was used to evaluate the obtained experimental data. The evaluation included the feed channel pressure drop, biomass accumulation and velocity distribution profile, with and without feed spacer. The feed spacer geometry used in practice was applied in the mathematical model (**Chapter 9**).

Despite an extensive biofilm literature, systematic studies on the influence of factors such as substrate load and linear flow velocity on the development of biofouling in spiral wound membrane systems are lacking. Therefore, the influence of substrate concentration, linear flow velocity, substrate load and flow direction on pressure drop development and biofilm accumulation was studied with biofouling monitors (**Chapter 10**).

Increased shear has been proposed as a method to control biofouling. The effect of flow regime on biofilm accumulation and morphology in monitors was studied with single phase flow (water) and two phase flow (water with air sparging: bubble flow) to determine potential biofouling control measures (**Chapter 11**).

Observations at a full-scale installation without biofouling problems and with stable performance led to the study of the influence of phosphate limitation to control biofouling. Monitors were used for a proof of principle experiment, evaluation of phosphonate-based and phosphonate-free antiscalants, and several treatment stages (**Chapter 12**).

Based on new insights derived from this book, an overview is given of several potential complimentary perspectives to solve or control biofouling (**Chapter 13**).

Problem Analysis

2

Reverse Osmosis and Nanofiltration: Biofouling and Membrane Performance

INTRODUCTION

Membrane processes like reverse osmosis (RO) and nanofiltration (NF) can produce (drinking) water of high quality, free of pathogenic microorganisms and (in)organic pollutants. The demand for drinking water in the world is increasing and regulations on drinking water quality become stricter (Shannon *et al.*, 2008). The effect of these developments and decreasing costs of membrane processes are illustrated by the role of membrane filtration processes in advanced water treatment practice (Mallevialle *et al.*, 1996; Patching and Fleming, 2003 and Shannon *et al.*, 2008).

A potential drawback of NF and RO is membrane fouling, resulting in an increase of normalized pressure drop over the feed-concentrate channel (NPD) and/or a decrease of normalized flux (MTC) and/or increase in salt passage. Such problems may increase the costs of plant operation strongly, and may be prohibitive for the application of membrane filtration in (drinking) water treatment. The major fouling mechanisms of NF and RO membranes are scaling, particulate and organic fouling and biofouling. Different types of fouling may occur simultaneously and can influence each other (Flemming, 1993). Scaling by inorganic compounds is usually controlled using a scale inhibitor, such as a polymer or an acid. Particulate fouling can be controlled by pretreatment, such as ultrafiltration. Thus, all types of fouling except biofouling and organic fouling – likely related types of fouling – are

---

*This chapter is based, with permission from the copyright holder, on a paper previously published in Water Research Vol. 42 No. 19 pp. 4856–4868 doi: 10.1016/j.watres.2008.09.002

controllable. Numerous authors describe biofouling problems in membrane installations (Paul, 1991, 1996; Flemming, 1993; Tasaka *et al.*, 1994; Ridgway and Flemming, 1996; Baker and Dudley, 1998; Huisman and Feng Kong, 2004; Schneider *et al.*, 2005 and Karime *et al.*, 2008). From 70 surveyed U.S. reverse osmosis membrane installations, 58 reported fouling problems, with biofouling as the most common operational problem (Paul, 1991). In the Middle East, about 70% of the seawater RO membrane installations suffers from biofouling problems (Gamal Khedr, 2000, 2002).

Microbiological studies on membranes showed no univocal data on the concentration of microorganisms. Some authors diagnosed biofouling by the presence of specific microorganisms on the membrane, without quantification (Gamal Khedr, 1998) of the methods (Baker and Dudley, 1998 and Van Hoof *et al.*, 2002) or without specifying the biological parameters (Hamida and Moch, 1996). In many studies it was considered that appearance of microorganisms was an indication of (advent) biofouling on the membrane surfaces. However, microorganisms are present on all surfaces in contact with water and their presence is not indicative for biofouling. Other studies showed that microorganisms are commonly observed on water-exposed surfaces, even in pure water systems (Mittelman, 1991). The presence of microorganisms was already observed on membranes after a short contact time, in the order of minutes (Ridgway and Flemming, 1996 and Schaule, 1992).

Biofouling is difficult to quantify because no univocal quantification methods linking biofouling and operational problems are described. Pressure drop is generally used as a good parameter for evaluating biofouling. An increase of pressure drop is however not conclusively linked to biofouling, since other factors may influence the pressure drop as well. Additionally, the pressure drop measurement may not be sensitive enough for (early) detection of biofouling. It is essential to have unambiguous definitions for biofilm formation, biofouling and operational problems in order to quantify the problem. The following definitions are proposed:

- *Biofilm formation*: the accumulation of microorganisms, including extra-cellular compounds, on a surface due to either deposition or growth (or both) (Hamilton, 1985; Costerton *et al.*, 1987 and Characklis and Marshall, 1990);
- *Biofouling*: biofilm formation causing unacceptable (operational) problems. (Characklis and Marshall, 1990);
- *Operational problem*: an operational problem of a membrane installation is at hand when the NPD increase and/or the MTC decrease exceed 15% of the start-up values. At variations larger than 15%, corrective actions are recommended and guarantees are restricted by the manufacturers of membrane elements.

The diagnosis 'biofouling' is only justified when a correlation exists between the encountered operational problem(s) and biomass accumulation as determined by ade-quate parameters. A range of biomass parameters and analytical tools are available (Ridgway and Flemming, 1996 and Speth *et al.*, 2000). However, studies on quantitative relationships between the NPD increase over a membrane installation and biomass accumulation are lacking. In this study, a suitable biomass parameter to quantify biofouling is proposed.

The main objectives of this study were:

1. To select a sensitive process parameter to characterize biofouling for process control;
2. To determine relationships between concentrations of selected biomass parameters in membrane elements and the NPD increase over membrane installations. The biomass parameters applied were adenosinetriphosphate (ATP), total direct cell counts (TDC) and heterotrophic plate counts (HPC);
3. To evaluate feed and concentrate water quality parameters ATP, TDC, Assimilable Organic Carbon (AOC) and the Biofilm Formation Rate in a glass-ring biofilm monitor (BFR) as indicators for biofouling.

MATERIALS AND METHODS

Test rig experiments

Pressure drop increase and biomass parameters were monitored in time in two parallel test rigs, with and without dosage of $10\,\mu g\ C\ L^{-1}$ of acetate, an easily biodegradable compound. The operation of the rigs was identical with exception of the acetate dosage (Vrouwenvelder *et al.*, 2000). Each test rig was loaded with one 4" RO membrane element and operated at a recovery of 10%. Ground water treated by aeration, rapid sand filtration followed by cartridge filtration was used as feed water. The water had a constant temperature of 13°C. Three days after starting up the test rigs, acetate ($10\,\mu g\ C\ L^{-1}$) was added to the feed water (DOC concentration of $1{,}900\,\mu g\ C\ L^{-1}$) of one test rig. Depending on the development of the pressure drop, the membrane element was taken from the test rig to determine biomass parameters by a destructive membrane study (autopsy).

Full-scale experiments

The 15 membrane filtration installations, different in staging of pressure vessels and number of membrane elements per pressure vessel, loaded with spiral wound 4" or 8" NF or RO membrane elements, were investigated. This study included polyamide and polysulfone thin film composite membranes from different suppliers. No disinfectant residual was present in the feed water. The types of water source and pretreatment prior to membrane filtration are shown in Table 2.1. Membrane elements were taken from different plants varying in ΔNPD to determine biomass parameters by autopsy.

Table 2.1 Raw water source, water treatment prior to membrane filtration (NF or RO), pressure drop increase (ΔNPD), feed water quality parameters (TOC, ATP and AOC) and total and 7 of the metals with the highest concentration in the membrane element

Water source[1]	Pretreatment[2]	Plant performance	Water quality parameters			Inorganic elements in membrane elements mg m^{-2}							
		ΔNPD%	TOC mg C L^{-1}	ATP ng L^{-1}	AOC μg AOC-C L^{-1}	Ca	Fe	Si	Mg	Mn	Al	Ba	Total[3]
G	A/RSF (a)	450	1.3–1.6	4–20	10	101,000	7,70	1,400	430	117	72	61	111,000
W	UF	300	6	40–130	23–750	224	155	35	39	2	16	≤1	580
S + W	CS/RSF/CS/UF (b)	224	7–13	≤1–140	28	1,250	470	3,400	143	510	2,300	47	8,700
S + G	RBF/A/DMF/A/RSF	150	6	4–15	7–27	n.d.	200	n.d.	n.d.	3	n.d.	n.d.	200
S	CS/RSF/O3/GACF/SSF	100	1–2	n.d.	90	19	≤10	5	n.d.	≤0.5	8	≤1	≤100
S	CS/RSF/CS/UF	90	5–10	4–370	n.d.	20	63	8	43	15	6	≤1	185
G	A/RSF	70	7–9	5	n.d.	n.d.	275	n.d.	n.d.	1.8	n.d.	n.d.	277
S	UF	20	n.d.	n.d.	n.d.	15	20	n.d.	0.4	≤0.5	n.d.	≤1	≤100
S	CS/SSF	≤10	1.8	1–6	4	n.d.	≤10	n.d.	n.d.	≤0.5	n.d.	n.d.	≤100
S	CS/RSF/UF	≤10	2.8	2–4	7	n.d.	≤10	n.d.	n.d.	≤0.5	n.d.	n.d.	n.d.
G	A/RSF/A/S/RSF/UF	≤10	8	2–6	11	13	≤10	5	2	15	1	≤1	≤100
S	CS/RSF/O3/GACF/SSF	≤10	1.4	≤1–6	5–6	n.d.	≤10	n.d.	n.d.	≤0.5	n.d.	n.d.	≤100
S	CS/RSF/UF	≤10	n.d.	≤1–4	n.d.	n.d.	≤10	n.d.	n.d.	≤0.5	n.d.	n.d.	n.d.
S	CS/RSF/GACF/UF	≤10	2.0	2–4	n.d.	n.d.	≤10	n.d.	n.d.	≤0.5	n.d.	n.d.	≤100

a, b: Case studies labeled with a and b, which are discussed in the text and presented in figures with these labels. n.d.: not determined.
[1]Water source: G, anaerobic groundwater; S, surface water; W, (pretreated) waste water.
[2]Water treatment: A, aeration; RBF, river bank filtration; DMF, dual media filtration; CS, coagulation sedimentation; GACF, granular activated carbon filtration; O3, ozonation; RSF, rapid sand filtration; S, softening; SSF, slow sand filtration; UF, ultrafiltration.
[3]total of inorganic metals analyzed or derived from ashes of material scraped from feed side of membrane.

Normalized pressure drop

The plants were operated at a constant flux by increasing the feed pressure. The term 'biofouling' in this paper is related to a feed channel NPD increase $\geqslant$15%, supported by a biomass parameter. The feed channel pressure drop is also named feed-brine pressure drop. To compare the effect of membrane fouling the pressure drop data were normalized for temperature (10°C) and flow (reference capacity related to membrane element type and element number in pressure vessel) according to Huiting *et al.* (1999). To compare data of the individual membrane installations, the increase of the NPD at the time of autopsy is expressed as a percentage of the starting value of the installation with new membranes, i.e. a ΔNPD value of 300% indicates that the NPD has increased by 300% since the start-up of the plant (equation 2.1). The average NPD value of five consecutive days after the start up of the plant (NPD_{t0} in equation 2.1) and five consecutive days prior to the membrane sampling (NPD_{t1}) was used for determining the ΔNPD.

$$\Delta NPD\ (\%) = \frac{NPD_{t1} - NPD_{t0}}{NPD_{t0}} \times 100\% \quad (2.1)$$

Sampling and study of membranes

Lead elements from the feed side were taken from test rigs and membrane installations. No cleaning was applied in the NF or RO installations for at least 10 days prior to membrane sampling. The membrane elements were packed in plastic bags and stored at 2°C until the destructive study, which was performed within 24 hours after removal of the membrane elements from the installation. Membrane elements were opened lengthwise by cutting the glass fiber casing. Subsequently, the membrane material was unrolled and 10 small sections ($7\,cm^2$) over the length of the membrane element were cut and placed in 10 mL of autoclaved tap water in a capped tube (diameter of 2.0 cm, height of 25 cm). The samples included the feed spacer, the membrane and the product spacer. Additionally, in the middle of the element the three membrane layers were sampled separately to determine the distribution of biomass over the individual layers. The tubes with membrane sections were stored at 2°C until analysis. The tubes with the membrane sections were placed in an ultrasonic cleaning bath (Bransonic, model 5510E-DTH, output 135 Watts, 42 kHz). The low energy ultrasonic treatment (2 minutes) followed by mixing on a Vortex (few seconds) was repeated three times. Next, volumes of water were collected from the tubes to determine biomass parameters (ATP, TDC and HPC). Additionally, samples of membrane and spacer of 15 by 15 cm were taken about 20 cm before the effluent side of the membrane for analysis of the concentration of accumulated metals by quantitative Inductively Coupled Plasma Mass Spectrometry (samples were destructed by pretreatment with aqua regia in a microwave oven). Also, material scrapings of the feed side of the membranes were collected after removal from the feed spacer. The volume and weight of the collected material and scraped surface area were registered. The biofilm thickness was derived from the volume of material collected and surface area sampled. The scraped material was analysed for the concentrations of total organic carbon (TOC; non-purgeable organic carbon by infra-red gas analysis), dry weight (gravimetrical method, residue after 105°C), ashes (gravimetrical method, residue after incineration at 550°C), proteins (Lowry *et al.*, 1951) and carbohydrates (Dubois *et al.*, 1956).

Biomass in membrane elements

Active biomass was determined in duplicate by measuring the adenosinetriphosphate (ATP) concentration from 100 μL samples (Holm-Hansen and Booth, 1966). The luminometer (Celcis, model Advance) added 100 μL of LuminEX-B reagent (Celcis) to a sample to release ATP from the bacterial cells. Subsequently, 100 μL of LumATE-PM (Celcis) was added for light production. The amount of light produced was measured with the luminometer (relative light units). The concentration of ATP was derived from the relative light units value using the conversion factors of the linear relationship between relative light units values and reference ATP concentrations. The ATP concentrations of autoclaved drinking water (<1 ng ATP L^{-1}) and two control solutions (2 and 100 ng ATP L^{-1}) were determined as quality control.

Total Direct Cell counts (TDC values, cells cm^{-2}) were performed with epifluorescence microscopy using acridine orange as fluorochrome by applying a slightly adapted method to eliminate fading (Hobbie *et al.*, 1977). The slides were examined with a Leica DM RXA microscope equipped with a 100 W power supply and phase and filter sets (I2/3) appropriate for acridine orange. Microorganisms were enumerated through a 100x/1.30 PL Fluotar oil immersion objective. Non-fluorescence immersion oil (Olympus, code 35505) was used because of low background fluorescence. A total of 10 random fields were counted.

Heterotrophic Plate Counts (HPC values, CFU cm^{-2}) were assessed by spreading 0.05 mL volumes of water, either directly or from decimal dilutions, in triplicate on plates with R_2A medium (Reasoner and Geldreich, 1985). Colonies were enumerated after 10 days of incubation at 25°C.

Biological parameters of feed water

Parameters to determine the concentration of microorganisms in water (ATP and TDC) were measured during the membrane study as well. Assimilable Organic Carbon (AOC) and the Biofilm Formation Rate (BFR) were used to assess the concentration of growth promoting substances in water. The AOC test is a bioassay with two well-defined pure cultures. From the maximum growth level of the two individual strains, the AOC concentration was calculated and expressed as μg of acetate-C equivalents L^{-1} (μg C L^{-1}; Van der Kooij, 1992). The BFR value was determined with an on-line operated biofilm monitor at a continuous linear flow rate of 0.2 m s^{-1}. The accumulation of active biomass (ATP) on the surface of glass rings in this monitor was determined as a function of time and the BFR value is expressed as pg ATP cm^{-2} day^{-1} (Van der Kooij *et al.*, 1995).

Effect of cleaning

Biomass parameters were determined in a membrane element taken from a membrane installation directly prior to and after applying a set of successive cleanings with industrial chemicals according to supplier prescriptions. The cleanings were applied for biofouling control. Both autopsied membranes were lead elements with the same operation time from pressure tubes installed in parallel. The time between the two membrane studies was about 24 hours.

Statistical evaluation

Statistical analysis of log transformed (log10) data was performed with SPSS, software version 11 for Windows.

RESULTS

For biofouling quantification, biomass parameters in membrane elements were investigated during membrane autopsies and compared with the increase of normalized feed channel pressure drop and feed water qualities in test rigs and 15 full-scale investigations (Figure 2.1, Table 2.1).

Biomass in membrane elements

The concentration of active biomass usually decreased during passage of the pressure tubes (unpublished data). Thus, the highest concentrations of active biomass were generally found in lead elements. The distribution of biomass over the length of lead elements differed. Severely fouled membranes showed high biomass concentrations that decreased over the element length, but less fouled membranes had lower biomass concentrations that were equally distributed over the element (Figure 2.2).

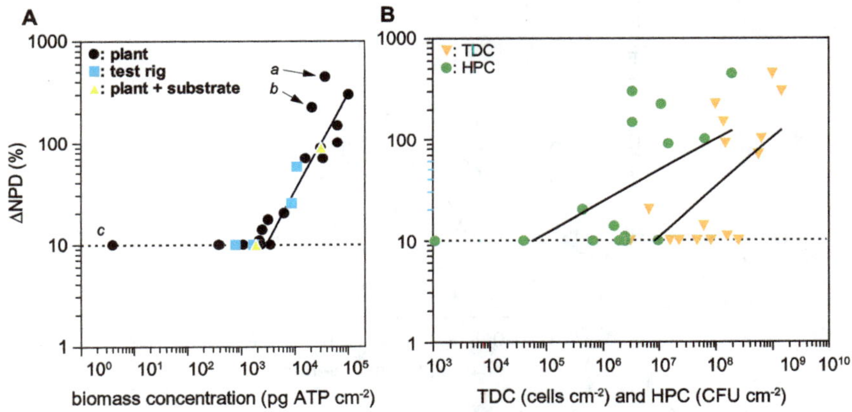

Figure 2.1 Relationships between the ΔNPD over membrane installations and ATP concentrations (**A**) and TDC and HPC values on R_2A media (**B**) in lead elements from membrane installations. (A) shows the data from full-scale and test rig studies, with and without dosage of a substrate to the feed water. (B) shows data from plant studies. The lines represent the calculated relationships; dotted lines represent the ΔNPD threshold. Details of case studies a, b and c are presented in the text

Most of the biomass was found on the feed side of the membrane and the feed spacer (Table 2.2). Some biomass was observed on the product side of the membrane and on the product spacer. Microscopic observations showed that cells on the product side were smaller than cells on the feed side of the membrane. In general, on the feed side of the membrane, the TDC analysis showed large clusters of up to thousands of bacterial cells. These clusters of cells are responsible for the relative large standard deviation of the TDC value.

The highest concentrations of biomass have the strongest effect on the feed channel pressure drop and ΔNPD, since the space for feed water flow is most limited at the location with a high biomass concentration. The maximum values for biomass parameters in lead membrane elements were used in subsequent discussions.

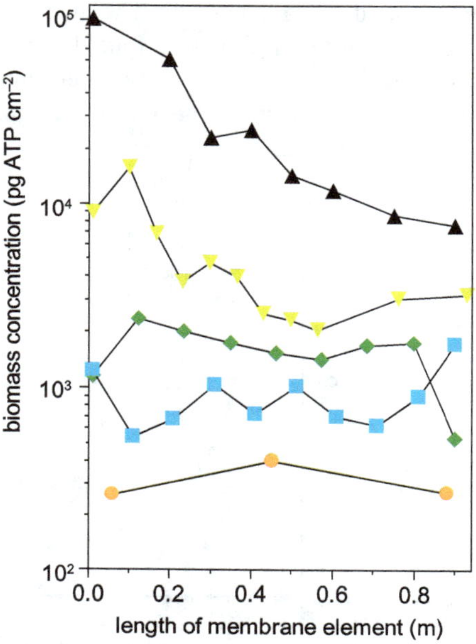

Figure 2.2 Distribution of ATP amount over the length of lead membrane elements taken from membrane installations with differences in pressure drop increase (▲ = 300%; ▼ = 100%; ◆, ■, ● = ≤10%). Each marker type represents a lead membrane element from one of the tested membrane installations. Severely fouled lead membranes showed high biomass concentrations that decreased over the element length and less fouled lead membranes had lower biomass concentrations that were equally distributed over the membrane element

Table 2.2 Distribution of biomass (minimum and maximum percentage of total for ATP, TDC and HPC) over the individual membrane layers of all membranes investigated

Membrane layer	ATP %	TDC %	HPC %
feed spacer	5–62	⩽1–95	⩽1–53
membrane	38–90	5–100	40–100
product spacer	⩽1–6	⩽1–4	⩽1–8

Dose-effect studies

In the membrane element (in test rig) without acetate dosage, no increase of ΔNPD was observed during the first 29 days, while the concentration of active biomass increased strongly (Figure 2.3). Dosage of 10 μg C L^{-1} as acetate

(0.5% of the feed water TOC concentration) to the feed water resulted in a rapid increase of ΔNPD and active biomass concentration in a parallel installed test rig operated under identical conditions (Table 2.3 and Figure 2.3). The dosage of this low amount of acetate to a test rig resulted in an at least four times higher ΔNPD value and a six times higher active biomass concentration than in the parallel test rig without dosage, with the same running time (29 days). In the test rig studies, all biomass parameters (ATP, TDC and HPC) increased in time with and without dosage of a substrate (Table 2.3). The gradual increase in time of active biomass and ΔNPD in the test rig studies without substrate dosage showed that the feed water (tap water) contains substrate.

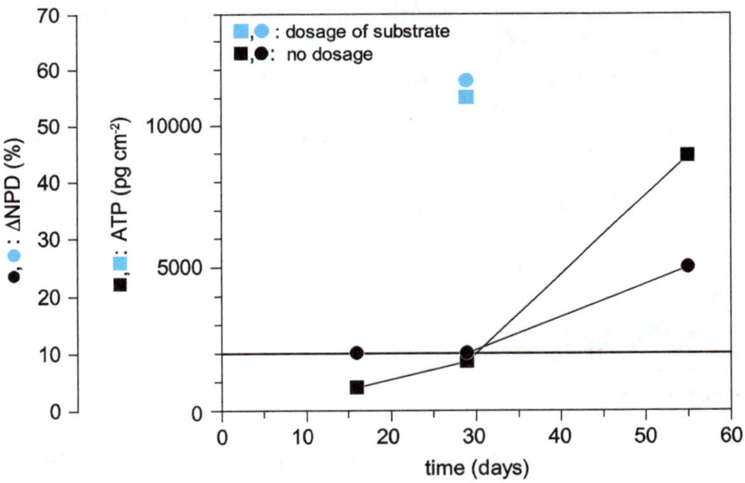

Figure 2.3 Development of ΔNPD and ATP in a RO membrane element in a test rig in time with and without dosage of 10 μg acetate-C L^{-1}. The dotted line represents the ΔNPD threshold

One full-scale membrane study was performed before and during an increase of ΔNPD and can therefore be interpreted as a dose-effect experiment (Table 2.3). This plant showed excellent performance for 2.5 years, i.e. a low and stable ΔNPD (≤10%) without cleanings and concentrations of active biomass of 1,000 pg ATP cm^{-2}. However, a sudden increase of ΔNPD (90%) was observed and autopsy showed an active biomass concentration of 31,000 pg ATP cm^{-2}. The only difference in time was the type of scale inhibitor dosed. Therefore, the bacterial growth potential of the scale inhibitors was determined. AOC analysis of the feed water with the new and original scale inhibitor demonstrated that the AOC of the feed water with the new scale inhibitor was 90 μg C L^{-1}, whereas the feed water AOC with the original scale inhibitor was <10 μg C L^{-1}. The AOC of the feed water without scale inhibitor was <10 μg C L^{-1}.

Table 2.3 Development of pressure drop (ΔNPD) and biomass parameters (ATP, TDC and HPC) in time and during dosage experiments of a biodegradable compound to the feed water of a test rig or pilot plant loaded with RO membrane elements

Study	Comment	ΔNPD (%)	ATP pg cm^{-2}	TDC cells cm^{-2}	HPC cfu cm^{-2}
test rig – no dosage	16 days running time	⩽10	7.9×10^2	4.0×10^7	1.7×10^4
	29 days running time	⩽10	1.7×10^3	n.d.	6.1×10^5
	55 days running time	25	8.9×10^3	7.8×10^7	8.0×10^4
test rig – with dosage	29 days with dosage	58	1.1×10^4	n.d.	1.4×10^7
plant + substrate	original scale inhibitor	⩽10	1.9×10^3	8.4×10^7	2.1×10^6
	new scale inhibitor	90	3.1×10^4	1.5×10^8	1.5×10^7

n.d. not determined.

All dose-effect studies showed that biodegradable compounds in the feed water lead to accumulation of biomass in the membrane element, causing an increase of ΔNPD. The biomass parameters (viz. ATP) seem to be more sensitive than ΔNPD. The concentration of biodegradable compounds in the water supplied to the membrane installation is of importance in controlling biofouling.

Fouling of membrane plants: use of biomass parameters

Characteristics of water source and pretreatment of the membrane installations are summarized in Table 2.1. The ranking of membrane installations in Table 2.1 is based on the ΔNPD value, i.e. a higher position in the table means a higher ΔNPD. In general, the Modified Fouling Index (MFI 0.45 μm) values and turbidity values of the feed water of the individual membrane installations were <1 s L^{-2} and <1 NTU.

The temperature of the feed water varied between 1 and 30°C; one plant operated at 60°C. The running time of the plants before sampling of a membrane element for autopsy varied between approximately 90 and 1,000 days. The ΔNPD values ranged from ≤10% to 450%. The accuracy of the ΔNPD value was about 10%.

The relationship between ΔNPD over the membrane filtration installation and the concentration of ATP, TDC and HPC in membrane elements is shown in Figure 2.1. At low level of ATP no ΔNPD is observed (Figure 2.1). The ATP values ranged from 4 to 102,000 pg ATP cm^{-2}. The ATP data from the test rig studies fit well within the data of the plant studies (Figure 2.1A). The TDC values ranged between 3×10^6 and 1×10^9 cells cm^{-2}. The standard deviation of the TDC value is relatively large (up to 100% of the TDC value) due to the presence of large cell clusters. The HPC values varied between 1.1×10^3 and 2.0×10^8 CFU cm^{-2} (0.04% and 20% of the total number of cells). In general, the HPC values are a small fraction of the total number of cells present (Figure 2.1B).

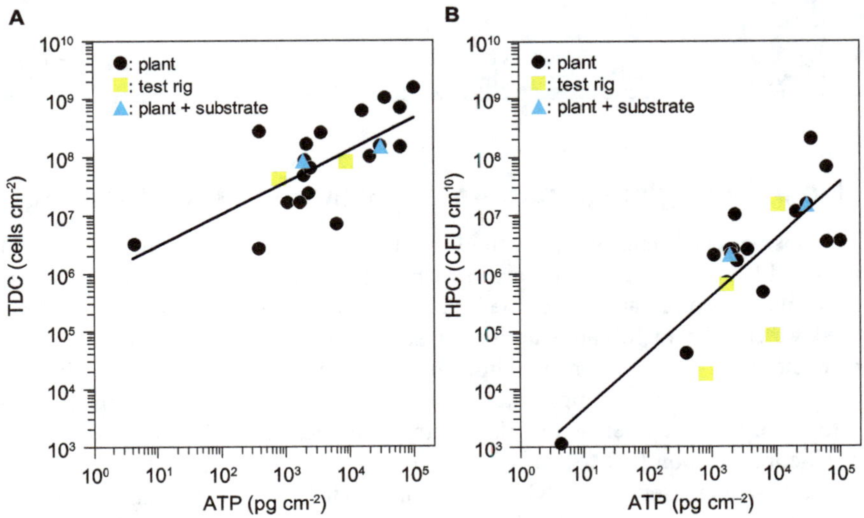

Figure 2.4 Relationships between ATP and TDC (**A**) and HPC values on R_2A media (**B**). The lines represent the calculated relationships.Weak but significant correlations between the ΔNPD and TDC ($p = 0.003$, $r^2 = 0.42$) and HPC ($p = 0.015$, $r^2 = 0.29$) were found (Figure 2.1B). More significant relationships were found between ATP and TDC ($p = 0.0001$, $r^2 = 0.51$) and HPC ($p = 7 \times 10^{-6}$, $r^2 = 0.62$), respectively (Figure 2.4). The higher significance of the correlations between ATP and the other biomass parameters, TDC and HPC, when compared to the correlations observed between TDC and HPC with ΔNPD illustrate the relative low sensitivity of ΔNPD

A highly significant relationship was found between ATP and ΔNPD ($p = 7 \times 10^{-11}$, $r^2 = 0.95$), without taking into account the corresponding data with ATP concentrations below 1,000 pg ATP cm^{-2}, since in that range ΔNPD values were not representative for the corresponding ATP data (Figure 2.1A). Also, data from the case studies a and b were excluded (see discussion).

Parameters of material scraped from the feed side of the membrane such as biofilm thickness, TOC, dry weight, ashes, proteins and carbohydrates from 6 locations showed differences (Table 2.4). However, the data of the individual parameters were limited and showed no relationship with ΔNPD (data not shown).

Table 2.4 Range of parameters in membranes taken from 6 plants

Parameter	Unit	Minimum	Maximum
thickness	µm	3	91
TOC	µg cm^{-2}	0.7	13
dry weight	µg cm^{-2}	5	73
ashes	µg cm^{-2}	<1	10
proteins	µg cm^{-2}	0.8	47
carbohydrates	µg cm^{-2}	0.5	26

Use of biological parameters of water to predict fouling

The dose-effect studies demonstrated that the feed water quality plays a role in the ΔNPD. This suggests the use of a biological parameter of the feed water as a relative simple and direct method to assess the biofouling potential of the feed water of a membrane filtration installation. Table 2.1 shows feed water characteristics like the concentration of organic carbon (TOC), active biomass (ATP) and the growth potential (AOC). TOC concentrations ranged from 1 to 13 mg C L^{-1}, whereas the plant with the highest ΔNPD had a relative low TOC-concentration (1.3 to 1.6 mg C L^{-1}). In the test rig study, 0.010 mg C L^{-1} as acetate increased the ΔNPD indicating that TOC concentrations as low as 1 mg C L^{-1} may cause biofouling. ATP concentrations ranged from <1 to 400 ng ATP L^{-1}. ATP concentrations higher than 10 ng L^{-1} were periodically observed only at plants with ΔNPD values exceeding 15%. However, lower ATP values were observed as well at elevated ΔNPD values, illustrating that the ATP concentration does not consistently correlate with ΔNPD. Note that ATP is a parameter describing the concentration of active biomass and not the substrate concentration present in the water. High concentrations of substrate can result in higher growth of biomass in the water.

The growth potential of water was assessed with a biofilm monitor (Table 2.5, BFR) and AOC analysis (Table 2.1). BFR values of feed water ranged from ≤1 to 1,400 pg ATP cm^{-2} day^{-1}. Low BFR values, ≤1 pg ATP cm^{-2} day^{-1} were observed at locations with extensive pretreatment and long running times at ΔNPD ≤10% without cleanings. Higher BFR values, ranging from 120 to 1,400 pg ATP cm^{-2} day^{-1}, were found at locations dosing a chemical containing biodegradable compounds or with restricted pretreatment resulting in reduced running times. At locations with BFR temporarily ≥290 pg ATP cm^{-2} day^{-1}, autopsies showed high active biomass concentrations indicating biofouling (Table 2.5). Biomass data from autopsies for BFR values in the range from 1 to 120 pg ATP cm^{-2} day^{-1} were not available. The concentrations of AOC of the feed water ranged from a few μg to 750 μg C L^{-1}. AOC concentrations ≥10 μg C L^{-1} were observed at locations with ΔNPD >15%.

Table 2.5 Biofilm formation rate in the biofilm monitor and running time to meet cleaning criteria

BFR (pg ATP cm^{-2} day^{-1})	Running time days (days)	Biofilm (pg ATP cm^{-2})	Pretreatment[a]
1,400	7	40,000	CS/RSF/CS/UF
310	30	21,000	RBF/A/DMF/A/RSF
290	40	30,000	UF
120	50	n.d.	treated waste water after MF
0.84	>365	1,800	CS/RSF/O3/GACF/SSF
0.82	>365	1,100	CS/RSF/SSF
0.32	>1,000	<900	CS/RSF/GACF/UF

[a]*Abbreviations used are explained in Table 2.1; MF, microfiltration; n.d., not determined.*

For high ATP and AOC values, biofouling may be expected but low values do not guarantee absence of biofouling. ATP and AOC analysis can be useful for rapid screening of factors influencing ΔNPD, as described under dose-effect studies. This rapid screening makes it possible to take directed actions. The ATP, AOC and BFR values are applicable for spotting of biofouling but are not suitable for process control, since low values do not guarantee absence of biofouling.

Pretreatment and cleaning

Membrane filtration installations with extended and robust pretreatment involving biological activity, such as granular activated carbon filtration and slow sand filtration experienced stable performance at low ΔNPD (Table 2.1). Membrane filtration installations with restricted pretreatment and high concentrations of biodegradable compounds (high BFR value) in the feed water experienced biofouling as determined with a destructive membrane study (Table 2.5) and high ΔNPD values. Exploratory studies on membrane cleanings showed that biomass could be (partly) inactivated, but that biomass was not removed from the membrane elements (Figure 2.5). Removal of biomass is essential for biofouling control, since the remaining inactivated biomass serves as food for microrganisms surviving the cleaning and supplied by the feed water, leading to rapid regrowth.

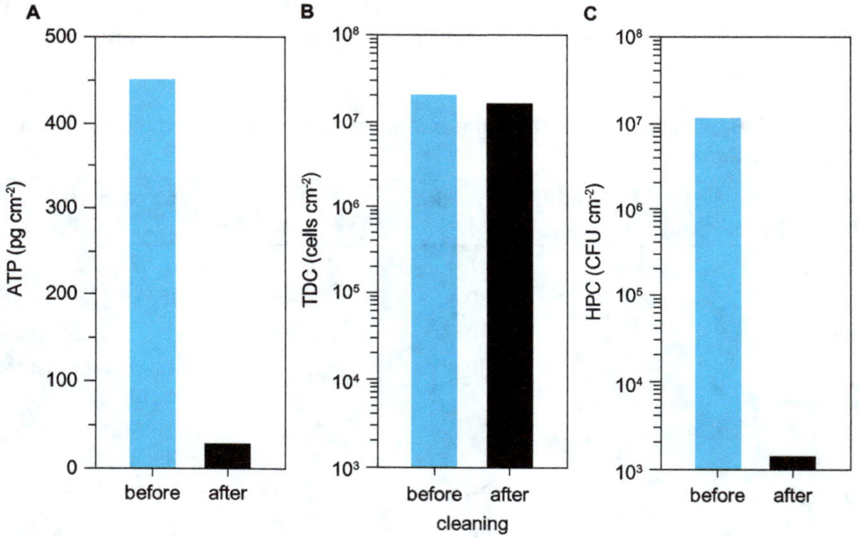

Figure 2.5 Effect of cleaning on the biomass parameters ATP (A), TDC (B) and HPC on R_2A media (C) in a membrane element. One lead element was taken from the plant directly prior to cleaning and another lead element from a parallel pressure tube was taken directly after the cleaning. Adapted from Vrouwenvelder *et al.*, 2001

DISCUSSION

Selection of a suitable parameter for biofouling

It is important to have an appropriate process parameter to characterize (bio)fouling in membrane installations. ΔNPD is the most logical process parameter. Usually, in practice applied pressure meter sensors do not allow sensitive accurate detection of ΔNPD and can therefore not be used as an (early) warning system enabling control actions such as preventive cleanings. The accumulation of biomass in the membrane elements was not detected with the ΔNPD measurement while the ATP concentration showed a clear increase during the first 30 days of the test rig study without substrate dosage (Table 2.3, Figure 2.3). Also in practice, the ΔNPD measurement was not able to discriminate plants containing membrane elements with clearly different active biomass concentrations ranging from about 3 to 3,000 pg ATP cm^{-2} (Figure 2.1A). Therefore, biomass parameters were evaluated as an alternative for ΔNPD.

The ATP concentration in a membrane element is a suitable parameter, since the highly significant relationship with ΔNPD ($p = 7 \times 10^{-11}$, $r^2 = 0.95$). The high sensitivity and accuracy makes ATP more distinctive than ΔNPD (Figure 2.1A). ATP can be used for accurate determination of the development of biofouling in a membrane filtration installation and may therefore be used as an (early) warning system for control actions like preventive cleanings (Figure 2.1A). The relationship between ATP and ΔNPD has predictive value for diagnosis of biofouling. The estimated value for ATP at ΔNPD of 15% was 3,100 pg ATP cm^{-2}.

The ΔNPD during plant studies was determined over the total installation, consisting of several staged pressure vessels with each pressure vessel containing multiple membrane elements, while during dose-effect studies ΔNPD was determined over one membrane element. Nevertheless, a good similarity between the plant studies and the dose-effect experiments was found with respect to the relationship between ΔNPD and ATP, measured in the lead element of the first stage during plants studies (Figure 2.1A).

TDC seems to be a suitable parameter to evaluate the effectiveness of cleanings. ATP can be affected by cleaning chemicals while TDC values are not (Figure 2.5). By TDC measurement the total number of cells, including active, inactive and dead cells is determined. So, TDC values can be used for evaluation of cleaning strategies on the removal efficiency of biomass from the membrane element. The large scattering of the TDC counts (Figure 2.1B), caused by the presence of clusters of cells, makes the parameter less appropriate than ATP for use as parameter for biofouling.

HPC is not an appropriate parameter to evaluate biofilm concentrations and biofouling. Comparison of TDC and HPC counts on R_2A media of biofilms in membrane elements (Figure 2.1B) clearly show that HPC values are relatively low in comparison with the TDC values. The 50-percentile value of HPC values was <2.5% of the TDC values. The relative substrate poor media and the long incubation time used for HPC values on R_2A media explains the relative high yield of colony counts compared to other media for colony counts. In addition, the standard deviation of the HPC values is wide spread (Figure 2.1B).

Extrapolation of the trend of the relationship between ΔNPD and ATP (Figure 2.6) shows that low ΔNPD values cannot be detected and discriminated on basis of current ΔNPD measurements but can be detected and discriminated on basis of ATP analysis, with an accuracy of 0.5 pg ATP cm^{-2}. The ΔNPD data and HPC and TDC counts show a similar trend, however, with a larger standard deviation (Figure 2.1B).

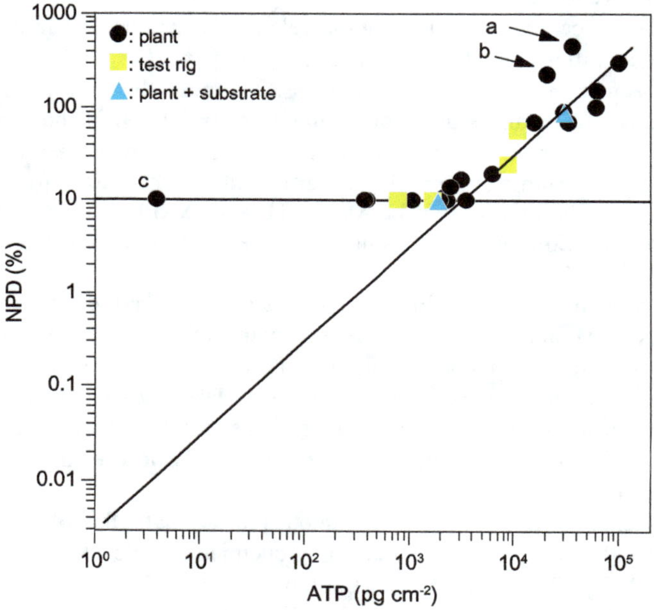

Figure 2.6 Hypothetical correlation between ΔNPD and ATP concentrations in full-scale installations and test rigs. The line represents the calculated relationship including the extrapolation to low ΔNPD and ATP values; the dotted line represents the ΔNPD threshold. The correlation suggests that fouling accumulation can be detected earlier and more accurately with proper pressure drop measurements

Comparison of the results of our study with literature data shows, that the ATP concentrations in this study on thin film composite membranes (4 to 102,000 pg ATP cm^{-2}) were lower than the values reported by Ridgway *et al.*

(1983) on cellulose based membranes (880,000 and 3,340,000 pg ATP cm^{-2}). Possibly, the cellulose acetate based membranes, which are susceptible for biological degradation, and the water quality may explain the relative high biomass concentrations found over the whole membrane installation by Ridgway. Griebe and Flemming (1998) reported TDC values of 6×10^6 and 1×10^8 cells cm^{-2} which covered approximately the range of our observations. In addition, concentrations of proteins, carbohydrates and the thickness reported in that study were in the same range as observed during our study (Table 2.4). Geesey and Bryers (2000) reported biofilms thicknesses varying between 10–100 μm on RO membranes, similar to the findings of our study (Table 2.4). The average ATP concentration per cell, which is a measure for metabolic activity, ranged from 9×10^{-7} to 7×10^{-9} ng ATP $cell^{-1}$ for the membranes from the plant studies (Table 2.1). These values were similar as observed during studies on biomass on granular activated carbon used in water treatment (Magic-Knezev and Van der Kooij, 2004) and ground water bacteria (Metge *et al.*, 1993).

The combination of ATP and TDC values in a membrane element gives valuable information in diagnosing biofouling and can be useful as process parameter for control actions. The results from this study on biomass parameters agreed well with data found in literature.

Use of feed water parameters as process parameters

The use of ultrafiltration, which is efficient for biomass removal from the water, is not sufficient to guarantee absence of biofouling in the subsequent NF or RO treatment (Table 2.1). Pretreatment with ultraviolet (UV) light to inactivate biomass in feed water most likely will not prevent biofouling as well. The TDC values in feed water, ranging from about $<10^2$ to 10^6 cells mL^{-1}, showed no relationship with biofouling. ATP concentrations periodically higher than 10 ng L^{-1} were observed only at plants with ΔNPD values exceeding 15% (Table 2.1). Possibly, the concentration of biomass in the feed water is of minor influence on the occurrence of biofouling. Biodegradable compounds in feed water seem to be stronger related to biofouling.

Monitoring the biodegradable compounds in feed water can be performed by assessment of the BFR determined with the biofilm monitor, combined with AOC analysis. The BFR varied between <1 and 1,400 pg ATP cm^{-2} day^{-1}. High BFR values ($\geq$290 pg ATP cm^{-2} day^{-1}) were observed at plants with biofouling and relative low BFR values ($\leq$1 pg ATP cm^{-2} day^{-1}) were observed at plants without biofouling (Table 2.5). A significant linear relationship between the concentration of acetate and the BFR has been reported for low concentrations of acetate added to water in a biofilm monitor, with a BFR of 35 pg ATP cm^{-2} day^{-1} for 1 μg L^{-1}

AOC (Van der Kooij *et al.*, 1997). Relatively low concentrations of substrate can lead to serious biofouling (Figure 2.3), which illustrate that the consequence of low AOC values on biofouling is not clear. However, the BFR values in the biofilm monitor determined of the feed water were a factor 5 to 100 lower than the estimated accumulation rates of biofilm formation in membrane elements from plants. This difference in BFR between the biofilm monitor and the membrane element was also found during the test rig studies with tap water. The BFR in the biofilm monitor was <2 pg ATP cm^{-2} day^{-1} but the BFR values in three elements, with a running time varying from 16 to 55 days were <30 pg ATP cm^{-2} day^{-1}. The maximum values for biomass concentration were clearly higher in the membrane elements than in the biofilm monitor. Possible causes for the disparity in BFR values and biomass concentrations between the biofilm monitor and the membrane elements can be differences in hydrodynamics (less efficient mass transfer in biofilm monitor), material type and roughness (the biofilm monitor contained glass rings with a smooth surface) and influence of filtration of water in the membrane elements or combinations thereof. A drawback of the biofilm monitor is that the development of the biofilm in time is assessed by periodic sampling and analysis by skilled personnel. The sampling frequency determines the availability of data. Despite these drawbacks, the biofilm monitor proved to be valuable in diagnosis of biofouling (Table 2.5).

High BFR and AOC values are indicative for biofouling. However, low values for these parameters may be sometimes difficult to relate to membrane performance, since biofouling has been observed with feed water that had low BFR and low AOC values. Thus, the parameters BFR and AOC are applicable for spotting of biofouling but are not suitable for process control.

Practical implications

General trends

Membrane filtration installations with extended and robust pretreatment involving biological activity experienced stable performance (relative low ΔNPD values) for running times up to 1,000 consecutive days without cleanings. Pretreatment steps with biological activity are rapid sand filtration, dual media filtration, granular activated carbon filtration and slow sand filtration (Table 2.1). Griebe and Flemming (1998) reported that the use of biofiltration as pretreatment resulted in a reduction of the concentration of biomass in membrane elements and an improvement of membrane performance, supporting the findings of this study. However, use of biological pretreatment before membrane processes is no guarantee for absence of biofouling. Even when biological pretreatment is used, chemicals dosed (such as scale inhibitors) and seasonal variations of the water quality need attention (Van der Hoek *et al.*, 2000; Vrouwenvelder *et al.*, 1998 and Carnahan *et al.*, 1995). Testing of 14 commercially available scale inhibitors based on polymers and dose-effect studies with test rigs demonstrated that chemicals used to prevent scaling differ greatly in their ability to promote growth of microorganisms and confirmed experiences in practice (Vrouwenvelder *et al.*, 2000).

Balancing of prevention (pretreatment and preventive cleanings) and control actions (cleanings) may be the most cost effective way to operate membrane installations. However, exploratory studies on membrane cleaning showed that biomass can be (partly) inactivated, but not removed out of the membrane elements (Figure 2.5). Schaule *et al.* (1993) reported that the fouling deposit in membranes, from a plant with an elevated feed channel pressure, before and after cleaning contained identical TDC values ($1 \times 10^8 - 2 \times 10^8$ cells cm^{-2}), supporting findings of our study. Kappelhof *et al.* (2003) showed that cleaning reduced the microbial activity of the biomass in the membrane element but within three weeks the original microbial activity was fully restored. This suggests that inactivated biomass is rapidly utilized by cryptic growth of biomass surviving the cleaning and biomass supplied with the feed water. Therefore, it is essential that cleanings remove biomass from the membranes.

Membrane suppliers have developed membranes that are supposed to be less susceptible for fouling, which may be another approach to control biofouling. However, the use of these low fouling membranes is no guarantee for stable plant performance. In our study, the plant characterized by the highest ΔNPD and highest biomass parameters was operated with low fouling membranes.

Membranes properties such as material charge, roughness, molecular weight cut off and hydrophilicity and variation in properties during operation may affect membrane process performance and fouling. However, irrespective of the (variety in) membrane properties used in the study, a significant correlation was found between ATP and ΔNPD (Figure 2.1A), suggesting that a variety of membrane properties is of minor importance for biofouling control.

Our studies at plants with ΔNPD $\geq$ 50% demonstrated that curative cleanings were not effective in restoring the original ΔNPD and membrane elements had to be replaced prematurely in all cases. Prevention of fouling by proper pretreatment is preferred over curative actions like cleanings. Probably, future systematic studies on membrane cleaning using appropriate parameters enable the evaluation and selection of chemical cleanings (procedures and chemicals). Limited systematic studies on cleanings were found (Whittaker *et al.*, 1984 and Ebrahim, 1994). Al-Amoudi and Lovitt (2007) reported that very few papers on cleaning of membrane to regenerate membrane performance have been published and there is an urgent need for extensive research. A combination of pretreatment and cleaning may be the way to restrict biofouling at minimum cost.

A 15% ΔNPD increase may be too high for corrective actions. Apparently, the definition for operational problem is arbitrary. Preferred is to take corrective actions at a lower ΔNPD increase, which is difficult to determine with the pressure drop measurements as applied in practice.

The relationship between ATP and ΔNPD is only correct in case of extensive pretreatment (i.e. ultrafiltration) removing other constituents, such as particles, that can cause a ΔNPD as well.

Special case studies

Low concentrations of inorganic elements in membrane elements were generally measured at plants with low ΔNPD values (ashes $\leq$100 mg m^{-2}, Tables 2.1 and 2.4).

For case studies a and b (Figure 2.1A), the ATP values cannot fully explain the observed ΔNPD increase. In case study a, besides high concentrations of biomass also relatively high concentrations of the inorganic elements calcium (100,000 mg m^{-2}) and iron (8,000 mg m^{-2}) have been found on the feed side of the membrane (Table 2.1). Calcium was present as $CaCO_3$. The sum of inorganic compounds corresponded well with the weight increase of the 8 inch element by 13 kg. Cleanings with a subsequent series of different chemicals about a month prior to the autopsy of the membranes showed no clear effect on the ΔNPD and the applied scale-inhibitor (dosage) was apparently not effective. In case study b, high concentrations of inorganic compounds have been found as well. Major inorganic compounds found were silica, aluminum, calcium, manganese and iron salts (Table 2.1). In conclusion, the high concentrations of inorganic compounds contributed to the ΔNPD increase and most likely explain the deviation of the results of case studies a and b from the general trend between ΔNPD and ATP as observed in Figure 2.1A.

Case study c also deviates from the relationship between ΔNPD and ATP. At case c both the concentrations of ATP and metals were low. Microscopic observations of material collected from the feed side from the membrane showed similar shaped and sized wire-like structures. The membrane installation in case study c was used in an industrial laundry and the pretreatment was not effective in removing all fibre pieces from the fabrics washed. Chemicals to control biofouling were not effective. Our membrane study showed that biofouling was not responsible for the operational problem and the recommended relative inexpensive adaptation of pretreatment solved the problem.

Case studies a, b and c show that it is important to study fouling in an integral way. Therefore, the concentration of inorganic compounds accumulated on the membrane must be determined and the scrapings of the membranes have to be examined microscopically.

Recommendations

For diagnosis of the type of fouling hampering the membrane installation or for determining the risk of biofouling, a destructive study is proposed of the lead element from the first stage. It is essential that the elements are processed as described and that no cleanings are applied for at least 10 days prior to membrane sampling. The diagnosis can be obtained within 8 hours (Vrouwenvelder *et al.*, 2003b).

Destructive membrane studies are expensive and laborious; therefore, a tool is needed for monitoring the development of fouling and evaluation of control actions. The ideal tool for monitoring of fouling has to meet a large number of requirements (Flemming *et al.*, 1998, 2003 and Patching and Fleming, 2003). However, no available monitor was fulfilling the combination of described requirements. Therefore, a tool was developed for the validation of membrane fouling: the Membrane Fouling Simulator (MFS, Vrouwenvelder *et al.*, 2006b, 2007a). The major advantages of the MFS are representativeness of spiral wound membranes and the small size requiring small amounts of water and chemicals. Using the MFS, fouling can be monitored by (1) operational parameters like pressure drop, (2) non-destructive (visual, microscopic) observations using the sight glass and (3) analysis of coupons sampled from the membrane and spacer sheet in the MFS.

Systematic research is needed, requiring a practical tool and a sensitive pressure analyser. A sensitive and robust pressure analyzer will provide proper pressure drop data enabling early detection of fouling (Figure 2.6).

SUMMARY

Biofilm accumulation in nanofiltration and reverse osmosis membrane elements results in a relative increase of normalized pressure drop (ΔNPD). However, an increase in ΔNPD is not exclusively linked to biofouling. In order to quantify biofouling, the biomass parameters adenosinetriphosphate (ATP), total cell count and heterotrophic plate count in membrane elements were investigated during membrane autopsies and compared with ΔNPD in test rigs and 15 full-scale investigations with different types of feed water. The combination of biomass related parameters ATP and total cell count in membrane elements seem to be suitable parameters for diagnosis of biofouling, whereas plate counts were not appropriate to assess biofouling. The applied ΔNPD measurement was too insensitive for early detection of fouling. Measurements of biological parameters in the water were shown to be not appropriate in quantifying biofouling. Evidently, there is a need for a practical tool, sensitive pressure drop data and systematic research.

Development of a Methodology

3

Development and Testing of a Membrane Fouling Simulator

INTRODUCTION

High pressure membrane filtration processes, reverse osmosis (RO) and nanofiltration (NF), produce drinking water of high quality, virtually free of pathogenic micro-organisms and (in)organic compounds. The global drinking water demand is increasing and also regulations on drinking water quality become stricter. Because of the decreasing costs of RO applications, an increasingly important role is predicted for this membrane technology in the future (Mallevialle *et al.*, 1996).

A serious problem in RO/NF is membrane fouling, resulting in an increase of the normalized feed channel pressure drop and/or reduction of normalized flux and/or increase of salt passage. Such problems cause a significant increase of operational costs – up to 50% of the total costs (Ridgway, 2003) – and may endanger the continuity of (drinking) water supply. Moreover, fouling will increase the use of cleaning chemicals and consequently introduce a waste problem. The major fouling mechanisms of RO and NF membranes are scaling, particulate and organic fouling and biofouling. Scaling by inorganic compounds is predictable and is usually controlled using an anti-scalant and/or an acid. Particulate fouling can be controlled by pretreatment, e.g. ultrafiltration. Thus, all types of fouling except biofouling and organic fouling – possibly related types of fouling – are controllable. Biofouling is considered as a major problem in RO and NF at the moment (Flemming, 1997a

---

*This chapter is based, with permission from the copyright holders, on two papers previously published in Journal of Membrane Science Vol. 281 No. 1–2 pp. 316–324 doi: 10.1016/j.memsci.2006.03.046 and Water Science & Technology Vol. 55 No. 8–9 pp. 197–205 doi: 10.2166/wst.2007.259

and Paul, 1991, 1996). In practice, the fouling problem may be even more complex because of possible interactions of different fouling types.

Prevention of membrane fouling requires in many situations intensive pretreatment, increasing the costs of water treatment. Consequently, research into the membrane fouling potential of raw water can be cost-effective. But also membrane selection, use of chemicals and cleaning strategies can help to control fouling problems.

Fouling studies are predominantly based on a trial and error approach with pilot and industrial installations. Fouling is diagnosed on the basis of parameters such as pressure drop and on membrane autopsy studies afterwards. Evidently, there is a need for better tools and measurements in which visual and microscopic observations of the membrane and spacer surface are possible, in-situ, real-time and non destructively (Flemming *et al.*, 1998, 2003).

Table 3.1 Requirements for a membrane fouling monitor and summarized results of testing of the prototype monitor and the Membrane Fouling Simulator (MFS)

Requirements	Description	Prototype	MFS
Representativeness	– spatial dimensions (spacer channel height)	+	+
	– materials (spacer and membranes)	+	+
	– linear velocity	+	+
	– flow distribution	+	+
	– relationship linear velocity and pressure drop	–	+
	– development of fouling		+
Membrane performance loss assessment	– feed channel pressure drop increase		+
	– flux reduction		
	– salt passage increase		
Fouling assessment	– non-destructive in-situ, real time inspection		+
	– analysis of coupons of membrane		+
Sensitive and reproducible			+
User friendly	– easy to handle, small, simple and robust		+
Low costs	– minimum of water and chemicals (and time)		+

Results of testing were considered (+) = satisfactory; (–) = not satisfactory; ( ) = not tested or evaluated.

Vitens formulated the requirements for a (bio)fouling monitor (Table 3.1). First of all, the tool must be representative for spiral-wound membrane elements used in practice, e.g. the materials used (membranes and spacers), spatial dimensions (height of feed and product spacer channels) and hydraulics (pressure drop and linear flow velocity and flow distribution) must be identical. Secondly, the obtained data must be accurate and reproducible, enabling the management of a database facilitating the interpretation of data collected at different locations and/or under different conditions. Thirdly, the monitor must be able to monitor fouling quantitatively by (1) operational parameters like the pressure drop and by (2) in-situ, real-time and non destructive observations (visual, microscopic and other methods) and by (3) analysis of coupons taken from the monitor (see Table 3.1).

A first prototype for such a monitor, named the Flat Sheet Monitor (FSM), was developed by Vitens in collaboration with Kiwa. The FSM is made of stainless steel and Perspex (cover lid), the external dimensions are $0.32\,m \times 1.03\,m \times 0.07\,m$ (Figure 3.1). The FSM has an effective membrane length slightly shorter (0.90 m) than the membrane length in a membrane element, enabling the use of coupons taken from spiral-wound membrane elements. The pressure drop can be monitored over the total FSM membrane length (0.90 m) and over the first and second half of the FSM (Figure 3.1).

Comparison of the specifications of the prototype monitor with spiral-wound membrane elements showed the same spacer channel height (Table 3.1). The monitor also contains the same materials as membrane elements (membrane and feed spacer) and can be operated with the same linear flow velocity. Dosage of a $KMnO_4$ solution (purple colour) to the feed side of the monitor showed that the flow was equally distributed over the monitor width (Vrouwenvelder *et al.*, 2007a). The relationship between linear flow velocity and pressure drop was determined over a spiral-wound membrane element in a single element test rig and over the prototype monitor. The relationship was also calculated using the friction coefficient described by Schock and Miquel (1987). For the spiral-wound membrane element, measured data fitted very well with calculated data (Figure 3.2A). However, the prototype monitor showed a distinct difference between measured and calculated data (Figure 3.2B), indicating that the monitor is not representative for spiral-wound membrane elements. The relationship between linear flow velocity and pressure drop was determined with a series of measurements with increasing and decreasing flow rates, which provided the same data. Possible explanations for the differences between the measured and calculated data may be a short circuiting and/or variations of the spacer channel height in the prototype monitor at different linear flow velocities. Dosage of a biodegradable compound to the feed water of the prototype monitor resulted in a pressure drop increase over the monitor (Figure 3.3A). Measurements over the first and second half of the monitor (Figure 3.3B) showed that the pressure drop increased predominantly over the first half of the monitor (0 to 0.45 m). Analysis of membrane coupons from the monitor showed that most biomass was present in the first half of the monitor. A suitable monitor must be representative for spiral-wound membranes, requiring an identical relationship between linear velocity and pressure drop. Hydrodynamics influence the biofilm structure and behaviour (Purevdorj-Gage and Stoodley, 2004). Therefore, instead of the size (length), the hydrodynamics of the monitor must be identical to spiral-wound membrane elements. In practice, biofouling is usually observed on the feed side of the membrane installation. Because of the hydrodynamics and the localization of biofouling, a smaller monitor was developed.

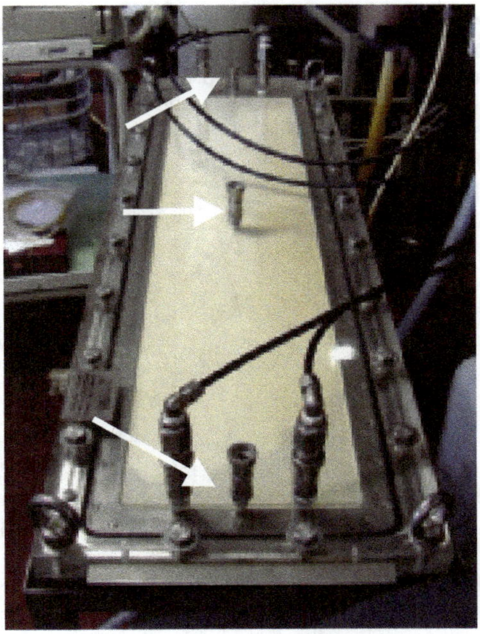

Figure 3.1 Side view of Flat Sheet Monitor (FSM, external dimensions of 0.32 m × 1.03 m × 0.07 m) with a transparent cover lid. The pressure drop can be measured over the length of the FSM (arrows) using quick connectors. Water was transported to and from the FSM by black tubes

The objective of this study was to develop a small all-round applicable tool for (bio)fouling prediction and control, the Membrane Fouling Simulator (MFS).

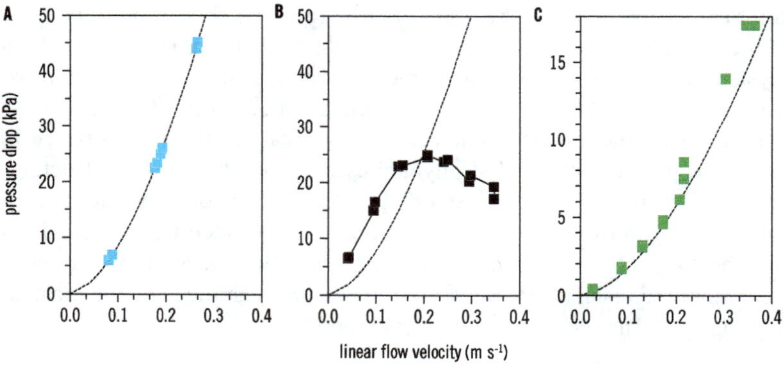

Figure 3.2 Relation between linear velocity (m s^{-1}) and pressure drop (kPa) in a spiral wound membrane element in a test rig (**A**), the FSM (**B**), and the MFS (**C**). The markers represent measured data and the dotted line represent calculated data using the function described by Schock and Miquel (1987)

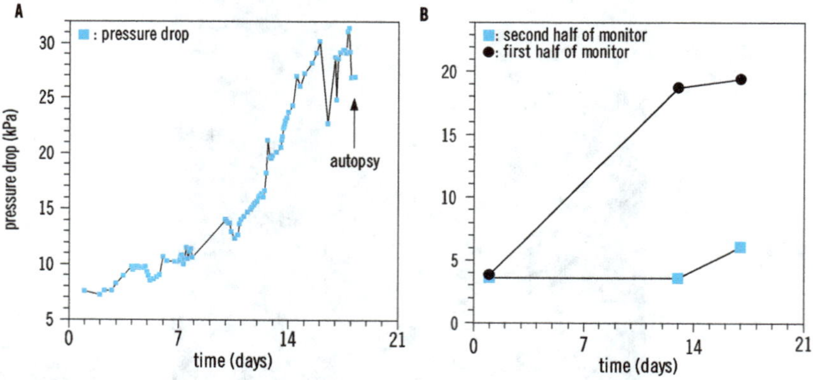

Figure 3.3 Development of pressure drop in time over the length of the Flat Sheet Monitor (A) and of the first and last half of the FSM (B)

MATERIALS AND METHODS

Membrane Fouling Simulator

Based on the experience obtained from the FSM, an adapted, smaller and cheaper monitor, the Membrane Fouling Simulator (MFS), was developed (Figure 3.4A). The external dimensions of the MFS are $0.07\,m \times 0.30\,m \times 0.04\,m$.

The MFS has an effective membrane length of 0.20 m, and a width of 0.04 m. So the MFS is simulating the fouling phenomena occurring in the first 0.20 m of membrane elements. Consequently, the water flow needed for testing (15–25 $L h^{-1}$) and chemical use is low. The MFS is constructed of two linked stainless steel templates containing coupons of membrane, feed and product spacer (Figure 3.4B). The height of the feed spacer channel and the product spacer channel is 0.80 mm and 0.25 mm, respectively. These dimensions are based on reported data (Schock and Miquel, 1987 and Van der Meer, 2003a) and analysis of spacers taken from membrane elements. The MFS is equipped with connectors for separate feed, concentrate and product flow. Also, connectors are present for measuring the feed pressure and differential pressure (pressure drop) over the length of the MFS on both the feed and product side of the membrane. A transparent window enables the performance of (microscopic) observations of the spacer and membrane during operation. The thickness of the window enables the use of a microscope for fouling thickness detection in the micrometer range. During operation, the glass window was covered with a light tight lid to prevent growth of phototrophic organisms.

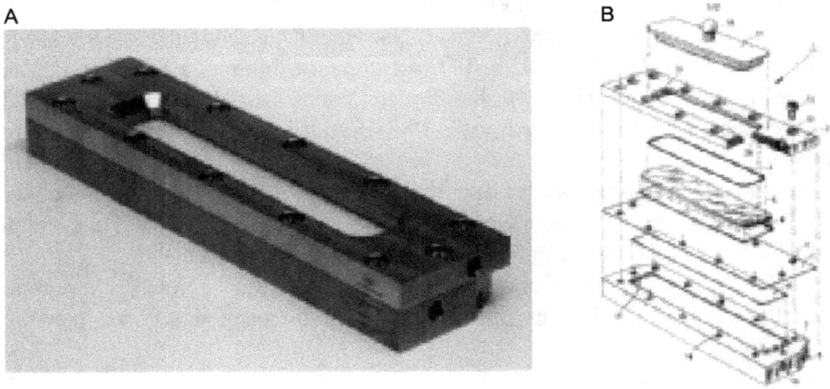

Figure 3.4 (**A**) Side view of the developed Membrane Fouling Simulator (MFS, external dimensions of $0.07 \times 0.30 \times 0.04$ m) with a transparent window. (**B**) Scheme of the MFS components (patent NL 1028474)

Membranes and spacers

New and unused spiral-wound nanofiltration membrane elements (Trisep TS80, 4 inch diameter) were used for the test rig studies and for preparation of membrane and spacer coupons for the MFS. The full-scale installation contained RO membrane elements (DOW Filmtec BWROLE-440, 8 inch diameter).

Experimental set-up for operation/monitoring of MFS and test rig

The MFS installation was fed with drinking water and subsequently consisted of two cartridge filters (10 and 1 μm pore size), two pressure reducing valves, manometer, dosage point (for biodegradable compound), MFS and flow controller. The feed flow in the MFSs was maintained at $16\,L\,h^{-1}$ (linear flow velocity of $0.16\,m\,s^{-1}$) during the experiments with and without dosage of biodegradable compounds. Note: In practice, in lead membrane elements linear flow velocities are applied in the range of 0.07 to $0.20\,m\,s^{-1}$ (Vrouwenvelder *et al.*, 2009a). The MFS was operated at a pressure of ~120 kPa to avoid degassing. The MFS was operated in cross flow, without permeate production. Drinking water prepared from ground water (treated by aeration, rapid sand filtration) was used as feed water of the MFS installation and the single element test rig. The water had a DOC concentration of $1.9\,mg\,L^{-1}$ and a constant temperature of 13°C.

Chemicals were dosed to the MFS feed water to enhance biomass growth. From a sterile vessel, containing a 5 L solution of concentrated substrate, substrate was dosed into the feed water prior to the MFS by a peristaltic pump (Masterflex) at a flow of $0.03\,L\,h^{-1}$. The dosage of substrate was checked periodically by measuring the weight of the dosing bottle. The chemicals $NaCH_3COO$, $NaNO_3$ and NaH_2PO_4 were dosed with a ratio C:N:P of 100:20:10. C was the growth limiting compound. N and P were dosed to eliminate growth limiting conditions caused by potentially low N and P concentrations in the feed water. The substrates were dissolved in ultrapure water. To restrict bacterial growth in the substrate dosage bottle, pH was set to 10.5 by dosing NaOH. Dosage bottles were replaced every 5 days. The substrate dosage ($0.03\,L\,h^{-1}$) was low compared to the feed water flow rate ($16\,L\,h^{-1}$). Thus, the pH of the feed water was not measurably influenced by substrate dosage.

During assessment of the relationship between linear flow velocity and pressure drop, the membrane element in the test rig was operated without permeate production.

A full-scale installation with persistent fouling problems (elevated pressure drop) was selected for a comparison study. In parallel to the full-scale installation, the MFS and a test rig were fed with surface water after pretreatment with ultrafiltration and dosage of a scale-inhibitor ($3\,mg\,L^{-1}$). No biodegradable compounds were dosed to the water. During these comparison studies, the membrane elements in the test rig and the full-scale installation were operated with production of permeate (recovery of 8%). The MFS and the test rig were operated with the same linear flow velocity.

A differential pressure transmitter (Labom) was used to determine the pressure difference over the feed spacer channel of the MFS and the membrane element in the test rig. The differential pressure transmitter was connected to the

MFS and test rig with quick connectors. Using the transparent window of the MFS, the development of the fouling in time was observed with a microscope (Leica Wild M10) equipped with a digital camera (Canon S70).

Sampling and analysis of membrane coupons

To determine the accumulated fouling, sections of membrane, feed and product spacers were taken from the MFS and membrane elements. The sections ($4\,cm^2$) were placed in 10 mL of autoclaved tap water in a capped tube. To determine the amount of biomass, the tubes with the membrane sections were placed in an ultrasonic cleaning bath (Bransonic, model 5510E-DTH, output 135 Watts, 42 kHz). The low energy sonic treatment (2 minutes) followed by mixing on a Vortex (few seconds) was repeated four times. Next, volumes of water collected from the tubes were used to determine biomass parameters (ATP and TDC).

Active biomass was determined in duplicate by measuring the adenosinetriphosphate (ATP) concentration from 100 μL samples (Holm-Hanson and Booth, 1966). The luminometer (Celcis, model Advance) added 100 μL of LuminEX-B reagent (Celcis) to a sample to release ATP from the bacterial cells. Subsequently, 100 μL of LumATE-PM (Celcis) was added for light production. The amount of light produced was measured with the luminometer (relative light units, RLU). The concentration of ATP was derived from the RLU values using the conversion factors of the linear relationship between RLU values and reference ATP concentrations. The ATP concentrations of autoclaved drinking water ($<1\,pg\ ATP\ mL^{-1}$) and two control solutions (2 and $100\,pg\ ATP\ mL^{-1}$) were determined as quality control.

Total Direct Cell counts (TDC values, cells cm^{-2}) were performed with epi-fluorescence microscopy using acridine orange as fluorochrome by applying a slightly adapted method to eliminate fading (Hobbie *et al.*, 1977). The slides were examined with a Leica DM RXA microscope equipped with a 100 W power supply and phase and filter sets (I2/3) appropriate for acridine orange. Microorganisms were enumerated through a 100x/1.30 PL Fluotar oil immersion objective. Non-fluorescence immersion oil (Olympus, code 35505) was used because of low background fluorescence. A total of 10 random fields were counted.

RESULTS

Relationship between linear velocity and pressure drop

The relationship between linear flow velocity and pressure drop over a single element test rig containing a spiral-wound membrane element and a MFS containing coupons of membrane and spacers taken from the same type of membrane element was calculated, using the methods described by Schock and Miquel (1987). Mathematically, the pressure drop is expressed as

$$\Delta p = \lambda \cdot \frac{\rho \cdot v^2}{2} \cdot \frac{L}{d_h} \tag{3.1}$$

where λ is the friction coefficient, ρ the specific liquid density, v the linear velocity, L the length of the membrane or MFS and d_h the hydraulic diameter. The length is only difference between the membrane element and the MFS with respect to equation 3.1. The friction coefficient as given by the correlation function by Schock and Miquel (1987):

$$\lambda = 6.23 \cdot \mathrm{Re}^{-0.3} \tag{3.2}$$

Where Re is the Reynolds number.

The measured linear flow velocity and pressure drop for both the membrane element and the MFS fitted very well with the calculated data (Figure 3.2AC). The sudden small jump in pressure drop in the MFS near a flow velocity of $0.22\,\mathrm{m\,s^{-1}}$ is caused by the use of another – less accurate – type of flow meter at flow velocities exceeding $0.20\,\mathrm{m\,s^{-1}}$. Evidently, the MFS and the membrane element had similar spatial dimensions (height of the feed spacer channel), resulting in an identical relation between linear flow velocity and pressure drop.

Flow distribution

The flow distribution in the MFS was determined by dosage of a solution containing $KMnO_4$ to the feed water and observations using the sight glass of the MFS. The front of the purple coloured $KMnO_4$-solution was equally distributed over the width of the MFS during transport of water trough the MFS, illustrating a plug flow regime. A plug flow regime was also observed in spacer filled feed channels of spiral-wound membrane elements by residence time distribution measurements using a step change in saline concentration (Van Gauwbergen and Bayens, 1997).

Sensitivity for fouling detection

In practice, biodegradable compounds in feed water of membrane installations have caused biofouling (Griebe and Flemming, 1998 and Vrouwenvelder *et al.*,

2000). Therefore, a biodegradable compound (acetate) was dosed to the feed water of the MFS to enhance biofouling. A concentration of 0.100 mg acetate-C L^{-1} in the feed water of a MFS resulted in a strong increase of the pressure drop over the MFS within 6 days, whilst the pressure drop over two MFSs without dosage remained constant (Figure 3.5A). In-situ visual observations made during operation of the MFS showed clear differences in accumulation of fouling in the MFSs with and without dosage of acetate (Figure 3.6). At the end of the research period the MFS units were opened for analyses of the accumulated material. On the membrane and feed spacer taken from the MFS with substrate dosage, higher concentrations of active biomass (ATP) and numbers of cells (TDC) were found when compared to the biomass concentrations on samples from the MFSs without dosage (Figure 3.5B, ATP data).

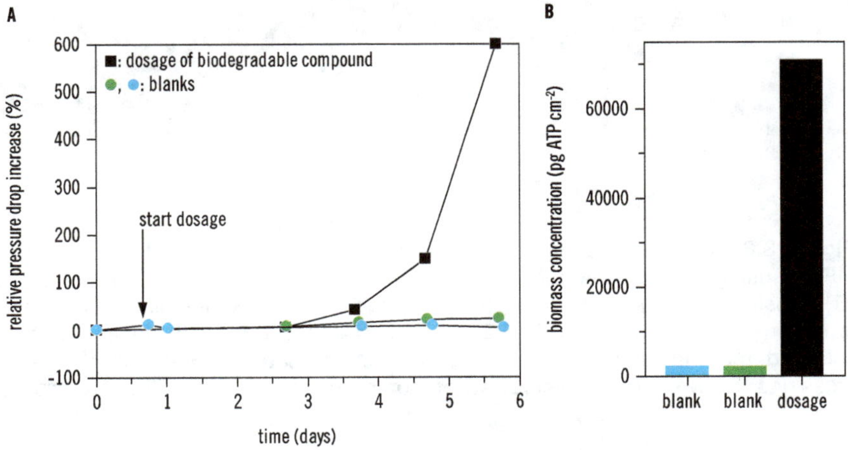

Figure 3.5 Pressure drop in time (**A**) over the MFS with and without dosage of a biodegradable compound (0.100 mg acetate-C L^{-1}) to the feed water of the MFS. Biomass concentration (**B**) on the membrane and spacer in the MFS after 6 days of MFS operation

This experiment illustrates that the MFS visualized fouling very well, in a distinctive way. In the MFS with dosage of a substrate, the pressure drop, the visual observations and the biomass concentration during analysis of the accumulated material univocally indicated a higher degree of fouling when compared to the MFSs without dosage.

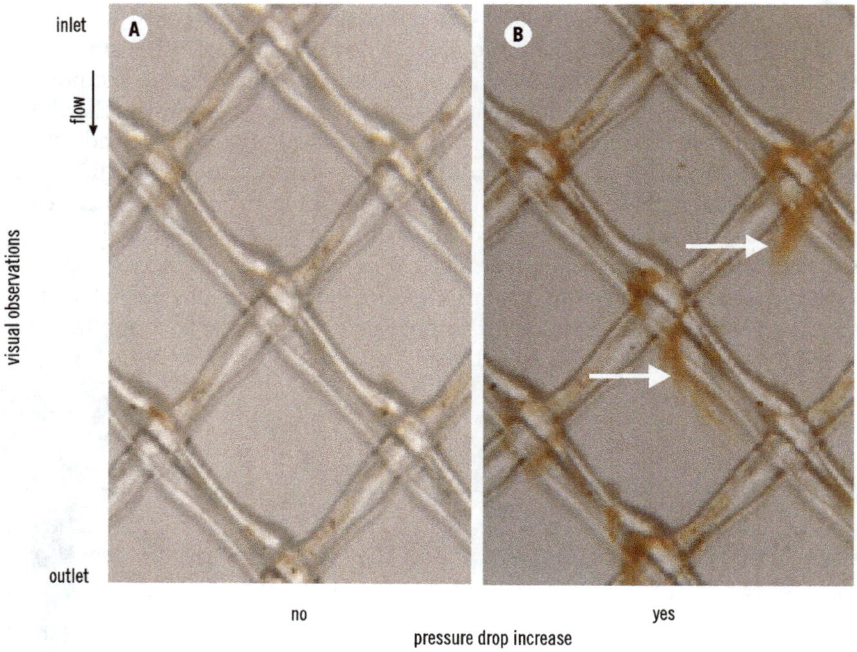

Figure 3.6 Visual on-line observations of the membrane and feed spacer in the MFS – using the transparent window during operation – with (**A**) and without (**B**) substrate dosage (after run time of 6 days). The flow direction is from top to bottom. The original scale is 6.2 × 10.7 mm. In the MFS with dosage, darker coloured material and the presence of clusters material in the feed spacer crossings (arrows) were observed when compared to the MFS without substrate dosage

Reproducibility of MFS experiments

In a parallel study with three MFSs, a biodegradable compound (0.200 mg acetate-C L^{-1}) was dosed to the feed water of each MFS. The pressure drop over the three MFSs increased within 4 days (Figure 3.7A). Comparison of the three MFSs showed a similar pressure drop at the start of the experiment and a similar development of the pressure drop in time. The biomass concentrations determined in the MFSs at the end of the research period were not significantly different (Figure 3.7B). A similar pressure drop at start-up of the experiment illustrates that the pressure drop was not affected by the individual coupons of the membrane and spacer in the MFSs. A parallel study without dosage of a biodegradable compound also showed a similar development of the pressure drop in time (Figure 3.5A, blanks) and similar biomass concentrations at the end of the research period (Figure 3.5B) in the MFSs. The biomass concentrations were equally distributed over the width of the MFS.

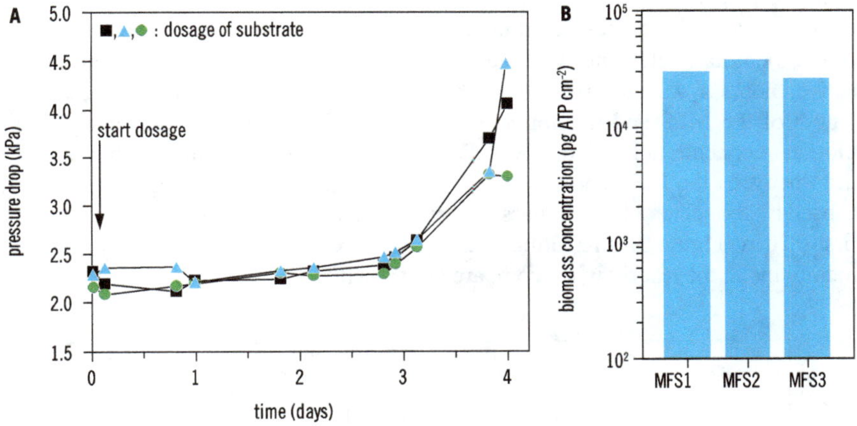

Figure 3.7 Pressure drop in time (**A**) over three MFSs operated in parallel with dosage of a biodegradable compound (0.200 mg acetate-C L^{-1}) to the feed water of each MFS. Biomass concentration (**B**) on the membrane and spacer in the MFSs determined after 4 days of MFS operation

It can be concluded that the MFS can provide reproducible data concerning pressure drop and biomass concentration.

Comparison of fouling in MFS and membrane elements

The biomass concentrations found during the analysis of the membranes from the MFS (Figures 3.5B and 3.7B) are within the range as observed with destructive studies on membrane elements taken from pilot and full-scale installations with and without operational problems (Ridgway *et al.*, 1983 and Vrouwenvelder and Van der Kooij, 2003a). However for evaluation of the representativeness of the MFS for fouling development in membrane elements, comparative studies are needed.

An MFS and a test rig were operated in parallel with a full-scale installation. The feed channel pressure drop over the MFS was lower than the pressure drop over the membrane element at the start of the study (Figure 3.8A). The pressure drop over the MFS increased more rapidly in time than over the element (Figure 3.8A). In order to compare the pressure drop over the MFS and the membrane element in the test rig, the pressure drop over the membrane element was corrected for the MFS length, using a factor 4.8 for the difference in length (length of element = 0.96 m/length of MFS = 0.20 m). The pressure drop over the MFS and the membrane element in the test rig corresponded closely for 17 days (Figure 3.8B). Thereafter, a more rapid increase of the pressure drop was observed over the MFS when compared to the membrane element. At the end of the research period, the concentration of active biomass was determined in the MFS and membrane element. Over the length of the MFS and membrane element, a decline was observed of the active biomass concentration (Figure 3.9). The difference in development of the pressure drop between the MFS and the membrane element from the test rig is most likely caused by the relative low biomass concentration in the membrane element after the first 0.20 m (Figure 3.9) resulting in a relative low pressure drop increase over the membrane in the test rig when compared to the MFS.

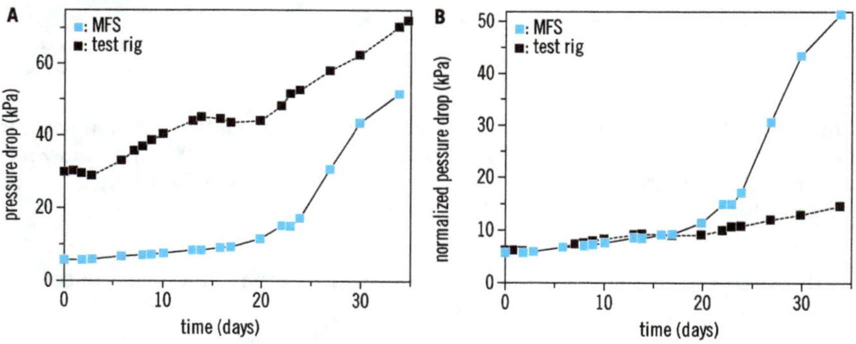

Figure 3.8 Pressure drop in time (**A**) over the MFS and membrane element (test rig) that were operated in parallel. The normalized pressure drop in time (**B**) over the membrane element corrected for the MFS length

A lead membrane element taken from a full-scale installation – supplied with the same water as the MFS and test rig – contained about the same concentration and distribution of active biomass over the length of the element as the MFS and the element from the test rig (Figure 3.9). Since the pressure vessel of the full-scale installation contained six membranes in series, no data were available on the pressure drop over the lead membrane element. The full-scale installation suffered from fouling, which was detected with the MFS within 30 days (Figure 3.8).

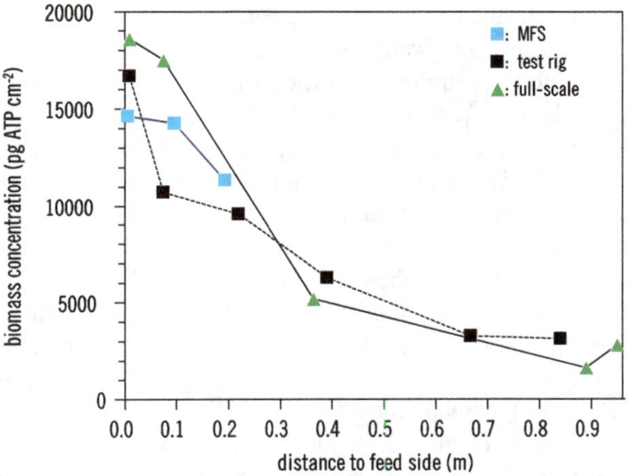

Figure 3.9 Biomass concentration (pg ATP cm^{-2}) over the length of the MFS and membrane modules from test rig and full-scale installation

Evidently, the MFS showed the same development of pressure drop in time and the same concentration of biomass as in spiral-wound membrane elements. The MFS enables to detect biofouling at an early stage, since biomass predominantly accumulated on the first 0.20 m of the lead membrane elements (Figure 3.9).

DISCUSSION

Evaluation of the Membrane Fouling Simulator

The MFS, a new tool for fouling prediction and control, is presented. The MFS is representative for spiral-wound membrane elements used in practice with regard to the materials used (membranes and spacers), spatial dimensions (height of feed and product spacer channels) and the hydraulics (pressure drop and flow distribution). As clearly demonstrated using the MFS, fouling can be quantified and characterized by (1) the pressure drop, (2) in-situ, real-time and non-destructive (visual, microscopic) observations using a transparent window and (3) analysis of membrane coupons sampled from the MFS. A comparison study of the MFS and membrane elements showed the same development of pressure drop in time and the same biomass accumulation. The small scale of the

MFS makes it easy to handle and minimizes the costs of investment, operation and use of chemicals and water. Comparison of the characteristics of the developed MFS with the requirements for an ideal monitor (Table 3.1) indicates that the MFS fulfils most requirements, illustrating the potential to be a suitable tool for fouling prediction and control. The characteristics of the MFS clearly discriminate the MFS from other available tools (Flemming, 2003).

The biomass concentrations found during the analysis of the membranes and spacers from the MFS (Figures 3.5B, 3.7B and 3.9) are within the range observed by destructive studies on membrane elements taken from pilot and full-scale installations with and without operational problems. The comparative study of the MFS and membrane elements in a test rig and a full-scale installation showed the same fouling (Figure 3.9), but a more rapid development of the pressure drop in the MFS (Figure 3.8A), explained by the distribution of biomass over the length of the membrane elements (Figure 3.9). The MFS was able to detect fouling earlier when compared to the pressure drop development over the membrane elements/ installation. All data collected during the studies with the MFS and membrane elements were in agreement with each other and clearly indicate the potential of the MFS to be a suited tool for fouling prediction and control.

More parallel studies of the MFS and spiral-wound systems are needed for evaluation of the representativeness of the fouling development in the MFS. The maximal feed pressure of the MFS is about ~150 kPa, because of the presence of the transparent window. A flux can be obtained in the MFS by a difference in osmotic pressure over the membrane, which requires the flow of a saline solution along the product side of the membrane. Adaptation of the MFS with a thicker transparent window, enables to operate the MFS at higher pressures facilitating a pressure driven flux and the possibility to determine the salt passage. This monitor is under development and is expected to satisfy all requirements (Table 3.1) of a monitor for fouling control.

Water types with a high membrane element fouling potential where rapidly and easily identified with the MFS. The duration of the studies varied between 4 to 35 days (Figures 3.5A, 3.7A and 3.8).

The sensitive pressure drop measurement as well as the observations using the transparent window seems appropriate parameters for fouling monitoring. With the MFS, in-situ identification of the type and quantification of the extent of fouling may be possible by for example dosage of a specific (colouring) tracer to the feed water of the MFS and the use of the transparent window. Obviously, the MFS can also be combined with other fouling detection methods.

The supplementary material at the end of this chapter contains an update (September 2009) on monitor development and use.

Potential fields of application for the Membrane Fouling Simulator

Model studies using the MFS may lead to the development of membrane systems less susceptible to fouling. The effects of parameters that may influence fouling can be studied systematically enabling the selection of key parameters related to fouling. Important parameters are e.g. substrates and micro-organisms in the feed water, hydrodynamic conditions (velocities), flux, spacers and membranes. The research can subsequently focus on the main important parameter(s) to control fouling. Additionally, newly developed spacers and membranes can be tested and alternative set-ups for membrane installations and methods of operation can be evaluated and optimized.

The fouling potential of water can be determined and evaluated with the MFS. Moreover, the MFS is suited for evaluation and comparison of different pretreatment schemes (Figure 3.10A) and evaluation and comparison of the fouling potential of chemicals dosed to the water prior to membrane filtration (Figure 3.10B). These studies can be used for selection or adaptation of the water source, pretreatment and chemicals dosed in order to minimize the risk of fouling. Scale-inhibitors can differ strongly in bacterial growth potential, the MFS can be used to select the chemical with a low fouling potential. The MFS can be connected to the concentrate, as a scaling monitor. Also, the MFS can be connected to the product for quality control. Parallel studies with MFSs containing different types of membranes and/or spacers may be suited to select membrane elements less susceptible for fouling.

Cleaning in combination with pretreatment may be the – cost effective – way to control fouling. The MFS can be used to evaluate the effects of biocide dosage, preventive and curative cleanings and to develop strategies to control fouling at low costs with a low impact on the environment (e.g. no overdose of biocide, discharge regulations). Identical fouling is obtained in parallel operated MFSs (Figures 3.5B and 3.7B). Evidently, the influence of different fouling control actions can be assessed by using individual MFSs (Figure 3.10C).

Several fouling mechanisms can be studied individually, simultaneously and systematically with the MFS. The obtained information can be used for a better selection of pretreatment to overcome (site specific) fouling. Also, the contribution and interactions of individual fouling types in relation to practical fouling problem can be unravelled. The effect of pretreatment on biofouling, particulate and organic fouling (also named NOM-fouling) can be determined with one set-up, using four MFSs in parallel (Table 3.2, Figure 3.10C). Differences between the individual MFSs with respect to production, use of membrane and biocide dosage enable the discrimination of individual fouling types.

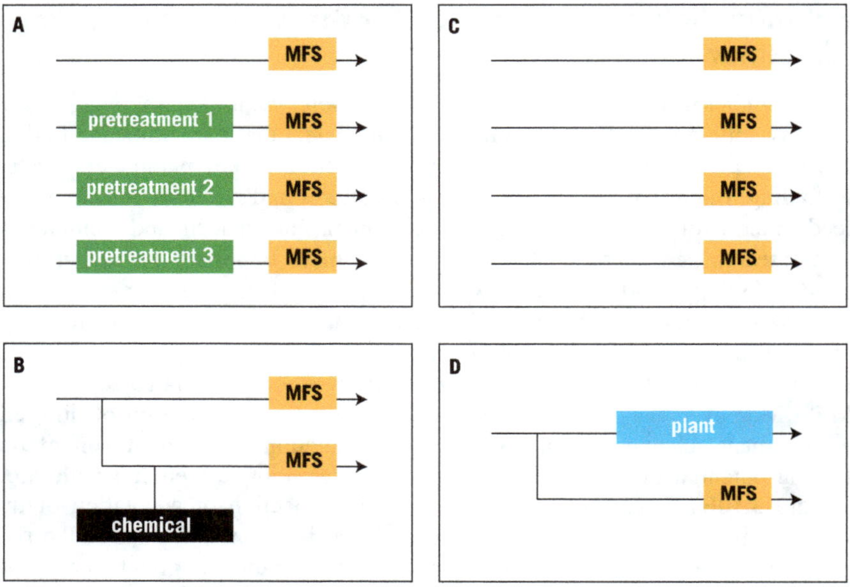

Figure 3.10 Schematic overview of potential use of MFS, **A** = fouling potential of water, evaluation and selection of pretreatment, **B** = chemical dosage, **C** = individual fouling studies, fouling control and cleaning studies, membrane selection and model studies, **D** = early warning system

Table 3.2 Setup for assessment of the contribution of individual fouling mechanisms using four MFSs in parallel

MFS	Fouling	Pretreatment	Production	MFS content	Dosage of biocide	Analysis
1	biofouling	UF	no	spacer	no	ΔP
2	particulate fouling	no	no	spacer	yes	ΔP
3	NOM	ILC/UF	yes	spacer + RO membrane	yes	MTC
4	all fouling types	no	yes	spacer + RO membrane	no	ΔP + MTC

UF = ultrafiltration; ΔP = pressure drop; ILC = in line coagulation; MTC = normalized flux.

An early warning system for fouling in spiral-wound membrane system requires early detection of fouling which enables to take timely and directed actions. The MFS can be connected to the feed water of a system (Figure 3.10D). The MFS has multiple methods to detect fouling quickly and accurately. For example, measuring a parameter like the pressure drop – with a sensitive pressure analyzer – over the MFS may provide a simple early warning system that can give an alarm when the pressure drop exceeds a value. The use of the transparent window and the destructive study of membrane coupons from the MFS also facilitate early detection of fouling.

DEVELOPMENT OF A SET OF NEW MONITORS

In practice, a clean water flux is determined for evaluating the effect of cleanings, indicating the need to develop a monitor with pressure driven permeate production. Recent research by Herzberg and Elimelech (2007) and Chong *et al.* (2008) suggest that biomass enhanced osmotic pressure may influence membrane performance by reducing the flux and increasing the salt passage. The effect of biomass enhanced osmotic pressure is reported to be stronger than the reduction in trans membrane frictional resistance. In their studies, flow cells were used containing membranes but no feed spacers. It is likely that their findings would have been different when feed spacers had been present. This underlines the need for a *representative* monitor able to study in RO/NF systems the influence of fouling on *all* performance indicators: (i) increase in pressure drop across membrane elements (feed-concentrate channel), (ii) decrease in membrane permeability, (iii) increase in salt passage. Therefore, a monitor with pressure driven permeate production named the HP-MFS is being developed (Figure 3.11E). This high pressure monitor can also be used as a portable scaleguard. The HP-MFS has no sight glass due to pressure operation up to 6,000 kPa (60 bar).

The metal MFS made of 316 stainless steel with a product spacer channel (Figures 3.4 and 3.12B), used in the studies described in this chapter, has been adapted. The adapted metal MFS is made of SMO254 steel to enable seawater biofouling studies and the product spacer channel was removed; the SMO254 MFS scheme is shown in Figure 3.12C.

Nuclear Magnetic Resonance (NMR) is a powerful tool to study membrane biofouling, since it is a quantitative non-invasive measurement technique. Magnetic materials can not be used since magnetic material influence the NMR signal. Because of the interference of the metal MFS with the NMR signal, a PVC version of the MFS had to be developed. In order to fit in a 200 MHz super wide-bore magnet, the MFS dimensions had to be scaled down. The small

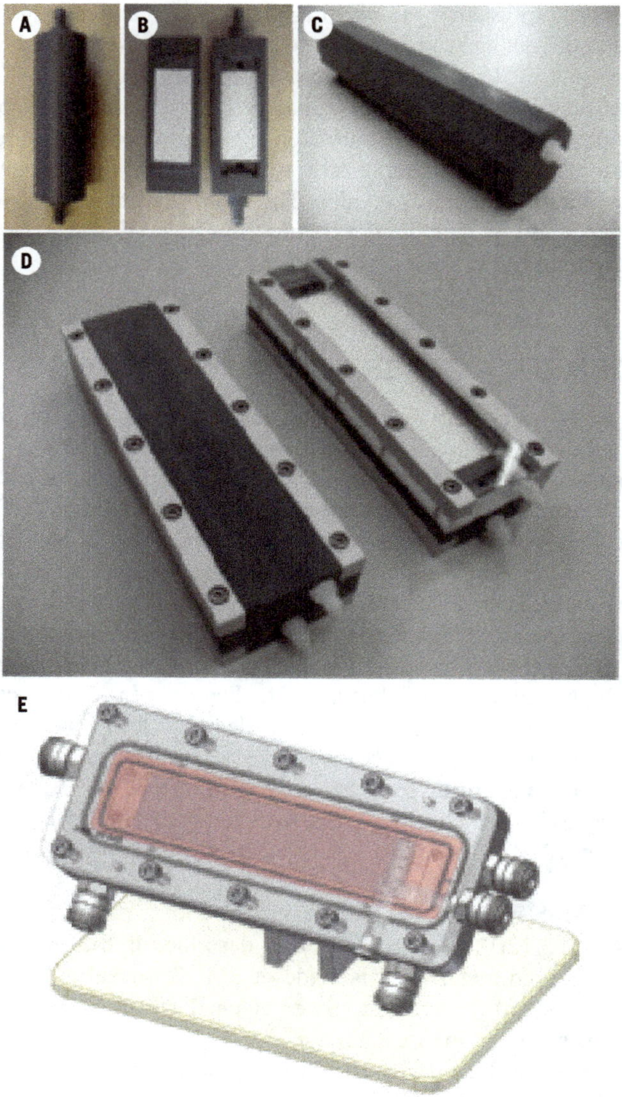

Figure 3.11 Selection of MFSs recently developed and under development: Small PVC monitor for NMR studies (**A,B**), PVC monitors with similar dimensions as the steel MFS (**C,D**), Steel monitor suitable for pressure driven permeate production at feed pressures up to 6,000 kPa (**E**). The spacer and membrane sheet sizes are 1.6 cm × 4.0 cm for (A,B) and 4.0 cm × 20.0 cm for (C,D) and about 4.0 cm × 20.0 cm for (E).

MFS has feed channel dimensions of 0.00077 m × 0.016 m × 0.040 m (Figure 3.11AB). In the plastic MFS, the water flows through the spacer placed between two membrane sheets. This MFS was produced with a manufacturing accuracy of 20 μm. The PVC MFS was found suitable for biofouling studies (Graf von der Schulenburg *et al.*, 2008 and Vrouwenvelder *et al.*, 2009c). Additionally, a larger PVC version of the MFS was made for NMR studies in larger diameter magnets (Figure 3.11C), with feed spacer and membrane sheet dimensions of 0.04 m × 0.200 m; the same dimensions as the metal MFS (Figure 3.4). The PVC MFSs suitable for NMR studies are closed using glue and therefore single use only (Figure 3.11ABC).

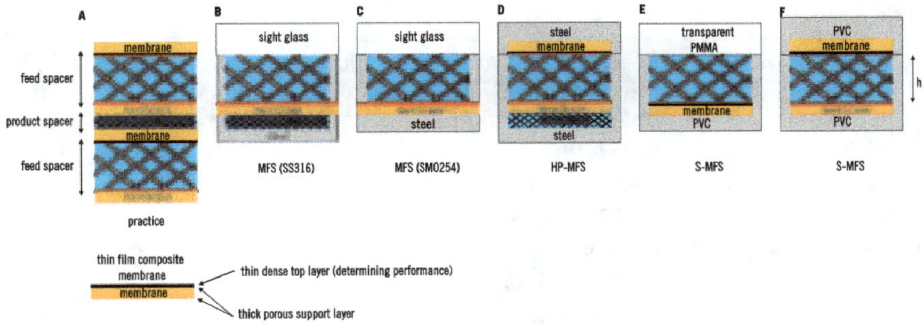

Figure 3.12 Scheme of monitors illustrating differences. Spiral wound membrane module scheme (**A**) is shown for comparison. MFS (**B**, Figure 3.3) with sight window and HP-MFS (**D**, Figure 3.11E) for pressure driven permeate production. Plastic S-MFS with transparent (**D**) and PVC-top (**E**, Figure 3.11D). S-MFS (**D** and **E**) can be made quickly with varying flow channel heights (h). The flow cells shown in Figure 3.11ABC are variations on the S-MFS (**E**), used for NMR studies.

Metal MFS (Figure 3.4) operation caused in some cases sweating of water at the edge of the top and bottom MFS halves, were the membrane is localized (Figure 3.12BC). Membranes should be placed with the thin top layer at the feed spacer side in the MFS. Generally, placement of a new membrane and spacer sheet and/or use of new screws and/or O-rings in the MFS solved the leakage problem. O-ring replacement is recommended when water sweats near the MFS sight window. These sweating problems were observed at the start of metal MFS studies.

A plastic MFS, named spacer-MFS (S-MFS), was developed to improve the metal MFS and to enable biofouling studies with varying spacer thicknesses.

The S-MFS can be used with and without sight window (Figures 3.11D and 3.12EF). The spacer and membrane sheet dimensions (0.040 × 0.200 m) and hydrodynamic behaviour are the same as the metal MFS (Figure 3.4). The S-MFSs has been in use for studies since early 2008 without any reported leakage problems. Several companies, research institutes and universities use the S-MFS. The S-MFS can be made quickly with varying flow channel heights.

An overview of the monitor types is given in Table 3.3 and a scheme of the individual monitor characteristics is presented in Figure 3.12.

Table 3.3 Monitor types and key characteristics[a]

Monitor name	Material	Key characteristics	Figure	Scheme
MFS	SS316	Feed spacer channel pressure drop. Flux obtained by osmotic pressure.	Figure 3.4	Figure 3.12B
MFS	SMO254	Feed spacer channel pressure drop.	Figure 3.4	Figure 3.12C
S-MFS	PVC	For spacer studies	Figure 3.11A–D[b]	Figure 3.12EF
HP-MFS[c]	SMO254	Pressure driven permeate production up to 6,000 kPa. Intended use is scaleguard.	Figure 3.11E	Figure 3.12D

[a]*All monitors are applicable for feed spacer channel pressure drop measurement;*
[b]*Some monitors are suitable for NMR studies (Figure 3.11ABC);*
[c]*Adjustment of flow channel height may require use of inserts. The HP-MFS is being developed.*

MFS operation

Thin film composite membranes are composed of multiple layers, consisting of a thin dense top layer and a thick porous support layer. The thin top layer should face the feed spacer in the MFS (Figure 3.12). The top layer membrane side is usually smoother and more shiny than the support layer side. The membrane top layer may show the feed spacer print when the membrane is taken from a membrane element.

A scheme of equipment suggested for operation of the (S-)MFS is presented in Figure 3.13. Use of a cartridge filter (1–10 μm pore size) to reduce fouling caused by the water may be considered for comparison studies in the laboratory.

Use of cartridge filters may result in a biofilm that is less visible during in situ sight window observations.

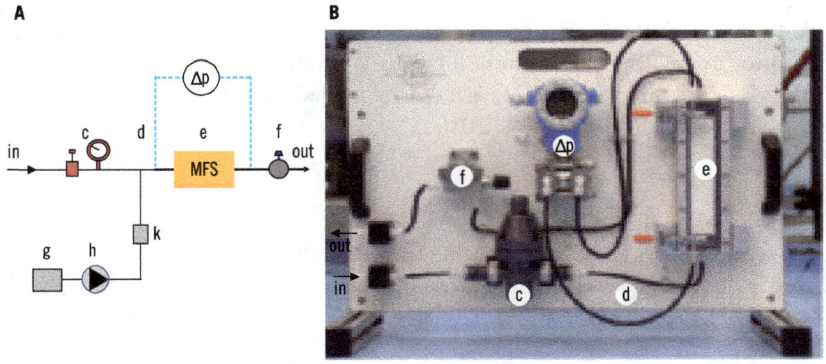

Figure 3.13 Installation for operation of the (S-)MFS. Scheme (**A**) and picture of (S-)MFS unit (**B**). The feed water (in) pressure is reduced using a pressure reducer (c). Chemicals can be dosed (d) to the feed water of the (S-)MFS. The pressure drop over the MFS (e) is measured with a sensitive accurate differential pressure transmitter (Δp). The flow is regulated using a flow controller (f). Chemical dosage requires a chemical reservoir (g), pump (h) and a pressure relief valve (k) to prevent back flow

The feed pressure for a steel MFS should be below ~150 kPa to reduce the risk of breakage of the glass window. A feed pressure of 200 kPa (2.0 bar) for the S-MFS had not caused leakage problems (the maximum feed pressure for the S-MFS is 400 kPa).

Dosing of chemicals to the MFS feed water require a pressure relief valve to avoid backflow to the dosage vessel and have accurate chemical dosing. Chemical dosage should be calibrated at the operating pressure and water flow rate. The pressure relief valve pressure should be set higher than the feed pressure and lower than the chemical dosage pump capacity.

Use of one pressure transmitter for each MFS is recommended over one pressure drop transmitter for several MFSs. When using several MFSs in combination with one differential pressure transmitter than a valve system for measuring the pressure drops is strongly preferred over using quick connectors. Connecting and disconnection quick connectors may affect the accumulated biomass in the MFS.

Pressure differential transmitter type PT4 in Table 4.1 in chapter 4 (shown in Figure 3.13B) has been found suitable for sensitive accurate biofouling monitoring.

Mixing of the dosed chemical with feed water may be a point of attention. The feed flow rate, chemical flow rate, pump, tube diameter, and use of static mixer should be considered.

The flow controller may foul, resulting in a reduction of the linear flow velocity in the (S-)MFS. Dosage of chlorine ($\sim$1 mg Cl_2 L^{-1}) to the MFS outlet water prior to the flow controller will reduce biofouling of the flow controller.

An MFS system with remote access to logged data like pressure drop has been used (Vrouwenvelder *et al.*, 2010c).

MFS use

Since developing the first MFS in 2006, over 183 (S-)MFSs are in use by a broad range of companies, research institutes and universities all over the world (Table 3.4). A selection of MFS systems during (on site) studies is shown in Figure 3.14. The use of the monitors enables systematic biofouling studies. Therefore, a breakthrough in biofouling control is anticipated for the near future.

Table 3.4 Overview of monitor use (incomplete)

Aspects	Details	No. of studies
Prevention and control	– early warning system	9
	– chemical dosage (antiscalants, biofouling inhibitors)	6
	– cleanings (strategies, efficiency and	6
	minimization of chemical use)	4
	– other	
Feed water	– monitoring	11
	– water fouling potential	10
	– pretreatment comparison/selection	7
Materials	– membrane and spacer tests	5
Model studies	– relation fouling accumulation and membrane performance	2
	– influence of operational/environmental parameters on membrane performance	3
	– spacer studies	5
	– mathematical modeling	1
Other studies than NF and RO	– monitor for drinking water distribution net	1
	– UF membrane fouling studies	3

Figure 3.14 Selection of steel MFSs in use. Model studies in laboratory (**A**). On site studies on influence of chemical dosages and development of water distribution net monitor (**B**). Anti-scalant and biofouling inhibitor testing (**C**). On site biofouling and cleaning tests at a full-scale installation suffering from biofouling (**D**)

SUMMARY

A tool is developed for the validation of membrane fouling: the Membrane Fouling Simulator (MFS). The MFS uses the same membranes and spacers as present in commercial spiral-wound reverse osmosis and nanofiltration membrane elements, has similar hydrodynamics and is equipped with a sight window. Using the MFS, fouling can be monitored by (1) operational parameters like feed channel pressure drop, (2) non-destructive (visual, microscopic) observations using the window and (3) analysis of coupons sampled from the membrane and spacer sheets in the MFS. The major advantages of the MFS are (i) representativeness of spiral-wound membranes and (ii) the small size requiring small amounts of water and chemicals, reducing research costs and enhancing the possibility to test several MFS units in parallel. The MFS can be applied for early warning, characterization of the fouling potential of feed water, comparison of different pretreatment schemes and for evaluation of fouling control applying different chemicals. The MFS can be used in model studies to develop integrated membrane systems less susceptible to fouling. The MFS is a suitable tool for testing newly developed membranes and spacers. The first MFS tests showed that the results are representative for membrane elements used in practice under the same operating conditions. A comparison study of the MFS and spiral-wound membrane elements showed the same pressure drop development in time and the same fouling accumulation. The MFS is an effective scaled-down version of a full-scale system and allows to study the biofouling process occurring in the first 0.20 m of reverse osmosis/nanofiltration elements.

4

Early Biofouling Detection and Pressure Drop Measurements

INTRODUCTION

The demand for drinking water in the world is increasing and regulations on drinking water quality become stricter (Shannon *et al.*, 2009). High pressure membrane filtration processes like reverse osmosis (RO) and nanofiltration (NF) can produce drinking water of high quality. Declining costs of membrane processes stimulate an increasing role for membrane filtration processes in advanced water treatment practice (Mallevialle *et al.*, 1996; Patching and Fleming, 2003 and Shannon *et al.*, 2009).

A potential drawback in membrane filtration applications is membrane fouling. The corresponding increase of plant operation costs may be prohibitive for the application of membrane filtration in (drinking) water treatment. One of the major types of fouling is caused by biofilm accumulation in membrane elements (Paul, 1991; Flemming, 1993; Tasaka *et al.*, 1994; Ridgway and Flemming, 1996; Baker and Dudley, 1998; Schneider *et al.*, 2005 and Karime *et al.*, 2008). This type of fouling, biofouling, increases the feed channel pressure drop (Characklis and Marshall, 1990), the pressure drop between feed and brine lines (over the feed spacer channel).

The feed channel pressure drop measurement currently used in practice is not sensitive enough to quantify biofouling in a membrane filtration installation timely and can therefore not be used as an (early) warning system enabling control actions such as preventive cleanings (Vrouwenvelder *et al.*, 2008b). Membrane

*This chapter is based, with permission from the copyright holder, on a paper previously published in Journal of Membrane Science Vol. 338 No. 1–2 pp. 92–99 doi: 10.1016/j.memsci.2009.04.016

manufacturers recommend to perform corrective cleanings at a total pressure drop increase of 15%. This total pressure drop increase may be too high for effective fouling control. It could be more efficient to take action at a lower pressure drop increase. However, this approach is impossible with pressure drop measurements as applied in practice at this moment. In practice, pressure is measured before and after stages of the membrane filtration installation, using pressure transmitters. Then, the pressure drop is derived by subtracting two measured pressures.

The objective of this study was to evaluate different types of pressure transmitters for pressure drop measurements for *accurate early* fouling detection in membrane filtration installations. The development of feed channel pressure drop was followed over individual membrane elements, individual stages and the total nanofiltration installation under field conditions. The relation between the pressure drop increase and accumulated fouling was determined, using membrane autopsies on elements differing in pressure drop increase. Nanofiltration membranes were used in the study because of potential practical application to improve removal of dissolved (in)organic and biodegradable compounds (Verliefde, 2008 and Van Dijk and Van der Kooij, 2005).

MATERIALS AND METHODS

Experimental set-up

Water Supply Company Vitens produces drinking water (25,000,000 m^3 $year^{-1}$) from anaerobic ground water at treatment plant Spannenburg in the Netherlands. Vitens applies subsequently aeration, rapid sand filtration, deacidification, softening and rapid sand filtration. The location was selected because of the potential use of membrane filtration (i) to soften the water and (ii) to improve the biological stability of the water (reduce aftergrowth) since the water is distributed without disinfectant. At the location TDS removal is not a primary aim for NF application. The TDS has little influence on biofouling and characterizing/quantifying of biofouling with differential pressure drop measurements. Drinking water from this treatment plant was treated additionally in a pilot plant consisting of UF and NF (Figure 4.1). The UF installation – consisting of two membrane elements (X-flow UFC M5) – was operated in a dead-end mode at constant flow with a flux of 70 L m^{-2} h^{-1}, a backwash twice an hour and three chemical cleanings a day.

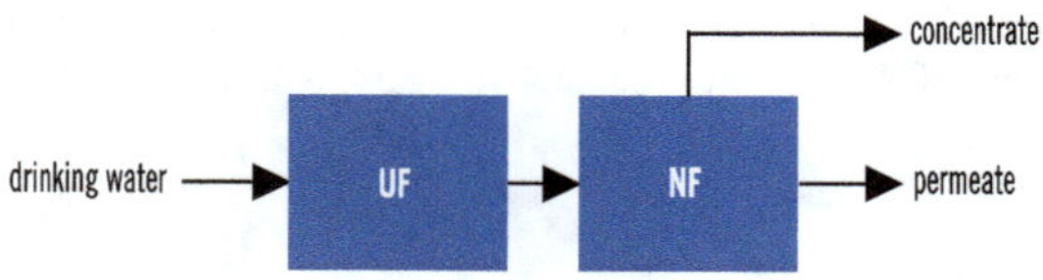

Figure 4.1 Experimental set-up pilot plant

The two stage NF installation (Figure 4.2), contained a total of 18 spiral wound Trisep TS80 TSF membrane elements, with a molecular weight cut-off of 200 Da. For this research project, a third stage single element (3.1 in Figure 4.2) was added to the installation. An antiscalant (7 mg L^{-1} 4Aqua OSM 92) was dosed to the UF permeate to prevent scaling. The antiscalant concentration was based on the projection program from the membrane supplier to avoid $CaCO_3$ scaling. The NF installation was operated continuously at a feed flow of 3.2 $m^3\ h^{-1}$ with a recovery of 75% for 146 days. The flux of the two stage NF installation was 20 L $m^{-2}\ h^{-1}$. The linear flow velocity in the lead membrane elements was 0.16 m s^{-1}. At the start of the study the feed pressure of the NF installation was about 600 kPa. The process parameters monitored were pressure drop, flow and conductivity. The flux was calculated (Verdouw and Folmer, 1997 and Huiting *et al.*, 1999). For assessment of the static pressure and the pressure drop over individual membrane elements, quick connectors on the NF pressure vessels were installed. Some elements in this installation were operated with and without permeate production. Previous work (Vrouwenvelder *et al.*, 2009b) showed that this was not effecting the development of biofouling and pressure drop. No chemical primary disinfection was applied and no disinfectant residual was present in the drinking water (The UF NF feed water) and no cleanings of the NF installation were applied during the study.

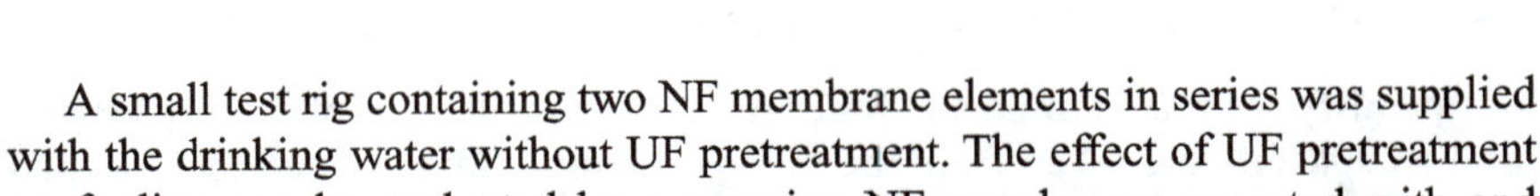

Figure 4.2 Scheme of nanofiltration installation with coding and order of membrane elements

A small test rig containing two NF membrane elements in series was supplied with the drinking water without UF pretreatment. The effect of UF pretreatment on fouling can be evaluated by comparing NF membranes operated with and without UF pretreatment. The membranes in the test rig were operated with the same conditions as the pilot NF installation fed with ultrafiltrate (Figure 4.2).

Pressure drop measurements

Feed channel pressure drop measurements were performed with three transmitter types, differing in accuracy and long term stability (Table 4.1). Pressure transmitters (PT1, Endress + Hauser, PMC131) were used for continuous monitoring of the feed pressure, concentrate pressure and permeate pressure for stage 1, stage 2 and the total NF installation. A digital precision manometer (PT2, Endress + Hauser, type PMC41) was used to assess the pressure before and after individual membrane elements. The calibrated measuring range for PT2 was 0 to 650 kPa. A differential pressure transmitter (PT3, J&M instruments, type JML1D2SH) was used to determine the pressure drop over individual membrane elements. The calibrated measuring range of PT3 was 0 to 16 kPa, which is within the upper range limit (0–100 kPa, Table 4.1). The digital precision manometer (PT2) and the differential pressure transmitter (PT3) were used for periodical (manual) measurements applying quick connectors on the pressure vessels. An alternative for PT3, with different specifications is PT4 (not tested). The feed channel pressure drop data were not normalized because of the highly stable temperature, pH-value and salinity of the water (Table 4.2).

During our studies short time fluctuations of the pressure drop reading of the differential transmitters were observed. For PT3 and PT4, pressure drop fluctuated with about 0.3 kPa during short time intervals. At different pressure drop levels the same fluctuation was observed. The variation of pressure drop was higher than the pressure transmitter manufacturers' specifications. The cause for these fluctuations was most likely the effect of pumping. In a set-up without pumping the short time pressure drop variation was much less and in the order of manufacturer specifications.

It is anticipated that the water salinity will not affect the accuracy of differential pressure measurements (using PT3 or PT4). In high salinity water it is essential that corrosion of the pressure transmitters is prevented. Part of our previous studies were performed with seawater desalination installations (Vrouwenvelder *et al.*, 2008b). No indications were found for an influence of salinity onto differential pressure measurements.

Sampling and study of membranes modules

To quantify the accumulated fouling, sections of membrane, feed and product spacer were taken from the membrane modules. The sections (4 cm^2) were placed in a capped tube filled with 10 mL of autoclaved tap water. To determine the amount of biomass, the tubes with the membrane sections were placed in an ultrasonic cleaning bath (Bransonic, model 5510E-DTH, output 135 Watts, 42 kHz). The low energy sonic treatment (2 minutes) followed by mixing on a Vortex (few seconds) was repeated three times. Next, water volumes collected from the tubes were used to determine biomass parameters (ATP and TDC).

Table 4.1 Pressure transmitter types, specifications and suitability for accurate early fouling detection

Code	Pressure transmitter	Type	Accuracy % of measuring range	URL	Long term stability	Maximum overload	Suitability
PT1	E + H Cerabar T: PMC131	PT	0.50%	0–400 kPa	±0.15 % of URL	1,670 kPa	0
PT2	E + H Cerabar M: PMC41	PT	0.20%	0–400 kPa	±0.10 % of URL	1,670 kPa	+
PT3	JM Instruments; JML1D2SH	PDT	0.10%	0–100 kPa	±0.10 % of URL	600 kPa	++
PT4	E + H Deltabar S: PMD70	PDT	0.05%	0–50 kPa	±0.05 % of URL	10,000 kPa	+++

URL = upper range limit; PT = pressure transmitter; PDT = pressure difference transmitter. The URL of pressure transmitters is applicable to the chosen pressure-cell in the transmitter. Pressure-cells with higher or lower ranges are available. The article code of PT4 is E + H PMD70-AAA7FKYAAA.

Table 4.2 Water quality parameters of UF and NF feed, permeate and concentrate

parameter	Unit	UF feed average ± sd	NF feed average ± sd	NF permeate average ± sd	NF concentrate average ± sd
temperature	°C	11.5 ± 0.3	11.5 ± 0.3	11.5 ± 0.3	11.5 ± 0.3
Susp. sol.	mg L^{-1}	⩽2	⩽2	⩽2	n.a.
TDS	mg L^{-1}	n.a.	460 ± 10	62 ± 15	1580 ± 130
DOC	mg C L^{-1}	7.4 ± 0.5	7.5 ± 0.4	<0.2 ± 0.3	32 ± 6
UV abs.	L m^{-1}	20 ± 1.1	19 ± 1.6	<0.5 ± <0.5	78 ± 11
Colour	mg Pt L^{-1}	16 ± 1.4	16 ± 1.4	<2 ± <2	80 ± 22
pH	$-\log [H^+]$	7.6 ± 0.1	7.6 ± 0.1	6.7 ± 0.3	7.9 ± 0.1
Conductivity	mS m^{-1}	50 ± 1.6	50 ± 1.1	8 ± 2	162 ± 14
HCO_3^-	mg L^{-1}	287 ± 9	286 ± 8	30 ± 10	1023 ± 39
Fe	mg L^{-1}	0.03 ± 0.02	⩽0.01 ± 0.01	⩽0.01 ± 0.01	⩽0.01 ± 0.01
Mn	mg L^{-1}	⩽0.01 ± 0.01	⩽0.01 ± 0.01	⩽0.01 ± 0.01	⩽0.01 ± 0.01
Na	mg L^{-1}	72 ± 6	73 ± 6	13 ± 0.7	242 ± 16
Cl	mg L^{-1}	30 ± 1	30 ± 1	6.2 ± 2.5	107 ± 12
Ca	mg L^{-1}	33 ± 4	32 ± 4	<5 ± <5	123 ± 24
Mg	mg L^{-1}	9.9 ± 0.9	9.8 ± 0.6	<0.5 ± <0.5	37 ± 1.8
Ka	mg L^{-1}	2.3 ± 0.1	2.4 ± 0.1	0.4 ± 0.1	8.3 ± 0.8
Ba	mg L^{-1}	0.032 ± 0.006	0.032 ± 0.006	0.006 ± 0.005	0.100 ± 0.011
Si	mg L^{-1}	n.a	15 ± 0.6	5 ± 0.4	34 ± 15
Sr	mg L^{-1}	n.a.	0.15 ± 0.02	0.002 ± 0.002	0.57 ± 0.08
F	mg L^{-1}	n.a.	0.08 ± 0.01	<0.05 ± <0.05	0.17 ± 0.11
As	µg L^{-1}	0.02 ± 0.008	<0.01	n.a.	n.a.
ATP	ng ATP L^{-1}	10.3 ± 3.9	3.4 ± 1.5	<1 ± <1	n.a.
TDC	cells mL^{-1}	$31 \times 10^4 \pm 2.8 \times 10^4$	$1.7 \times 10^4 \pm 0.6 \times 10^4$	$4.3 \times 10^4 \pm 2.8 \times 10^3$	n.a.

n.a. = not analyzed; sd = standard deviation.

Active biomass was determined in duplicate by measuring the adenosinetriphosphate (ATP) concentration from 100 μL samples (Holm-Hansen and Booth, 1966). The luminometer (Celcis, model Advance) added 100 μL of LuminEX-B reagent (Celcis) to a sample to release ATP from the bacterial cells. Subsequently, 100 μL of LumATE-PM (Celcis) was added for light production. The amount of light produced was measured with the luminometer (relative light units, RLU). The concentration of ATP was derived from the RLU values using the conversion factors of the linear relationship between RLU values and reference ATP concentrations. The ATP concentrations of autoclaved drinking water (<1 pg ATP mL^{-1}) and two control solutions (2 and 100 pg ATP mL^{-1}) were determined as quality control.

Total Direct Cell counts (TDC values, cells cm^{-2}) were performed with epi-fluorescence microscopy using acridine orange as fluorochrome by applying a slightly adapted method to eliminate fading (Hobbie *et al.*, 1997). The slides were examined with a Leica DM RXA microscope equipped with a 100 W power supply and phase and filter sets (I2/3) appropriate for acridine orange. Microorganisms were enumerated through a 100x/1.30 PL Fluotar oil immersion objective. Non-fluorescence immersion oil (Olympus, code 35505) was used because of low background fluorescence. A total of 10 random fields were counted.

Additionally, samples of membrane and spacer of 15 by 15 cm were taken about 20, 50 and 80 cm from the feed side of the membrane for analysis of the concentration of accumulated metals by quantitative Inductively Coupled Plasma Mass Spectrometry (ICP-MS, samples were destructed by pretreatment with aqua regia in a microwave oven).

Feed water

In addition to physicochemical water quality parameters (including TOC, pH, temperature and conductivity), the concentration of biomass (ATP, TDC) in the water (UF feed, NF feed, permeate and concentrate) was measured.

RESULTS

The development of feed channel pressure drop in the pilot nanofiltration installation was investigated over (i) individual membrane elements, (ii) individual stages and (iii) the total nanofiltration installation with different types of pressure transmitters (Figures 4.3 and 4.4). For relating pressure drop increase with accumulated fouling, membrane autopsies on elements differing in pressure drop increase were performed (Figure 4.5).

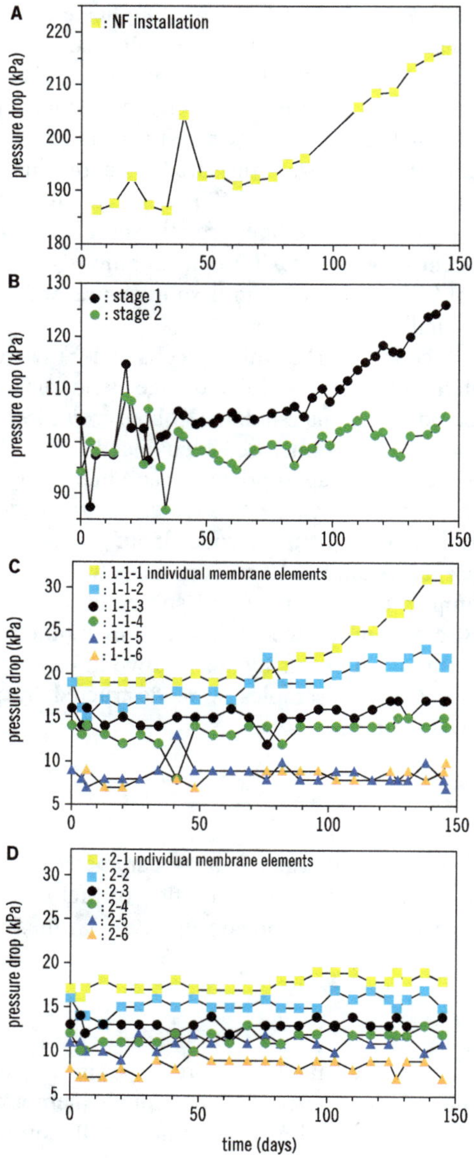

Figure 4.3 Pressure drop in time over the NF installation (**A**), stages (**B**) and individual membrane elements in stage 1 (**C**) and stage 2 (**D**) of the NF installation supplied with ultrafiltrate. Pressure transmitter PT1 was used for (A) and (B) and PT2 was used for (C) and (D)

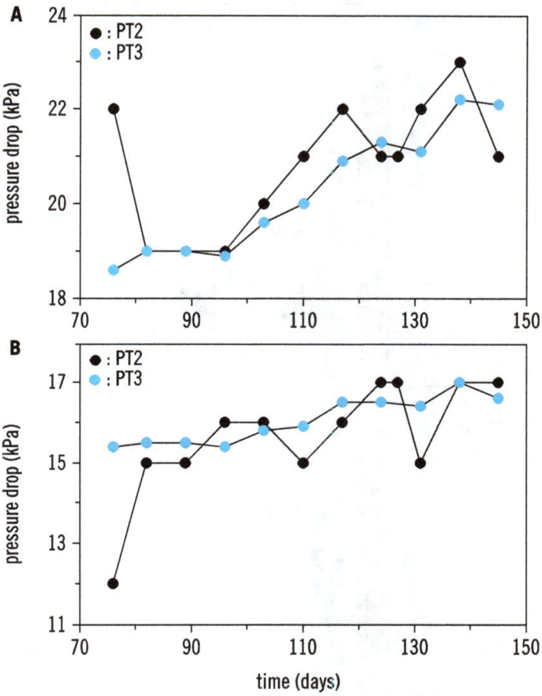

Figure 4.4 Pressure drop in time over membrane module 1-1-2 (**A**) and 1-1-3 (**B**) from the NF installation determined with PT2 and PT3. The standard deviation for PT2 (2.6 kPa) was larger than for PT3 (0.016 kPa). PT3 provided more sensitive data. The pressure drop measurements of individual membrane elements with PT3 were performed starting day 76. The pressure drop can fluctuate up to about 0.3 kPa for PT3 and PT4 during short time intervals. At different pressure drop levels the same fluctuation was observed

UF pretreatment

UF treatment of the drinking water had a limited effect on the water quality (Table 4.2). UF reduced the concentration of iron, active biomass (ATP) and the number of bacteria (TDC; Table 4.2). The presence of bacteria in the UF permeate is most likely related to growth of bacteria on surfaces of materials and not to UF leakage. The UF installation was operated with hardly any measurable permanent fouling, which was expected based on the relatively low flux ($70\,L\,m^{-2}\,h^{-1}$) and high cleaning frequency ($3\,d^{-1}$).

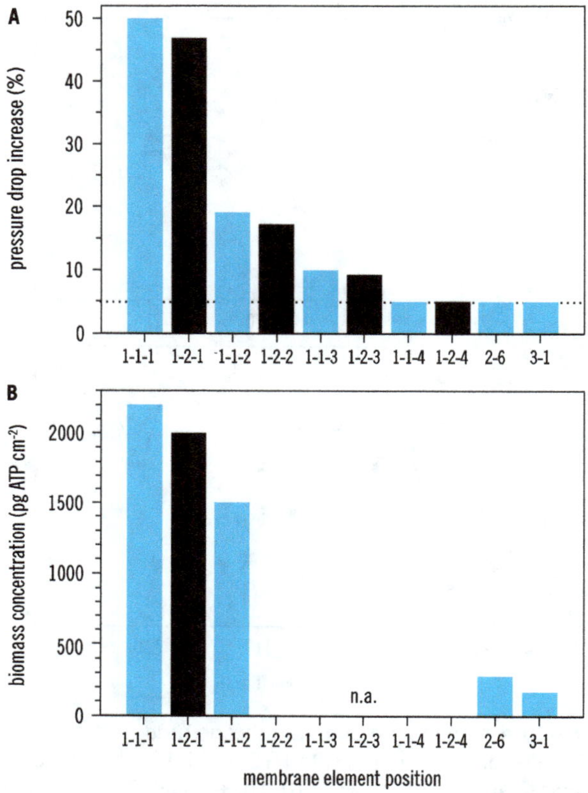

Figure 4.5 Pressure drop increase (**A**, PT3 data) and biomass concentrations (**B**) of selected membrane elements after 146 days of operation of the NF installation. The dotted line represents the estimated detection limit of the measurement. n.a.: not analyzed. In stage 1 two parallel pressure vessels are present. Each pressure vessel contains six elements. For example, membrane modules 1-1-1 and 1-2-1 are the lead membranes from the two parallel pressure vessels (Figure 4.2)

UF treatment had a significant effect on NF membrane fouling, with respect to both type and extent of fouling. *Without UF pretreatment*, within 38 days the pressure drop increased with $\geqslant$110% over a small test rig containing two NF elements in series. The exact pressure drop increase could not be determined, since the maximum signal for the pressure transmitter (PT1 type) was exceeded. In the lead module without UF pretreatment, much higher concentrations of biomass (7,800 pg ATP cm^{-2}) and inorganic elements (4,000 mg m^{-2}) were observed than on the lead element from the NF installation with UF pretreatment

(≤2,200 pg ATP cm^{-2} and ≤0.2 mg m^{-2}). Without UF pretreatment, the highest content of inorganics was found for calcium (2,000 mg m^{-2}), iron (1,500 mg m^{-2}), silicon (300 mg m^{-2}) and manganese (46 mg m^{-2}). The membrane module in position 2 of the test rig without UF pretreatment contained only 55% of the concentration of biomass and inorganic elements present in membrane module on position 1, illustrating a decline of fouling over the length of the test rig. The total inorganic content declined also over the length of the membrane elements, with the highest content on the first 20 cm of the lead element. Visual inspection of the NF membrane surfaces showed clear differences between the membrane modules as well (Figure 4.6). Figure 4.6 shows that without UF pretreatment, darker coloured and more material was observed on the membranes compared to the membranes operated with UF pretreatment. Without UF pretreatment, inorganic fouling and biofouling of NF membranes were observed. Most likely both types of fouling were contributing to the pressure drop increase over the test rig.

Figure 4.6 Membrane coupons from membrane elements, illustrating that the membranes from the NF test rig (foreground) were visually more fouled than the membranes from the NF installation (the center shows element 2-6 and 3-1 and the background element 1-1-1, 1-1-2 and 1-2-1)

With UF pretreatment, the pressure drop over the NF installation showed no increase within 38 days. In 146 days, the pressure drop increased gradually by about 15%. Inorganic fouling ($\leqslant$0.2 mg m^{-2}) was not observed but biofilm accumulation (2,200 pg ATP cm^{-2}) still occurred, although at a lower level than without UF pretreatment.

Evidently, UF pretreatment had great impact on NF performance: UF reduced the rate of pressure drop increase and reduced inorganic fouling and to a lesser extent biofilm development in NF membrane elements. The following sections present data obtained with the NF installation after UF pretreatment.

Development of pressure drop

A standard pressure transmitter (PT1) was used for pressure drop measurement of the installation and stages and a high precision pressure transmitter (PT2) were used for the individual membrane elements.

The pressure drop over the NF *installation* increased about 15% within 146 days after starting up (Figure 4.3A). The pressure drop measurements over the individual *stages* of the NF installation showed that the pressure drop increase over the first stage was higher than the second stage increase (Figure 4.3B). The variation of the pressure drop data during the first 40 days limits the accuracy of the pressure drop increase calculation over the two stages. The pressure drop measurements over *individual membrane elements* revealed the highest pressure drop increase (and pressure drop) of the lead membrane element from the first stage compared to membrane elements from other locations in the membrane installation (Figure 4.3CD). The two parallel first stages of the installation showed the same pressure drop for individual membrane elements at the same position in the pressure vessel (data not shown).

The pressure drop increase over the NF installation was predominantly caused by the pressure drop increase over the *lead membrane elements* in the first stage (Figure 4.3C), which had a pressure drop increase of about 50% in 146 days. The pressure drop increase was low ($\leqslant$5%) in all elements at position 4, 5 and 6 in the first stage, all elements in the second stage and the third stage single element (Figure 4.3).

At the start of the experiment, pressure drop declined with increasing distance to the feed side of the installation (Figure 4.3CD). This pressure drop decline is caused by the reduction of linear flow velocity due to permeate production in the membrane elements. For each element, permeate production is about 10% of the element feed flow.

The normalized flux of the NF installation did not decline during the research period (Verdouw and Folmer, 1997). So, based on the normalized flux no indications were found for reduced membrane performance.

Comparison of pressure drop measurements

Using pressure difference transmitter PT3, a distinct gradual increase of the pressure drop in time was observed over membrane elements 1-1-2 and 1-1-3, which was not observed with pressure transmitter PT2 (Figure 4.4). PT3 provided less scattering and more accurate pressure drop data than PT2. The data obtained with PT3 illustrate that the pressure drop increase declined over the first three elements of the first stage of the NF installation (Figure 4.5A). The other membrane elements in the first and second stage showed no detectable pressure drop increase.

Use of a pressure difference transmitter (PT3) enables more sensitive and accurate monitoring of the pressure drop, facilitating earlier detection of fouling than with standard (PT1) or even high precision pressure transmitters (PT2).

Fouling analysis

A pressure drop increase of 15% over the NF installation supplied with ultrafiltrate was selected for membrane autopsies. Because of clear differences in pressure drop increase, membrane elements 1-1-1, 1-1-2, 1-2-1, 2-6 and 3-1 were taken from the nanofiltration installation after 146 days continuous operation (Figure 4.5A).

The total concentration of inorganic compounds in all membrane elements was below the detection limit of the analysis performed ($\leq 0.2\,\mathrm{mg\,m^{-2}}$ membrane surface area). In practice, these low values ($\leq 0.2\,\mathrm{mg\,m^{-2}}$) were observed in installations without pressure drop problems, suggesting that the performance of the installation in this study was not influenced by accumulated inorganic compounds.

The biomass concentrations in the selected membrane modules showed significant differences (Figure 4.5B). The concentration of biomass decreased with increasing distance to the feed side of the membrane installation. The biomass concentrations (ATP) related well with the pressure drop increase, since high biomass concentrations where found in the membrane elements with a high pressure drop increase and relative low biomass concentrations where found in elements with a low pressure drop increase. The relationship between pressure drop increase and biomass concentration is not linear (Figure 4.5). The lead membrane elements from the first stage (1-1-1 and 1-2-1) had the highest pressure drop increase and the highest active biomass concentration (Figure 4.5). The standard and high precision pressure transmitters PT1 and PT2 were not sensitive enough to discriminate individual membrane elements by pressure drop increase as observed with pressure difference transmitter PT3 (Figure 4.5A).

DISCUSSION

Pretreatment effect

Without UF pretreatment, a strong pressure drop increase and fouling (biomass and inorganics) was observed in NF membrane elements. UF treatment reduced the pressure drop increase and fouling accumulation in the NF installation. With UF pretreatment, biofouling was still observed but inorganic fouling was absent. Dosage of chemicals such as scale inhibitors may contribute to fouling (Vrouwenvelder *et al.*, 2000). A scale-inhibitor was dosed to the water ($7\,mg\,L^{-1}$) after UF to prevent scaling of the NF installation. This scale-inhibitor had a low bacterial growth potential (Vrouwenvelder *et al.*, 2000) and is expected not to contribute to the biofouling development in this period.

After ultrafiltration, a drinking water can still cause biofouling by use as feed water for in a NF installation. The pretreatment of the water treatment plant includes two rapid sand filtration steps in which biodegradable compounds are removed extensively from the feed water by biological activity of biomass on the sand. Apparently, this pretreatment did not remove nutrients sufficiently to avoid biofouling in the NF installation. High requirements must be met for feed water of a membrane installation with spiral wound membranes to prevent biofouling. For stable operation of NF or RO plants without a pressure drop increase, extensive (biological) pretreatment can be an effective approach to reduce the nutrient concentration. For example, the RO installation Jan Lagrand of PWN in Heemskerk was operated for over 3 years without a significant pressure drop increase, without preventive cleaning (Galjaard *et al.*, 2005). This RO installation is supplied with water after extensive pretreatment consisting of open storage of lake water, microstraining, coagulation, sedimentation, dual media filtration, granular activated carbon filtration and ultrafiltration.

Biofouling mechanism in lead membrane elements

Recent investigations into the community structure analysis of biofilms in membrane modules from RO membrane plants suffering from reduced performance have been reported (Chen *et al.*, 2004 and Pang *et al.*, 2005, 2007). 'The bacterial community of the RO membrane biofilm was clearly different from the bacterial community present at other locations in the treatment plant, indicating the development of a specialized bacterial community on the RO membranes. The species in the biofilm were different from those in the feed water, indicating that the biofilm was actively formed on the RO membrane sheets and was not caused by filtration of bacteria present in the feed water' (Bereschenko *et al.*, 2008).

Experiments with membrane fouling simulators at the same location with the same pretreatment and operated without nutrient dosage showed no increase of feed channel pressure drop and no biomass accumulation while experiments with monitors with nutrient dosage showed an pressure drop increase over the feed spacer channel within a 10 day research period, indicating that the substrate concentration is an key parameter for biomass accumulation.

Bacteria accumulate on material surfaces by two processes: attachment and growth by cell multiplication. Attachment is a prerequisite for biofilm formation (Van Loosdrecht *et al.*, 1990). The presence of microorganisms on membranes was already observed after a short contact time with water containing microorganisms, in the order of minutes (Schaule, 1992 and Flemming, 1993). Microorganisms are commonly observed on water exposed surfaces, even in pure water systems (Mittelman, 1991). Initial adhesion of microorganisms to materials has shown not to be predictive for biofilm formation (Gjaltema *et al.*, 1997). The presence of attached microorganisms is not indicative for biofouling. Once attached, the cell may grow, reproduce and produce extracellular polymers, utilizing nutrients transported by the bulk fluid (Characklis and Marshall, 1990). Obviously, biofouling of nanofiltration and reverse osmosis systems is caused by biofilm growth. Thus, nutrients can be considered as biofouling potential. The dominance of biofouling in nanofiltration and reverse osmosis membrane installations with microfiltration/ultrafiltration pretreatment indicates that nutrients are not rejected by microfiltration and ultrafiltration membranes (Vrouwenvelder *et al.*, 2008b). The biofouling process is based on the transformation of dissolved into non-dissolved compounds (nutrients into biofilm).

Biomass accumulates fastest on locations were supply of easily degradable compounds is highest: the feed side of the installation. Studies with the Rototorque, an annular reactor designed as perfectly mixed reactor resulting in the same substrate concentration in the total reactor, showed that most biofilm accumulated at the inlet side of the feed water containing substrate (Gjaltema *et al.*, 1994), supporting the results of the biofouling studies. A flat sheet monitor with approximately the same membrane and spacer length and flow channel height as membrane modules was developed to study membrane fouling (Vrouwenvelder *et al.*, 2007a). Dosage of substrate to the feed water of the monitor resulted in a pressure drop increase over the monitor. Measurements over the first and second half of the monitor showed that the pressure drop increased predominantly over the first half of the monitor. Analysis of membrane coupons from the monitor showed that most biomass was present in the first half of the monitor (Vrouwenvelder *et al.*, 2007a). In all systems fed with low substrate concentrations the highest biomass accumulation is observed at the feed side.

Concentrating biodegradable compounds in the water over the membrane filtration installation as a consequence of water production is not resulting in higher substrate concentrations causing higher biomass concentrations in the membrane modules at the outlet side of the installation (Figure 4.6B, module 2-6 and 3-1 have low biomass amounts). The easily biodegradable substrates were already utilized in the first elements of the first stage.

Biofouling monitoring

Biomass accumulation in membrane modules increases the feed channel pressure drop before flux decline (Vrouwenvelder *et al.*, 2009c). By early warning of biofouling a signal for the feed channel pressure drop increase over a small part of the installation is given before the feed channel pressure drop increase over the total installation is increased followed by flux decline. In other words, an early warning signal is given before biofilm accumulation advances and leads to problems such as a feed channel pressure drop increase over the total installation and flux decline. Early detection of fouling requires monitoring where the fouling starts. In case of biofouling this is the *lead membrane module* of the first stage (Figures 4.6 and 4.7). Full-scale data from membrane installations suffering from elevated pressure drop showed that most biomass is present in lead membrane elements (Carnahan *et al.*, 1995 and Vrouwenvelder *et al.*, 1998, 2008b), supporting the findings of this study. For early detection of a pressure drop increase, monitoring of the lead membrane element is required (Figure 4.7).

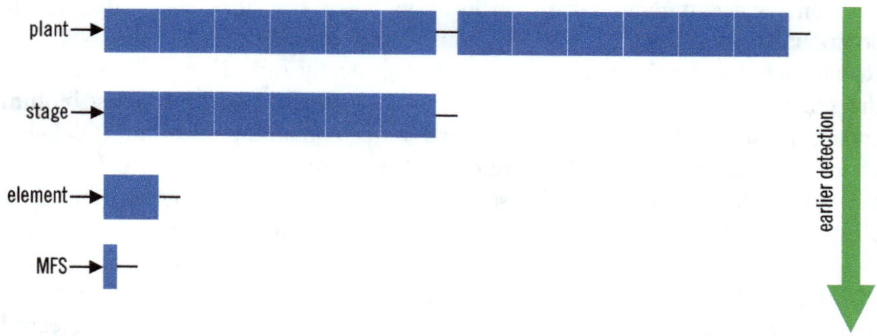

Figure 4.7 Biofouling is earlier detected when pressure drop monitoring is applied at the lead module at the feed side of membrane installation. An alternative for early biofouling detection is shown, a small tool named membrane fouling simulator (Vrouwenvelder *et al.*, 2006b, 2007a)

Sensitive pressure drop data should be preferred over water quality parameters, since the pressure drop increase is directly related to membrane biofouling of NF and RO membranes (Characklis and Marshall, 1990 and Vrouwenvelder *et al.*, 2008b), operated after microfiltration or ultrafiltration. Also, pressure drop measurement is less expensive than analysis and evaluation of water quality parameters, and can easily be applied on-line. In extensively pretreated water a pressure difference transmitter is the most simple and robust system to monitor biofouling.

In practice, monitoring of the pressure drop over lead membrane modules can be achieved by (i) using connectors on the pressure vessels before and after the lead membrane, (ii) using a single element test rig parallel to the installation operated under the same conditions, (iii) installing an additional stage with single membrane elements prior to the membrane installation and (iv) using the membrane fouling simulator (Vrouwenvelder *et al.*, 2006b, 2007a), a tool representative for membrane elements. For other types of feed water the pressure drop measurement can be a suitable tool to detect other types of fouling than biofouling in spiral wound NF and RO membranes. For example, it may be applicable for monitoring inorganic fouling and differentiate between particulate fouling and scaling.

Selection of pressure transmitter

Using a sensitive pressure difference transmitter provides accurate data with low scattering, enabling closer monitoring. Pressure transmitter types PT1 and PT2 are not suitable for accurate early biofouling detection. For assessment of the pressure drop, two pressure measurements are required. Obviously, two pressure measurements result in a lower accuracy of the pressure drop value than the accuracy of pressure transmitter specifications (Table 4.1). Also, the pressure drop is relatively small compared to the pressure, especially for high pressure membrane processes NF and RO. Since the two measured pressures are subtracted, a small error in the pressure measurement has a large impact on the pressure drop, explaining the lower accuracy of PT1 and PT2 (6.5 and 2.6 kPa standard deviation) compared to PT3 and PT4 (0.016 and 0.008 kPa).

This study has shown that pressure differential transmitters are suitable for pressure drop monitoring. Differential transmitters measure the pressure difference in one measurement. Therefore, the accuracy of PT3 and PT4 are much higher than standard (PT1) and high precision pressure transmitters (PT2, Table 4.1). PT3 provided the most accurate pressure drop data in this study. This transmitter contains two large diameter (8 cm) diaphragm seals to detect the pressure difference. These diaphragm seals are sensitive to damage by (transient) exceeding the maximum overload pressure. Although PT3 performed very well in our studies a more robust alternative is preferred for use in practice. The pressure difference transmitter PT4 has a higher accuracy, lower long term variability and due to the use of ceramic membranes a clearly higher maximum overload (10,000 kPa) than PT3. Both, PT3 and PT4 are available with different accuracy ranges, upper range limit and maximum overload.

Potential fouling control

A new and simple approach for biofouling control can be derived from the observations of this study: the pressure drop increase of the lead membrane (Figure 4.5) and the distribution of pressure drop over the modules in the NF installation (Figure 4.3CD). The use of an additional stage with e.g. large diameter (16 inch) membrane elements before the membrane installation improves (i) monitoring, (ii) control efficiency since less membranes are involved and (iii) flexibility of control measures (Wessels *et al.*, 2001 and Cornelissen *et al.*, 2007).

Another approach to reduce the effect of biofouling can be switching the position of lead biofouled elements with (clean) modules from the last position in the last stage or by changing the flow direction in the stages. The lead membrane has a higher linear flow velocity (and pressure drop) than the membrane element on the last position in the membrane installation. So, by changing the positions of the modules in the installation or changing the flow direction, the pressure drop over the installation will be reduced. The pressure drop of the lead biofouled membrane is reduced from 32 kPa to 10 kPa by changing (i) the membrane position in the installation (to position 6 in the stage) or (ii) the flow direction in the pressure vessel. This is based on calculations using the data of this study and methods described by Schock and Miquel (1987), assuming that only the friction coefficient is affected by the biofouling. The pressure drop of a clean membrane element on the last position in a stage (8 kPa on position 6, Figure 4.3D) is nearly the same as the biofouled membrane element (10 kPa) on the last position. If the stages contain more than six membrane elements in series, then the pressure drop of the lead membrane (32 kPa) will be reduced to even lower values (9 kPa on position 7 and 7 kPa on position 8). It is not clear yet, whether biofouling in the membrane element at the last position will decline in time due to nutrient limitation conditions; the concentration of easily biodegradable nutrients in the water and the linear water velocity are relatively low at this position.

Control of biofouling of NF/RO installations may be obtained by operating the membrane elements at relative low linear flow velocities. High biomass concentrations in the membrane elements seem to have little effect on the pressure drop increase at these low linear flow velocities. Studies to determine a hydraulic optimized design of staging and number of membrane modules per pressure vessel resulted in the development of a hydraulic optimized pressure vessel (Optiflux), which reduces pressure loss and increases membrane productivity (Van der Meer, 2003a,b). Using the optimized pressure vessel, the amount of lead modules is doubled and consequently the linear flow velocity is reduced (Van der Meer *et al.*, 2003a). Several full-scale installations using this optimized concept are in operation since 1999 to 2005 without problems (Van Paassen *et al.*, 2005). Combining low linear flow velocities with flow direction changes in the pressure vessels may be more effective for biofouling control.

SUMMARY

Accurate detection of biofouling in membrane filtration installations is assessed by monitoring the feed channel pressure drop over individual membrane elements of a full-scale installation using a sensitive pressure difference transmitter. Feed channel pressure drop increased mainly in the lead element of a nanofiltration installation. Also most biomass was present in the lead element, corresponding to membrane autopsy results obtained in several full-scale installations. The proposed pressure drop measurement over the lead element enables early detection of biofouling, facilitating timely and more effective control actions, with reduced environmental impact and cost.

5

Biofouling Studies and the Application of MRI

INTRODUCTION

The interplay of an increasing population, rising water pollution levels and stricter drinking water quality regulations in some parts of the world has lead to a significant rise in demand for ultrapure water. An increasingly important role in water purification is taken by membrane filtration technologies (Mallevialle *et al.*, 1996), such as reverse osmosis (RO) and nanofiltration (NF) membranes for desalination. However, membrane filtration suffers from susceptibility to biofouling: the accumulation of attached microorganisms to surfaces in the membrane unit. The negative consequences of biofouling are several-fold and well documented (Patching and Flemming, 2003). One of them is the increase of the feed channel pressure drop, resulting in higher pumping and cleaning costs eventually leading to early membrane replacement. Biofouling occurs widely in industry (Duranceau and Knippel, 1996; Chen *et al.*, 2004b; Belfer *et al.*, 2005; Pang *et al.*, 2005; Jarusutthirak and Amy, 2006 and Xu *et al.*, 2006) and provokes significant additional operational costs for membrane plants (Flemming *et al.*, 1997a,b and Ridgway, 2003). Biofouling is conventionally studied and monitored by macroscopic parameters such as pressure drop or by membrane 'autopsy' (Flemming *et al.*, 1997a), which is destructive opening and inspection. There is a substantial need for novel measurement techniques that enable non-invasive, real time and spatially resolved observation of such biofouling (Flemming *et al.*, 1997a,b, 1998).

---

*This chapter is based, with permission from the copyright holder, on a paper previously published in Journal of Membrane Science Vol. 323 No. 1 pp. 37–44 doi: 10.1016/j.memsci.2008.06.012

In this study we demonstrate the first application of nuclear magnetic resonance microscopy (NMR) to a RO membrane module. The NMR protocols used allow the extraction of (i) the spatial biofilm distribution in the membrane module, (ii) the velocity field and its evolution with biofouling and (iii) propagators, which are distributions of molecular displacement of a passive tracer (e.g. salts, organic molecules) in the membrane. In this study we employ ^{1}H NMR and the passive tracer signal is due to the ^{1}H of the water molecules.

Thus far NMR has been employed to study colloid deposition in cartridge filters (Dirckx *et al.*, 2000) or the flow profile in hollow-fibre arrays (Heath *et al.*, 1990 and Yao *et al.*, 1995). In this study two systems are investigated: (i) an industrial scale spiral wound reverse osmosis (RO) membrane (Hydranautics ESPA1-2540) and (ii) a flow cell filled with a feed spacer geometry typically used to separate different membrane sheets in industrial NF or RO membranes. In future the flow cell will be used for more fundamental studies of biofouling parameters, such as spacer designs. Spiral wound membranes consist of semi-permeable membranes separated by feed and permeate spacers, all wound around a hollow tube in a spiral fashion. The feed spacers separate the membranes and enhance tangential cross-flow and thus turbulence between membranes. The clean water solution permeates the membrane and leaves the membrane module through the hollow tube in the centre. Little information is currently available on where biofouling preferentially occurs within such a membrane module and how this influences the flow field through the module.

METHODOLOGY

Membranes systems

The two systems studied are as follows.

Membrane module

An industrial spiral wound RO membrane with an outer diameter of 2.5 in. (Hydranautics ESPA1-2540). A cross-section of the spiral wound membrane is shown in Figure 5.1.

Flow cell

The second system is a PVC flow cell with inner dimensions of 0.077 cm × 1.60 cm × 4.00 cm, designed to contain a typical membrane feed spacer of the same dimensions placed between two membrane sheets. The flow cell was produced with a manufacturing accuracy of 20 μm and water flow occurred between the membrane sheets through the spacer. It was shown previously that the design enables reproducible spacer fouling studies (Vrouwenvelder *et al.*, 2006b). The feed spacer porosity is 0.85. The use of PVC prevents any influence from the flow-cell material on the NMR signal. The flow-cell is shown in Figure 5.2A. Note the orientation of the Cartesian coordinates in both systems for future reference. Figure 5.2B shows a sketch of the spacer geometry.

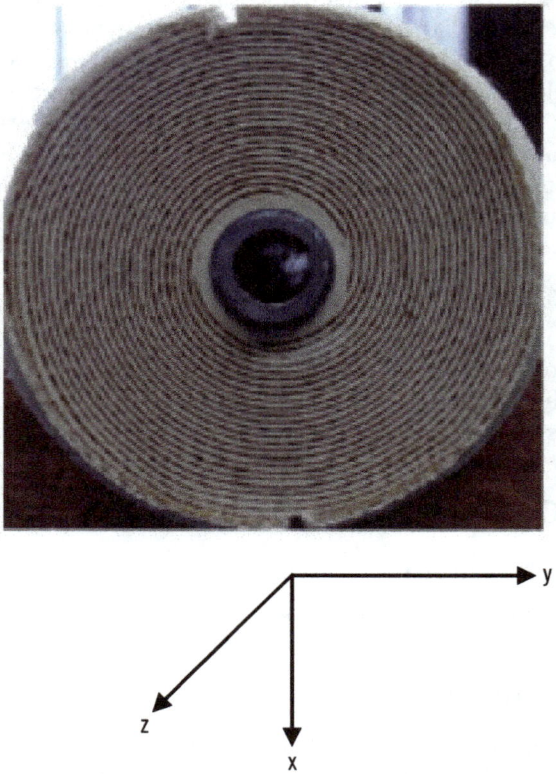

Figure 5.1 Cross-section photo of the RO membrane module

BIOFOULING PROCEDURE

Membrane module

The membrane module was supplied with Cambridge, U.K. tap water at a volumetric flow rate of $120\,\mathrm{L\,h^{-1}}$. This was supplemented with a feed stock nutrient solution containing CH_3COONa, $NaH_2PO_42H_2O$, $NaNO_3$ at $6\,mL\,h^{-1}$ for 2 weeks. The concentrations of CH_3COONa, $NaH_2PO_42H_2O$, $NaNO_3$ in the flow cell feed water were $1{,}000\,\mu g\,L^{-1}$, $150\,\mu g\,L^{-1}$ and $355\,\mu g\,L^{-1}$, respectively. The centre tube of the membrane was blocked during both fouling and NMR experiments.

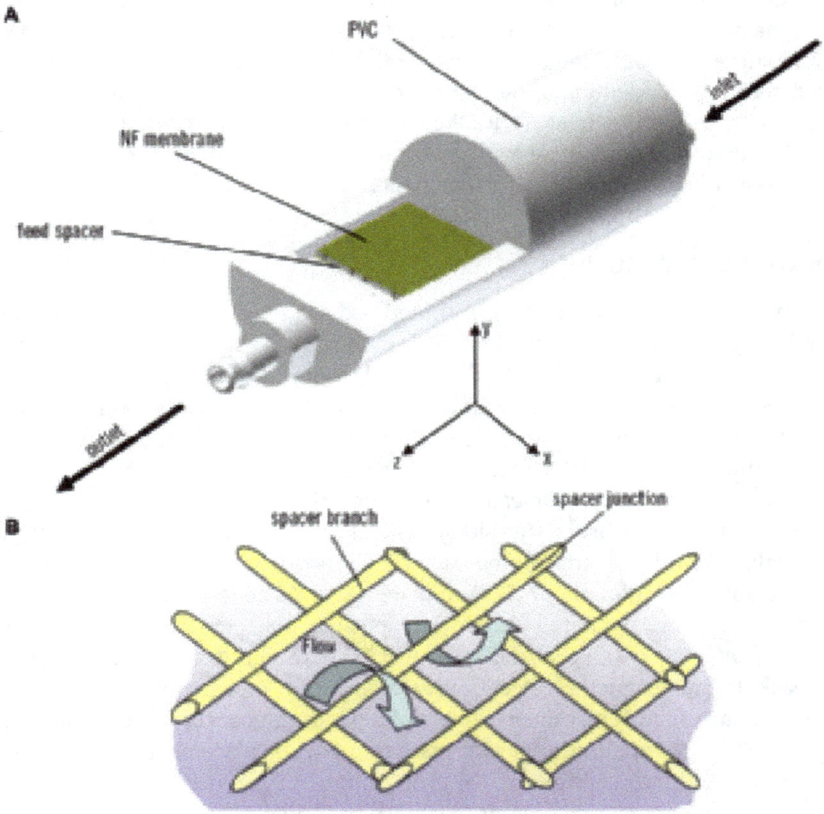

Figure 5.2 (**A**) Schematic of the PVC flow cell. Note the orientation of the Cartesian dimensions. The flow direction is in positive z direction. (**B**) Schematic of the feed spacer. The feed spacer branches have half the heights of the inner flow cell volume and flow can thus either pass above or below the spacer element

Flow cell

The flow cell was supplied with tap water at a volumetric flow rate of $6.4\,L\,h^{-1}$. This was supplemented with the same feed stock solution containing CH_3COONa, $NaH_2PO_42H_2O$, $NaNO_3$ as for the membrane module at $28\,mL\,h^{-1}$.

Nuclear magnetic resonance (NMR) microscopy

All NMR experiments were carried out on a 4.7 T superconducting super wide-bore magnet employing radio frequency (rf) coils with a diameter of 64 mm for the membrane module and 38 mm for the flow cell. ^1H detection (200 MHz) was used with signal resulting almost exclusively from the water content. Note that the ^1H NMR signal correlates with the ^1H density and thus high image intensities are obtained in regions filled with water and low signal intensities are obtained in regions filled with gas or other materials (e.g. feed spacer).

Membrane module

Two-dimensional (2-d) radial (x–y) and axial (x–z) images were acquired employing a Rapid Acquisition Relaxation Enhancement (RARE) pulse sequence (Hennig *et al.*, 1986) with a spatial resolution of 107 μm/pixel. The use of RARE enabled T_2 signal relaxation (Stejskal, 1965) weighted image contrast between biofilm-containing regions and free pore space; this procedure has been successfully employed in a number of previous studies (Hoskins *et al.*, 1999; Manz *et al.*, 2003; Seymour *et al.*, 2004 and Graf von der Schulenburg *et al.*, 2007). A standard phase shift velocity encoding pulse sequence was employed (Callaghan, 1991) to image the superficial flow component (z-component) with a spatial resolution of 220 μm/pixel (256 × 256 pixel). Z-displacement propagators were acquired employing pulsed field gradient (PFG) NMR (Hoskins *et al.*, 1999), specifically the APGSTE variant to reduce the influence of background magnetic susceptibility gradients. Propagators are probability distributions of molecular displacement of a passive tracer (in this case the water molecules). They are used to obtain mass transport displacement statistics. They sample the entire three-dimensional (3-d) volume of the membrane module and are significantly quicker to acquire than 2-d or 3-d velocity images. 32 gradient increments were acquired for observation times, Δ, of 100, 500 and 1,000 ms; gradient increments, ΔG, and pulse gradient durations, δ, were varied to avoid signal fold-over artefacts. The flow rate during velocity imaging and propagator measurement was $60\,ml\,min^{-1}$. More details on NMR microscopy methodology employed can be found elsewhere (Callaghan, 1991).

Flow cell

Structural two-dimensional images of 512 × 256 pixel of the flow cell were acquired employing a RARE pulse sequence with a slice thickness of 2 mm. The field of view was 50 mm × 25 mm resulting in an isotropic in-plane image resolution of 98 μm/pixel. A RARE factor of 64 was employed. The RARE factor indicates the relative increase in acquisition speed and provides the necessary T_2 relaxation weighting in the images required to identify regions of biofilm growth. Two-dimensional velocity images of the superficial flow component (z-component) were acquired employing a standard phase shift velocity encoding pulse sequence [more details with respect to this method can be sourced from Ref. Callaghan (1991)] with an image plane size of 256 × 128 pixel resulting in a spatial resolution of 210 μm/pixel. For phase shift velocity encoding, gradient increments, ΔG, and pulse gradient durations, δ, were decreased with increasing biofilm accumulation to increase the velocity imaging range and avoid fold-over artefacts. The flow rate during velocity imaging was 1.2 mL min^{-1}.

RESULTS AND DISCUSSION

Membrane

A high-resolution (107 μm/pixel) 2-d (x–y) image of a radial cross-section of the industrial scale membrane module (un-fouled) is shown in Figure 5.3. The light white rings filled with water contain regular black marks of lower ^1H density indicating the feed spacer elements (see red ring). The membrane envelopes and the product spacer appear in grey (see green box). This becomes more evident when zooming into a small region of 5.35 mm × 5.35 mm of the image. The black regions in the membrane module are regions of very low ^1H spin density and are air bubbles or tube material. The image allows the identification of irregularities in the spiral winding of the membranes. The dark regions at the bottom of the image (see blue circles) are due to inhomogeneities in the rf field. Figure 5.4 shows a centred axial image (x–z) of the un-fouled membrane module with a non-isotropic resolution of 156 μm/pixel in the z and 218 μm/pixel in the x dimension. The feed space elements are again visible.

A high resolution selected quadrant of a 2-d radial (x–y) image of the spiral wound membrane after biofouling is shown in Figure 5.5. Grey regions of biofilm accumulation (due to T_2 relaxation signal contrast) are visible (see red circles) and long streamer-like biofilm shapes are present (see green circle). Figure 5.6AB shows 2-d (x–y) radial images of the superficial velocity component (superficial direction of flow out of the page) before and after biofouling. Significant heterogeneity is introduced into the flow field by biofilm growth, which leads to a significant proportion of stagnant zones with velocities of $\sim 0\,\mathrm{m\,s^{-1}}$ (corresponds to blue in the image). This is compensated for by faster channels forming elsewhere. The velocity standard deviation within the 2-d slices shown in Figure 5.6AB increases from 0.000532 to $0.002021\,\mathrm{m\,s^{-1}}$ due to biofouling.

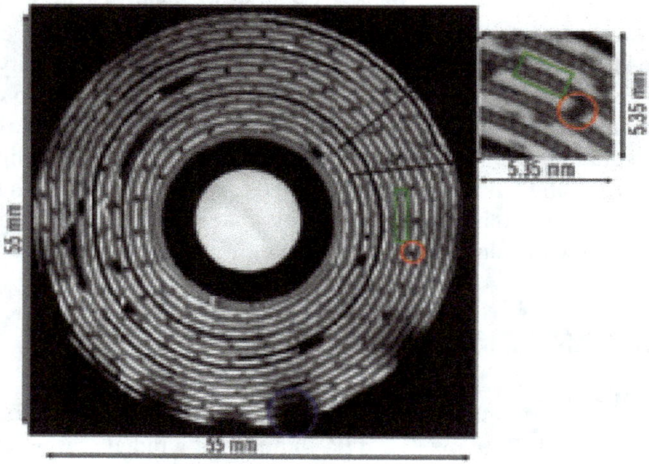

Figure 5.3 2-d radial (x–y) high-resolution (107 μm/pixel) image of the un-fouled membrane module. The light white rings filled with water contain regular black marks of lower ^{1}H density indicating the feed spacer elements (see red ring). The membrane envelopes and the product spacer appear in grey (see green box)

To quantify this impact of biofouling on the overall transport dynamics in the whole membrane module, propagator measurements were employed. The propagators, or distributions of water molecule displacement due to advection and diffusion, in the membrane module before biofouling for observation times, Δ, of 100, 500 and 1,000 ms are shown in Figure 5.7. The propagator for $\Delta = 100\,\mathrm{ms}$ is a skewed distribution with a stagnant peak. The stagnant peak is due to the stagnant water in the blocked central pipe and water molecules elsewhere in the module that have no net displacement due to diffusion and advection in the system. The stagnant peak is thus diffusion dominated. The high displacement tail of the distribution corresponds to the advected water molecules. For larger Δ, a portion of the stagnant peak decreases as more water molecules join advected channels and the advected part of the distribution shifts to larger mean displacements. These observations are typical for propagator measurements in complex porous media; such propagator measurements have been successfully applied to rocks (e.g. Ref. Packer and Tessier, 1996), chromatography columns (e.g. Refs. Lebon *et al.*, 1996 and Tallarek *et al.*, 1998) and sphere packings (e.g. Refs. Lebon *et al.*, 1996 and Amin *et al.*, 1997). The mean velocity of the water molecules stays constant (within 4%) for all three Δ and is in agreement with the expected mean velocity of $6.3 \times 10^{-4}\,\mathrm{m\,s^{-1}}$; indicating the quantitative nature of the propagator measurements. Figure 5.8 shows the propagator distributions before and after biofouling for $\Delta = 100\,\mathrm{ms}$ (a) and $\Delta = 1{,}000\,\mathrm{ms}$ (b), respectively. The biofouling introduces a longer tail for water following the faster advective channels. These observations are typical for biofilm growth in porous media (Seymour *et al.*, 2004, 2007 and Graf von der Schulenburg, 2007). The long tail of fast displacements in the propagator will result in a short residence time distribution contribution and the stagnant peak will result in a considerably longer residence time distribution contribution. They will thus result in an increase in residence time standard deviation in the system and a reduction in membrane performance.

Figure 5.4 2-d axial (x–z) image of the membrane module with an anisotropic resolution of 156 μm/pixel in the z dimension and 218 μm/pixel in the x dimension

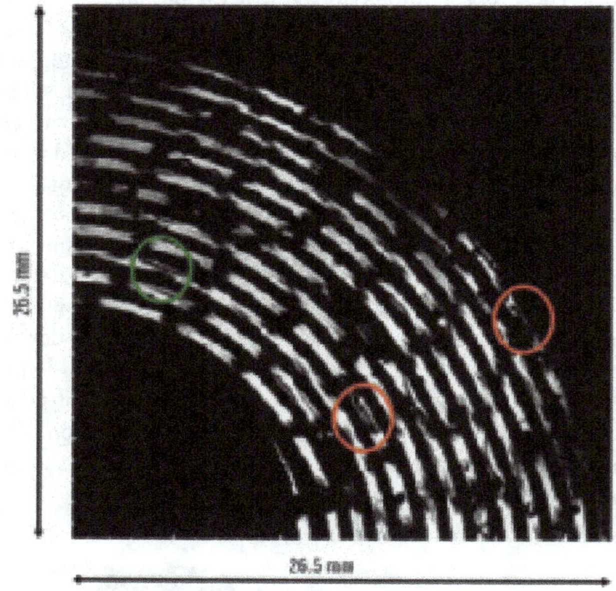

Figure 5.5 Enlarged view of a quadrant of a 2-d radial (x–y) high resolution (107 μm/pixel) image with T_2 weighting to image the spatial biofilm distribution after biofouling. Regions of biofilm growth are clearly visible and marked by the red circles. Long streamer-like biofilm shapes are also present (see green circle)

Flow cell

A 2-d axial x–z image of the flow cell before biofouling is shown in Figure 5.9A. The image intensity corresponds to the water density at each spatial position. The spacer junctions are clearly visible as black regions of low water intensity and the spacer branches are at intermediate image intensity as there is water either above or below them (as shown in Figure 5.2B). The darker area marked by the red circles is erroneously due to glue that entered the flow-cell during construction. Figures 5.9B-D are T_2-weighted images and show the same flow cell after 2, 3 and 4 days of nutrient supply. Figure 5.9B shows regions of biofilm growth as highlighted by the red circles. Note that the T_2-weighting also leads to smaller image intensities in the pixels adjacent the spacer branches. The biofilm growth seems to occur preferentially at the spacer junctions close to the inlet of the geometry. Also filament biofilm streamers are observed. These observations of biofilm location and shape are in good agreement with visual examinations of biofilm growth on membrane spacers (Dirckx *et al.*, 2000). Figure 5.9CD shows further biofilm accumulation and eventually apparent clogging of regions of the spacer geometry. The streamers reach lengths of up to ~6 mm.

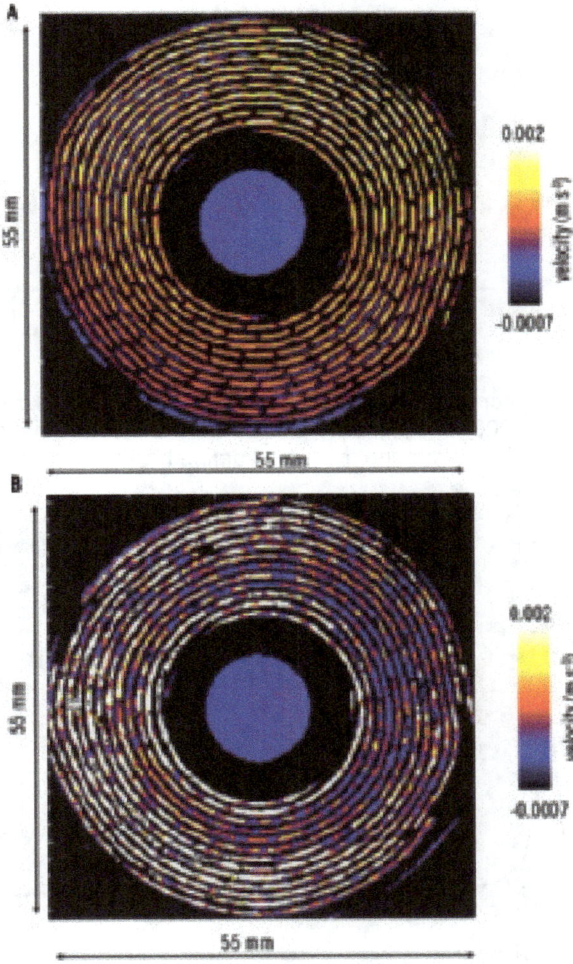

Figure 5.6 2-d radial (x–y) velocity images of the membrane module before (**A**) and after (**B**) biofouling. The image resolution is ~220 μm/pixel. The images show the superficial flow component (z component) on a colour scale from $-0.0007\,\mathrm{m\,s^{-1}}$ (black) to $0.002\,\mathrm{m\,s^{-1}}$ (light yellow). Flow is out of the page

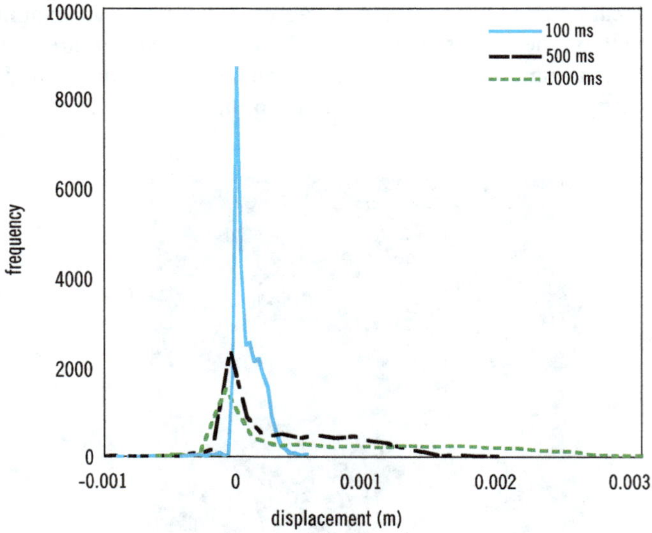

Figure 5.7 Propagator distributions for membrane model without biofouling for $\Delta = 100$, 500 and 1,000 ms. Frequency is in arbitrary units. Note that displacement is in the direction of the membrane axis (z, as defined in Figure 5.1)

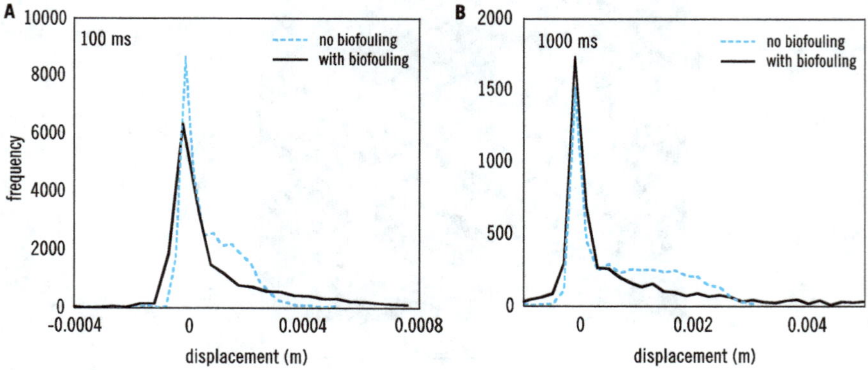

Figure 5.8 Propagator distribution for membrane model with and without biofouling for **(A)** $\Delta = 100$ ms and **(B)** $\Delta = 1{,}000$ ms. Frequency is in arbitrary units. Note that displacement is in the direction of the membrane axis (z, as defined in Figure 5.1)

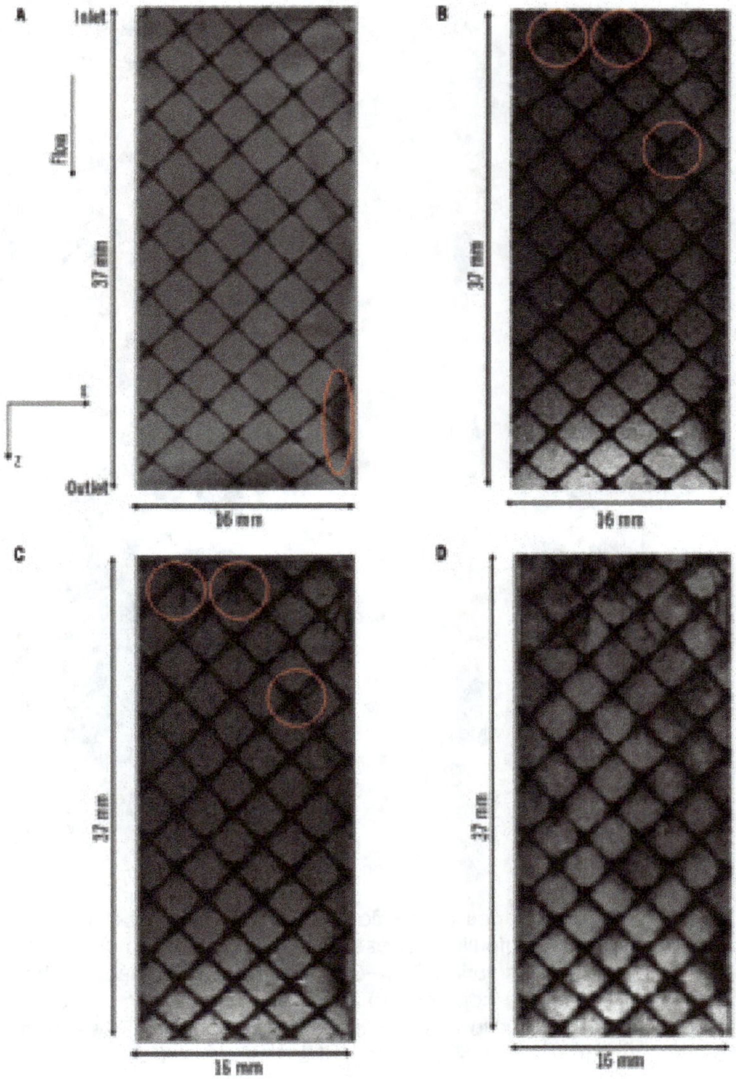

Figure 5.9 2-d images of the flow cell for 0 (**A**), 2 (**B**), 3 (**C**), 4 (**D**) days after nutrient supply for biofouling. The image resolution is ~98 μm/pixel. T_2 weighting is achieved via an echo time of 650 ms and biofilm appear as regions of low signal intensity. The darker area marked by the red circles in (A) is erroneously due to glue that entered the flow cell during construction. Biofilm growth preferably occurs at spacer junctions (red circles in (B and C))

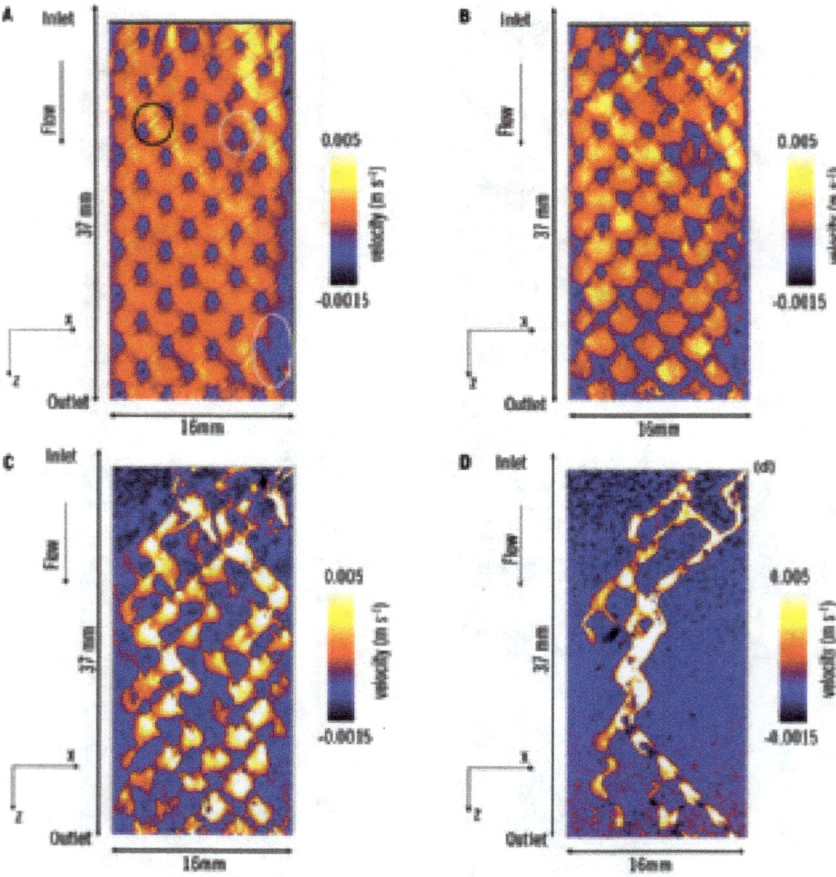

Figure 5.10 2-d velocity images of the flow cell 0 (**A**), 2 (**B**), 3 (**C**), 4 (**D**) days after nutrient supply for biofouling. These thus correspond to the images in Figure 5.9. The image resolution is $-210\,\mu m$/pixel. The images show the superficial flow component (z component) on a colour scale from $-0.0015\,m\,s^{-1}$ (black) to $0.005\,m\,s^{-1}$ (light yellow). Flow is from top to bottom. There is no flow at the spacer junctions and regions under the spacer branches tend to flow faster due to the local volume reduction at these positions (black circle). Some inhomogeneity is introduced by irregularities in the spacer material or the glue (white circles in (A))

Figure 5.10(A–D) shows the corresponding velocity images for 0, 2, 3 and 4 days of nutrient supply. The images show the superficial velocity component (z-component) on a colour scale, where white corresponds to high velocities and black to low velocities. Without biofouling the flow field is relatively homogenous with laminar parabolic flow profiles in the regions between spacer branches (Figure 5.10A). There is no flow at the spacer junctions and regions under the spacer branches tend to flow faster due to the local volume reduction at these positions (black circle). Some inhomogeneity is introduced by irregularities in the spacer material or the glue shown in Figure 5.9A (see white circles in Figure 5.10A). After 2 days, the biofilm growth at the spacer junctions shown in Figure 5.9B leads to a distortion of the flow field and regions of low velocities close to the spacer junctions (see Figure 5.10B). This is followed with time by blocking of entire regions (see Figure 5.10C) and a significant increase in flow field heterogeneity with large stagnant regions and fast flowing channels (see Figure 5.10D). This observed flow field heterogeneity can easily be quantified by extracting velocity distributions from the images.

Figure 5.11A shows the velocity distribution for 0 and 4 days of nutrient supply (corresponding to Figure 5.10AD). The initially narrow distribution (day 0) is centred around its mean velocity. Later the distribution shape changes to a strongly skewed distribution with a large peak in the low velocity region and a long tail of high velocities. This corresponds to the stagnant and channelling regions in the fouled geometry, respectively. An increase of velocity standard deviation of >5 times is observed over the biofouling period, as shown in Figure 5.11B. As expected, the mean velocity shown in Figure 5.11B for the four images stays constant (variation of $\sim 3\%$) and is in agreement with the expected mean velocity ($1.9 \times 10^{-3}\,\mathrm{m\,s^{-1}}$), indicating the quantitative nature of MRI velocimetry. This increase in velocity standard deviation implies an increase in residence time standard deviation which decreases the membranes performance. Further, the velocity images shown in Figure 5.10 allow the quantification of the fraction of membrane surface area that is still supplied with fresh inflow (i.e. does not present zero velocity (stagnant conditions)). This effective membrane surface area decreases with time drastically as shown in Figure 5.11C.

SUMMARY

There is a substantial need for novel measurement techniques that enable non-invasive spatially resolved observation of biofouling in nanofiltration and reverse osmosis membrane modules. Such measurements will enhance our understanding of the key design and operational parameters influencing biofilm

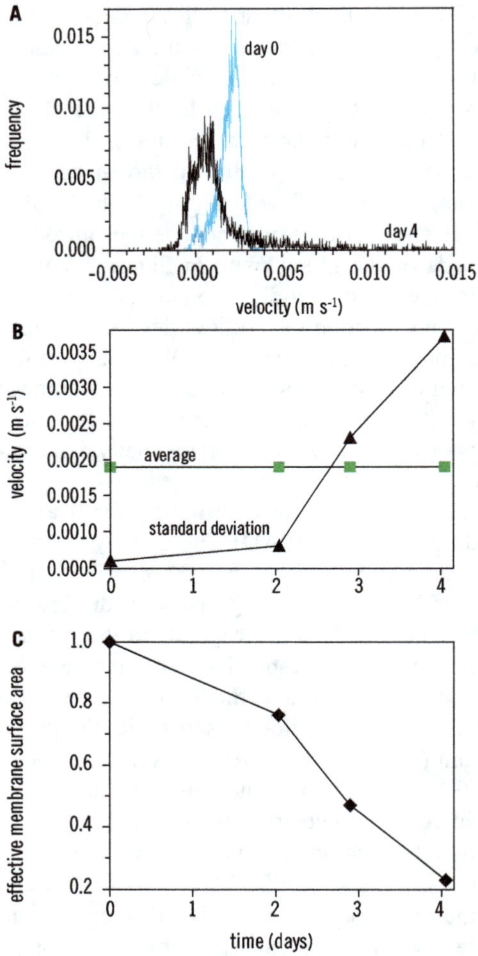

Figure 5.11 (**A**) Velocity distribution for flow cell after 0 and 4 days of nutrient supply for biofouling. The velocity distributions were extracted from the velocity images shown in Figure 5.10AD. (**B**) Average velocity and velocity standard deviation during 4 days of biofouling for the flow cell. The standard deviation increases significantly. The mean velocity is in good agreement with the expected mean velocity and stays constant in time. (**C**) Effective membrane surface area during 4 days of biofouling for the flow cell

fouling. In this study we demonstrate the first application of nuclear magnetic resonance microscopy (NMR) to a spiral wound reverse osmosis membrane module. The presented NMR protocols allow the extraction of the evolution with biofouling of (i) the spatial biofilm distribution in the membrane module, (ii) the spatially resolved velocity field and (iii) displacement propagators, which are distributions of molecular displacement of a passive tracer (in our case, water) in the membrane. From these measurements, the effective membrane surface area is quantified. Despite the opaque nature of membrane design, NMR microscopy is shown to be able to provide a non-invasive quantitative measurement of reverse osmosis membrane biofouling and its impact on hydrodynamics and mass transport. Minimal biofilm growth is observed to have a substantial impact on flow field homogeneity.

6

Developing 3D Computational Model

INTRODUCTION

Membrane filtration processes like reverse osmosis (RO) and nanofiltration (NF) produce high quality drinking water, virtually free of pathogenic micro-organisms and (in)organic pollutants (Mallevialle *et al.*, 1996 and Shannon *et al.*, 2008). A major problem in RO and NF applications is membrane fouling, resulting in a pressure drop increase, increasing the plant operation cost. Four major fouling mechanisms of RO and NF membranes can be discriminated: scaling, particulate and organic fouling and biofouling. Scaling by inorganic compounds is usually controlled using a scale inhibitor or an acid. Particulate fouling is controlled by extensive pretreatment (like ultrafiltration) removing the particulate matter. Thus, all types of fouling except biofouling and organic fouling – related types of fouling – are controllable. Biofouling is in practice the major fouling type in RO and NF membranes fed with extensively pretreated water. Biofouling is caused by growth of biomass i.e. biofilms in membrane modules (Ridgway and Flemming, 1996; Patching and Fleming, 2003 and Shannon *et al.*, 2008).

Insight in the factors influencing the development of biomass and pressure drop increase could be helpful in developing membrane systems less susceptible to biofouling. The relation between fouling accumulation and reduced membrane performance is complex. The linear flow velocity in membrane modules influences the substrate load, the substrate transport and spatial concentration distribution, biofilm growth, biofilm morphology and the effect of accumulated biofilm on

---

*This chapter is based, with permission from the copyright holder, on a paper previously published in Journal of Membrane Science Vol. 345 No. 1–2 pp. 340–354 doi: 10.1016/j.memsci.2009.09.024

the feed channel pressure drop increase (Vrouwenvelder *et al.*, 2009a). Biofilm accumulation can cause flow channel formation, reducing the water production flux (Vrouwenvelder *et al.*, 2009a). The effect of biofilms and particulate fouling on salt rejection (and on other solutes that can serve as nutrients to the biofilm), on concentration polarization and on the enhanced osmotic pressure are also important (Kim *et al.*, 2006; Herzberg and Elimelech, 2007 and Chong *et al.*, 2008). Furthermore, the concentration polarization may affect the biofilm cells physiology and viability by inducing higher substrate levels near the RO membrane (Herzberg and Elimelech, 2007). To unravel the influence of individual parameters on membrane performance a three dimensional mechanistic mathematical model is needed. Major achievements have been made on modeling the effect of spacer on mass transfer and fluid flow. However, fouling is a practical problem and a model coupling hydrodynamics and fouling is still lacking.

There are several important reasons to make this study necessary. First, although computational fluid dynamics models in membrane systems are becoming abundant in the literature (e.g., 3-d models (Karode and Kumar, 2001; Li *et al.*, 2002; Dendukuri *et al.*, 2005; Ranade and Kumar, 2006a,b; Koutsou *et al.*, 2007, 2009; Santos *et al.*, 2007; Shakaib *et al.*, 2007, 2009; Fimbres-Weihs and Wiley, 2007; Li and Tung, 2008b and Li *et al.*, 2009) and 2-d models (Pellerin *et al.*, 1995; Cao *et al.*, 2001; Geraldes *et al.*, 2002b; Schwinge *et al.*, 2002a,b, 2004; Koutsou *et al.*, 2004; Gimmelshtein and Semiat, 2005; Ahmad *et al.*, 2005, 2006; Subramani *et al.*, 2006; Shrivastava *et al.*, 2008; Fimbres-Weihs and Wiley, 2008; Wardeh and Morvan, 2008 and Guillen and Hoek, 2009)), none of these includes the biofilm growth in the feed channel with spacer separating the two membranes. Consequently, most of the obtained results cannot be applied to biofouling studies, and cannot be directly compared with experimental biofouling data.

Second, fully three-dimensional (3-d) models for flow and mass transfer coupled with biofilm growth are needed. Two-dimensional (2-d) models frequently reported in the literature are too simplified to actually represent in a correct way the hydrodynamics in a complicated geometry. 2-d models usually considered equally-spaced filaments with cylindrical, triangular or square section (in a cavity, zigzag or submerged configuration: Guillen and Hoek, 2009; Wardeh and Morvan, 2008; Ahmad *et al.*, 2005, 2006; Subramani *et al.*, 2006; Gimmelshtein and Semiat, 2005; Koutsou *et al.*, 2004; Schwinge *et al.*, 2002a,b, 2004 and Geraldes *et al.*, 2002b, but also other filament structures have been studied (Guillen and Hoek, 2009; Fimbres-Weihs and Wiley, 2008 and Shrivastava *et al.*, 2008). The potential impact of axially-orientated filaments parallel to the flow direction was not considered in 2-d numerical models. In other words, 3-d simulations are essential for a correct description of hydrodynamics. Until now, also the 3-d CFD studies have considered simplified spacer geometries, usually with one layer of straight cylindrical filaments crossed at an certain angle over another layer of straight cylindrical filaments (Koutsou *et al.*, 2007, 2009; Shakaib *et al.*, 2007, 2009; Fimbres-Weihs and Wiley, 2007; Ranade and Kumar; 2006a,b; Li *et al.*, 2002 and Karode and Kumar, 2001), see Figure 6.1C – the idealized 'diamond' geometry). The commercially available spacers are in reality more complex, with filaments of variable cross-section area and with an overlapping area at the filament crossings (Vrouwenvelder *et al.*, 2010a) and Figure 6.1AB). For this reason, we investigated in this study the extent that these geometry simplifications may have on the obtained flow patterns and on the biofouling development.

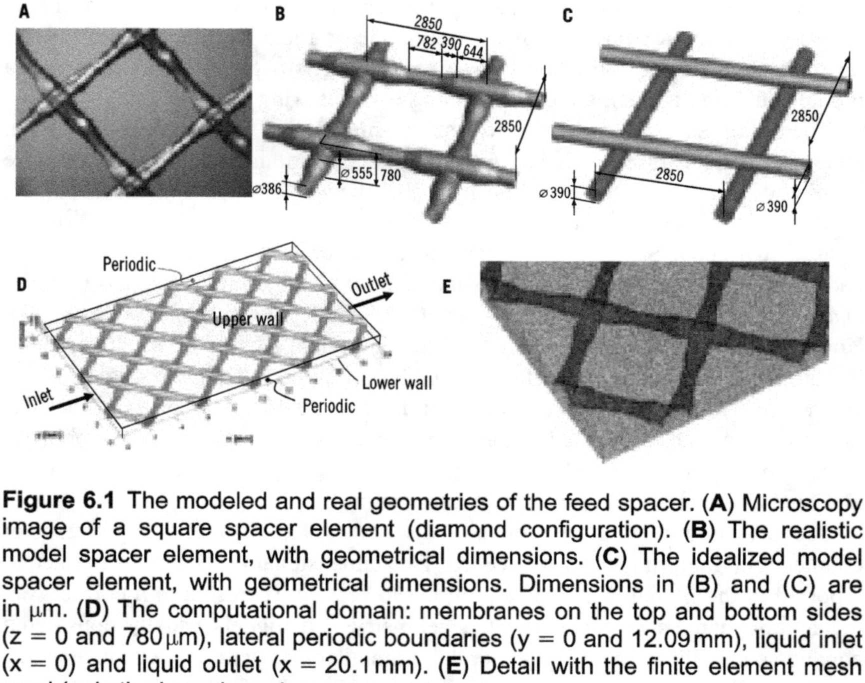

Figure 6.1 The modeled and real geometries of the feed spacer. (**A**) Microscopy image of a square spacer element (diamond configuration). (**B**) The realistic model spacer element, with geometrical dimensions. (**C**) The idealized model spacer element, with geometrical dimensions. Dimensions in (B) and (C) are in μm. (**D**) The computational domain: membranes on the top and bottom sides ($z = 0$ and $780\,\mu$m), lateral periodic boundaries ($y = 0$ and 12.09 mm), liquid inlet ($x = 0$) and liquid outlet ($x = 20.1$ mm). (**E**) Detail with the finite element mesh used (only the boundary elements are shown)

Third, most of the studies consider only a very small representative volume (Li *et al.*, 2009; Shakaib *et al.*, 2007, 2009; Fimbres-Weihs and Wiley, 2007; Koutsou *et al.*, 2007; Ranade and Kumar, 2006a,b and Li *et al.*, 2002). Usually, the size of such a computational domain is limited to one square formed by four crossing cylindrical spacer filaments. A few studies report simulations on larger domains (Koutsou *et al.*, 2009; Li and Tung, 2008b; Dendukuri *et al.*, 2005 and Karode and Kumar, 2001), for example with systems of approximately 3 by 5 square elements (Koutsou *et al.*, 2009). Without biomass growth their results show, as expected, a repetitive pattern in each square element once the flow is well established. Due to biofilm growth the flow will get disturbed in a heterogeneous manner, making it necessary to use larger computational domains.

A model including biofilm growth presents an increased computational difficulty compared with the traditional CFD in membrane devices. The modeling approach presented in this paper is based on previous work related with biofilm development in flow conditions, under mass transfer limitation and in complex geometries. The influence of flow on mass transfer and irregular biofilm surface formation on different support geometries has been studied with 2-d models (e.g., Picioreanu *et al.* (1999, 2000b) and in 3-d (Eberl *et al.*, 2000). Kapellos *et al.* (2007) developed an original 2-d simulator for biofilm development in granular porous materials, followed by the work of Graf von der Schulenburg *et al.* (2009) in three dimensions. One important characteristic of biofouling systems is that while the spacer geometry is fixed, the biofilm colonies will grow in time and the boundaries of the channel in which the water flows are continuously changing. We are dealing now with a problem with moving boundary. For this, new computational approaches had to be developed in this work so that the effect of the channel obstruction with biofilms on the flow pattern could be described with sufficient accuracy both over time and in space, while still being numerically efficient.

MODEL DESCRIPTION

Model geometry and computational domains

We present here a three-dimensional model describing the liquid flow, the mass transport of a soluble substrate and the biofilm development in the feed channel of spiral wound nanofiltration and reverse osmosis membrane devices. The complex geometry of the feed spacer creates an intricate flow pattern, further complicated by the non-uniform biofilm growth in the feed channel. Consequently, the geometry of the domain where the liquid is allowed to flow changes in time as the biofilm develops in the membrane module.

We mainly investigated in this study the spacer geometries as used in real applications. A photograph of one square element of the spacer is presented in Figure 6.1A. The spacer constitutes a solid obstacle within the rectangular flow channel, and keeps the two parallel membrane surfaces separated by 780 μm. The mean flow is sent at 45 degrees angle on the filaments ('diamond' configuration). While the majority of 3-d CFD studies focus on one square element formed by four crossing filaments, this work uses a larger computational domain of 5 by 3 squares. This is needed in order to describe the flow channeling effect created by clogging the flow channels with biofilm. Characteristic dimensions of the spacer are presented in Figure 6.1B. Secondly, the flow characteristics for the realistic spacer geometry are compared with those obtained by considering a simplified traditional arrangement of cylindrical filaments (diameter 390 μm) in two superposed layers crossing at 90 degrees. This idealized spacer geometry is shown in Figure 6.1C.

Momentum balance (hydrodynamics)

The water flow is modeled by the incompressible laminar Navier-Stokes equations (6.1):

$$\begin{cases} \rho(\mathbf{u} \cdot \nabla)\mathbf{u} + \nabla p = \nabla \cdot (\eta \nabla \mathbf{u}) \\ \nabla \mathbf{u} = 0 \end{cases} \tag{6.1}$$

where **u** is the vector of local liquid velocity (with components u, v and w on the three Cartesian directions x, y and z), p is the pressure, ρ is the liquid density and η the liquid dynamic viscosity. The assumption of laminar flow is justified at least for the initial (clean) module configuration. The Reynolds number is $Re = 127$ for an average inlet velocity $u_{in} = 0.163\,\mathrm{m\,s^{-1}}$, channel width $7.8 \times 10^{-4}\,\mathrm{m}$, water density $1{,}000\,\mathrm{kg\,m^{-3}}$ and dynamic viscosity $0.001\,\mathrm{Pa\,s}$. Although several computational studies reveal that in certain conditions time-dependent (unsteady) flow develops around the spacer (e.g., Schwinge *et al.*, 2002b and Ahmad and Lau, 2006), we chose here steady state flow. This assumption is, roughly, like using a solution where the unsteady patterns (i.e., vortexes, recirculation regions) do not exist because they were averaged in time. The steady state assumption is justified by the very long time interval over which we study the biofouling process (days) compared with the time scale of flow perturbations (seconds or less). It is also clear that this steady-state approximation of the flow pattern leads to less mixing in the channel than a truly transient solution would provide. Besides the large Re, we are aware that also other phenomena such as long and flexible vibrating biofilm structures ('streamers') can induce unsteady flow fields. Oscillating biofilm streamers have been experimentally observed (Stoodley *et al.*, 2008) and numerically modeled in a 2-d fluid-structure framework by (Taherzadeh *et al.*, 2009). We accept however the steady state compromise for the great simplification of flow calculations that it brings in an already very difficult system.

The boundary conditions for the fluid dynamics equations are presented in Figure 6.1D. The inflow plane at x = 0 is fragmented in four independent boundaries by the spacers. A fully developed laminar velocity profile with maximum velocity half way between the top and bottom membranes is assumed in the inlet ($x = 0$), with average velocity u_{in}. The liquid flow rate and consequently this velocity are kept constant in time. We adopted the constant flow rate condition because membrane filtration installation operators in practice want to maintain constant production of permeate (this is achieved by increasing the feed pressure and cleanings). It is also obvious that for constant pressure drop total clogging will occur instead of channeling. On the outlet boundary ($x = L_X$) zero-pressure (arbitrarily chosen as reference value) and zero-viscous-stress conditions are imposed. The top and bottom membrane surfaces ($z = 0$ and $z = L_Z$) as well as the whole spacer surface are no-slip boundaries, i.e., walls with zero liquid velocity. It was assumed that there is no water flow within the biofilm matrix because its permeability is shown to be negligible, in the order of $10^{-16}\,m^2\,s^{-1}$ (Fowler and Robertson, 1991 and McDonogh *et al.*, 1994). On the lateral boundaries ($y = 0$ and $y = L_Y$) periodic conditions are applied in order to minimize the unwanted edge effects of a limited computational domain, and mimic a system where the spacer would extend infinitely in the y direction. The periodic condition enforces identical distributions of the velocity on the plane at $y = 0$ with those on the plane at $y = L_Y$.

The solution of Navier-Stokes equations (6.1) provides the 3-d distributions of velocity $\mathbf{u} = (u_X, u_Y, u_Z)$ and liquid pressure p, and the viscous shear stress on the walls τ. The liquid velocity is further used to support the convective transport of solutes, while the value of shear stress determines the attachment of new biofilm cells on the solid surfaces.

Mass balance for soluble substrate

The biofilm cells grow as a function of the local concentrations of soluble substrates (e.g., organic sources of carbon and energy, oxygen, nitrate, carbon dioxide, etc.). For exemplification of the modeling approach it is sufficient to assume one limiting compound, for example dissolved oxygen. The 3-d distribution of substrate concentration C_S in the flowing liquid and in the biofilm results from the solution of a mass balance equation:

$$D_S \nabla^2 C_S - \mathbf{u} \nabla C_S + r_S = 0 \tag{6.2}$$

including transport terms by molecular diffusion with a diffusion coefficient D_S (for simplicity the same value in liquid as in the biofilm volume) and convection with the liquid velocity $\mathbf{u}$, balanced by the rate of substrate consumption r_S. To keep the model formulation simple, but realistic, traditional Monod kinetics with saturation is assumed for the only limiting substrate:

$$r_S = Y_{SX} \mu_m \frac{C_S}{K_S + C_S} C_X \tag{6.3}$$

Values for the reaction rate parameters are taken both from experiments in membrane modules (maximum specific rate $\mu_m = 2.8\ \text{day}^{-1}$ from Vrouwenvelder *et al.* (2010a) and according to established activated sludge models (ASM, see Henze *et al.*, 2000) for heterotrophic biomass with concentration C_X growing on a substrate with half-saturation coefficient $K_S = 0.063\ \text{mol m}^{-3}$ and yield $Y_{SX} = 0.52$ mol/C-mol biomass. For exemplification, in the current model simulations we assumed oxygen being the rate limiting reactant. Choosing the organic compound as limiting substrate will not change the general conclusions reached.

On the inlet boundary ($x = 0$) the water flow contains the limiting substrate with concentration $C_{S,in}$ constant in time. Only convective flux (no diffusion, $D_s(\partial C_s/\partial x) = 0$) applies on the outlet boundary ($x = L_X$). The top and bottom membrane surfaces ($z = 0$ and $z = L_Z$) as well as the whole spacer surface are impermeable walls (zero-flux boundaries). The flux of permeate through the membrane is only a very small fraction from the local convective flow in the feed spacer channel (Vrouwenvelder *et al.*, 2009b,c), therefore we neglected the flux here. Similar to the velocity field, on the lateral boundaries ($y = 0$ and $y = L_Y$) periodic conditions are applied for the oxygen concentration.

In the case of dynamic transport of a soluble conservative (non-reactive) tracer in the feed spacer channel, instead of the steady state mass balance equation (6.2) the time-dependent equation (6.4) was solved for the tracer concentration C_T:

$$\frac{\partial C_T}{\partial t} = D_T \nabla^2 C_T - \mathbf{u}\nabla C_T \tag{6.4}$$

with the same boundary conditions as for substrate. The only differences were that no reaction produces or consumes the tracer and that the inlet tracer concentration follows a function of time that approximates a pulse (duration 0.3 s):

$$C_{T,in}(t) = \frac{1}{1+e^{-300(t-t_1)}} - \frac{1}{1+e^{-300(t-t_2)}} \tag{6.5}$$

with the initial time of tracer injection $t_1 = 0.05$ s and final time $t_2 = 0.35$ s.

Mass balance for biomass

The development of a biofilm on the spacer and membranes is determined both by the substrate availability and by the hydrodynamic conditions (Van Loosdrecht *et al.*, 1995, 1997). On the one hand microbial cells attach to the solid surfaces, consume substrate, grow, divide, and produce extracellular polymers, thereby increasing the biofilm volume. On the other hand the liquid shear promotes biofilm detachment and therefore loss of biofilm volume. All these processes were integrated in previous multi-dimensional biofilm models, in different ways (Picioreanu *et al.*, 1998a,b, 2001, 2004; Xavier *et al.*, 2005 and Kapellos *et al.*, 2007).

We adapted here the discrete cellular automata (CA) biofilm model by Picioreanu *et al.* (1998a,b, 2001), which in principle corresponds to a biomass balance including terms for accumulation, transport (e.g., diffusion and convection), reaction (e.g., growth and decay) and biofilm/liquid mass transfer (e.g., attachment and detachment). In the CA approach, the whole computational space is divided into a mesh of cubic elements. In the biofilm region the CA cubes contain biomass with a concentration C_X, whereas in the liquid space $C_X = 0$. For simplicity, only one type of aerobic biomass was considered in this study.

In the cellular automaton representation the biomass growth is solved apart from the biomass transport and transfer, method which constitutes a simplified split-operator model.

Biomass growth – occurs in each biofilm cubic element with a rate given by:

$$\frac{dC_X}{dt} = r_X = \mu_m \frac{C_S}{K_S + C_S} C_X \tag{6.6}$$

where the left-hand side is the accumulation term and the right-hand side is the biomass growth rate, linked to the substrate consumption equation (6.3). In order to keep the analysis of model results as simple as possible we neglected here the biomass decay and detachment.

Biomass transport – takes place when the biomass concentration C_X reaches a maximum value $C_{X,m}$. The biomass redistribution in the neighboring cubes takes place according to the CA algorithm described in (Picioreanu *et al.*, 1998a) for biofilms on planar support, and adapted here for any arbitrary support geometry. Although empirical (not based on a rigorous theory, but being just a convenient algorithm), the CA model has proven to describe well experimental observations on formation of spatial structure of biofilms.

Biomass attachment – is included here with a constant rate of $r_{X,att} = 6 \times 10^{-4}$ (C-mol biomass) (m^{-2} module) day^{-1}. For the model solution this is equivalent to 200 cubic elements containing biomass newly added each day. The attachment place is chosen randomly, but is also a function of shear stress: it was considered that the most probable sites to be colonized are those exposed to low liquid shear. In this way, $r_{X,att}$ should be conceptually seen as a net attachment rate resulting from the difference between a real attachment rate and the rate of rapid detachment of this newly attached biomass before creating stable colonies. Consequently, attachment occurs only on surfaces where the viscous shear stress τ_{att} is less than 1.5, 10 or 100 Pa. These values of shear were chosen so that the biomass attaches: (1) only on the spacer filaments behind the crossings, (2) everywhere on the filaments, or (3) everywhere on the filaments and membranes, respectively.

Model solution

The model was implemented in a combination of: (i) MATLAB code (MATLAB 2007b, MathWorks, Natick, MA, www.mathworks.com) as the main algorithm script, (ii) COMSOL Multiphysics (COMSOL 3.5, Comsol Inc., Burlington, MA, www.comsol.com) finite element methods for solving the partial differential equations governing the flow field and the solute mass balances, and (iii) own Java code for the cellular automata biofilm model.

I. The main program code is a MATLAB script, which in a first section defines the:
   a. model input parameters – from Table 6.1;
   b. model geometry (the rectangular computational domain and the spacer geometry);
   c. three-dimensional COMSOL application modes for:
      1) momentum transfer – incompressible Navier-Stokes equations (6.1);
      2) mass transfer – diffusion, convection and reaction, equation (6.2), with the corresponding boundary conditions;
   d. 3-d mesh for finite element solution of the hydrodynamics and mass balances. We used a mesh of ~430,000 Lagrange P_2P_1 tetrahedral elements (quadratic accuracy) with a maximum size of 200 μm both in the volume and near the spacer boundaries (Figure 6.1E). This means ~2,100,000 equations simultaneously solved for hydrodynamics and ~663,000 for the solute mass balance.
   e. 3-d mesh for biomass and pseudo-viscosity (see step (f) in this algorithm) distributions. A mesh of $N_X \times N_Y \times N_Z = 250 \times 150 \times 10$ cubic elements was used (cube size ~80 μm).

II. The second section of the script solves the model equations in a time loop (time step $\Delta t = 6$ hours). A sequential approach was used, based on the assumption that characteristic time scales of hydrodynamics, mass transport and biofilm development are very different (Picioreanu *et al.*, 1999, 2000b). At any time t there are successively solved:
   f. hydrodynamics at steady state to get $\mathbf{u}$ and p (with COMSOL finite element methods), for a given geometry of the biofilm matrix and spacer. The spacer is explicitly defined as a solid object in COMSOL. The liquid and the biofilm however constitute the same sub-domain.

The biofilm matrix (assumed an impermeable and growing solid object) is implicitly defined by assigning very large arbitrary values of the liquid viscosity ($\eta = 10^4$ Pa s) in the computational volume occupied by biomass. This arbitrarily large water viscosity – only a numerical convenience, obviously ~ is termed 'pseudo-viscosity' in this study, in contrast with the real water viscosity ($\eta = 10^{-3}$ Pa s) applied in the regions without biofilm. A spatially variable viscosity function $\eta(x,y,z)$ is constructed on the tetrahedral finite element mesh by linear interpolation from the cubic CA mesh.

g. mass balances for the soluble substrate at steady state to get C_S (with COMSOL finite element methods) by using the calculated velocity field **u** and the given 3-d biomass distribution C_X. The 3-d distribution of all model state variables at this time t (i.e., **u**, p, C_S, and C_X) is saved after this step. Similar to the method applied in step (f) for the liquid viscosity, the 3-d biomass concentration $C_X(x,y,z)$ on the finite element mesh is found by linear interpolation from the CA mesh.

   To calculate the dynamics a soluble tracer transport in the feed spacer channel, instead of the steady state mass balance equation (6.2) a time-dependent equation was solved.

h. biomass balances including:
   1. attachment of new biomass using the distribution of shear t on the walls. The biomass attached only in places where $\tau < \tau_{att}$.
   2. growth of biofilm biomass to get the new C_X at $t + \Delta t$ from C_X at t and the previously calculated C_S,
   3. spreading (transport) of biomass using own Java algorithms as described in (Picioreanu *et al.*, 1998a,b).

Finally, the 3-d biomass distribution C_X (needed in the substrate mass balance) and the pseudo-viscosity distribution η (needed for the definition of solid walls in hydrodynamics) are updated on the cubic CA mesh. With the new biofilm geometry and biomass distribution so obtained a new time step starts with the hydrodynamic calculations (e).

MODEL RESULTS AND DISCUSSION

A series of case studies are presented here to exemplify the use of the model for investigation of: (1) three-dimensional interactions between biofilm and the flow pattern, and the importance of describing the real spacer geometry, (2) the effect of substrate mass transport limitations on biofouling and (3) the effect of biofouling on the solutes residence time distribution. Table 6.1 presents all model parameters used in the simulations.

Table 6.1 Parameters for the model of biofouling in feed spacer channels of membrane devices

Parameter	Description	Value	Units	Source
System dimensions				
L_X	Length of the computational domain along the main flow direction	20.1	mm	Chosen
L_Y	Width across the channel	12.1	mm	Chosen
L_Z	Height between the membranes Other dimensions of spacer filaments are presented in Figure 1.	0.78	mm	Measured spacer geometry
ε_L	Feed spacer channel porosity	0.84	m^3 liquid m^{-3} module	Calculated
a_S	Specific surface area feed spacer	7,700	m^2 spacer m^{-3} spacer	Calculated
a_V	Specific surface area feed spacer per module volume	1,230	m^2 spacer m^{-3} module	Calculated
Hydrodynamic parameters				
ρ	Water density	1,000	kg m^{-3}	
η	Water dynamic viscosity	10^{-3}	Pa s^{-1}	
u_{in}	Mean water inlet velocity	0.163	m s^{-1}	Chosen from practice (Vrouwenvelder *et al.*, 2009a)
u_{in}	Mean water inlet velocity in mass transfer cases only	0.0163	m s^{-1}	Chosen (for numerical stability lower than in practice)
Dissolved components				
$C_{S,in}$	Inlet concentration of substrate	0.25	mol m^{-3}	Chosen
$C_{T,in}$	Inlet concentration of tracer solute	1	mol m^{-3}	Chosen
D_S, D_T	Diffusion coefficients for substrate and tracer	2.5×10^{-9}	m^2 s^{-1}	Value for oxygen

(*Continued*)

Table 6.1 (*Continued*)

Parameter	Description	Value	Units	Source
Microbial growth and substrate consumption rate				
μ_m	Maximum specific biomass growth rate	2.8	day^{-1}	(Vrouwenvelder *et al.*, 2010a)
K_S	Monod half-saturation coefficient for substrate	0.063	mol m^{-3}	(Henze *et al.*, 2000)
Y_{SX}	Molar yield of substrate consumed per C-mol biomass	0.52	mol C-mol^{-1}	(Henze *et al.*, 2000)
$C_{X,m}$	Maximum biomass concentration in the biofilm	2,800	C-mol m^{-3}	Typical for a heterotrophic biofilm (Wanner *et al.*, 2006)
Microbial growth and substrate consumption rate				
$r_{X,att}$	Biomass attachment rate	6×10^{-4}	C-mol biomass m^{-2} module day^{-1}	Chosen
τ_{att}	Threshold viscous stress at the wall to allow biomass attachment	1.5, 10, 100	Pa	Chosen

Interaction between hydrodynamics and biofilm growth

(a) Importance of 3-d simulations

This case is intended to demonstrate why three-dimensional computations are needed for a good description of hydrodynamics in the feed spacer channels of membrane modules. Figure 6.2 presents two-dimensional sections (slices) through the computed 3-d field of total velocity, $U = |\mathbf{u}| = \sqrt{u_X^2 + u_Y^2 + u_Z^2}$. The velocity distributions from Figure 6.2 are calculated with a clean spacer, thus at day 0 of membrane module operation when no biofilm has formed yet. They will serve as reference for the following cases with biofilm growth.

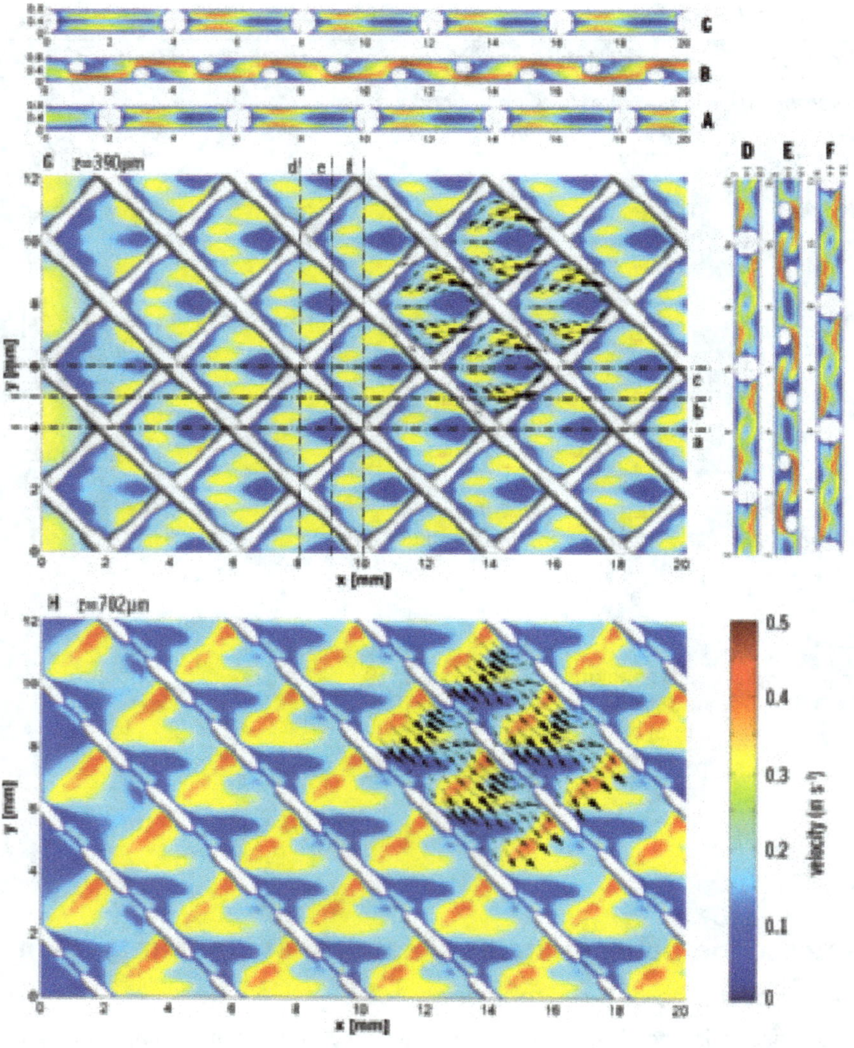

Figure 6.2 *Liquid velocity distribution in the feed channel without biofilm.* Two-dimensional distributions of liquid velocity are presented in different sections through the 3-d computational domain. (**A–C**) velocity in x-z planes at y = 4.03, 5.04 and 6.05 mm; (**D–F**) velocity in y-z planes at x = 8.06, 9.07 and 10.08 mm; (**G,H**) velocity in x-y planes at z = 390 and 702 μm. The spacer is shown as white crossing filaments. The colour scale indicates the liquid velocities (red – highest, blue – lowest). Arrows indicate the flow velocity direction and magnitude

It is apparent from Figure 6.2 that the flow pattern is mostly established after passing one square spacer element (about $x = 4$ mm). The flow velocity distribution in all remaining square elements is identical (from $x = 4$ to 20 mm). The periodic condition imposed on the y direction also ensures repeatability of the flow characteristics on this direction. It can be therefore seen that the velocity values on the side plane at $y = 0$ perfectly match those on the other side at $y = L_Y = 12$ mm (Figure 6.2GH). From these results it can be concluded that in the absence of a heterogeneous biofilm growth, the flow pattern in the membrane module could be described by computing the fluid dynamics in one square element (such as in previous studies by Shakaib *et al.*, 2007, 2009; Fimbres-Weihs and Wiley, 2007; Koutsou *et al.*, 2007 and Ranade and Kumar, 2006a,b).

The flow structure is however fully three-dimensional and follows a tortuous path not only up and down past the spacer filaments (Figure 6.2A–C), but also in the lateral direction y (Figure 6.2D–F). The highest flow velocities ($\sim 0.5\,\mathrm{m\,s^{-1}}$) appear, as expected, in the narrow section between the filaments and the membrane surface (Figure 6.2B). These high velocities can also be seen in the section in the x-y plane parallel with the membrane at height $z = 702\,\mu\mathrm{m}$ (Figure 6.2H), compared with a section through the middle plane at z = 390 μm (Figure 6.2G). No two-dimensional computations can describe this flow pattern, with changing the dominant flow direction from 0 degrees with the x axis in the middle ($z = 390\,\mu\mathrm{m}$) to 45° at $z = 702\,\mu\mathrm{m}$ or –45° at $z = 72\,\mu\mathrm{m}$ (see arrows on Figure 6.2GH and Figures 6.3, 6.4 and 6.5).

The pressure drop over the 20 mm long feed spacer channel simulated here is about 0.8 kPa in the case of a clean device (or $0.4\,\mathrm{kPa\,cm^{-1}}$). In an actual membrane module this value depends on the membrane position in the installation (Vrouwenvelder *et al.*, 2009d). Biofouling is usually observed in the lead module of the lead pressure vessel (Carnahan *et al.*, 1995 and Vrouwenvelder *et al.*, 1998, 2008b, 2009d). In practice in clean lead membrane modules values around $0.4\,\mathrm{kPa\,cm^{-1}}$ have been observed in (Vrouwenvelder *et al.*, 2010a), which agree well with the simulation results. Figure 6.6 presents a 2-d section through the pressure field, in the middle of the channel ($z = 390\,\mu\mathrm{m}$). It can be clearly seen that the pressure changes only slightly within a square spacer element, which means that the pressure drop due to membrane walls is very small. The main resistance, and thus pressure drop, is encountered by the flow passing over the spacer filaments. This is evident in Figure 6.6BC presenting the values of liquid pressure on different lines along the membrane module: the pressure falls abruptly in steps. The steps are higher at the filament crossings (Figure 6.6B – line B) than in the middle of the square sides (Figure 6.6B – line A). A section through the channel middle (Figure 6.6C) shows that the pressure even increases in front of the filament crosses.

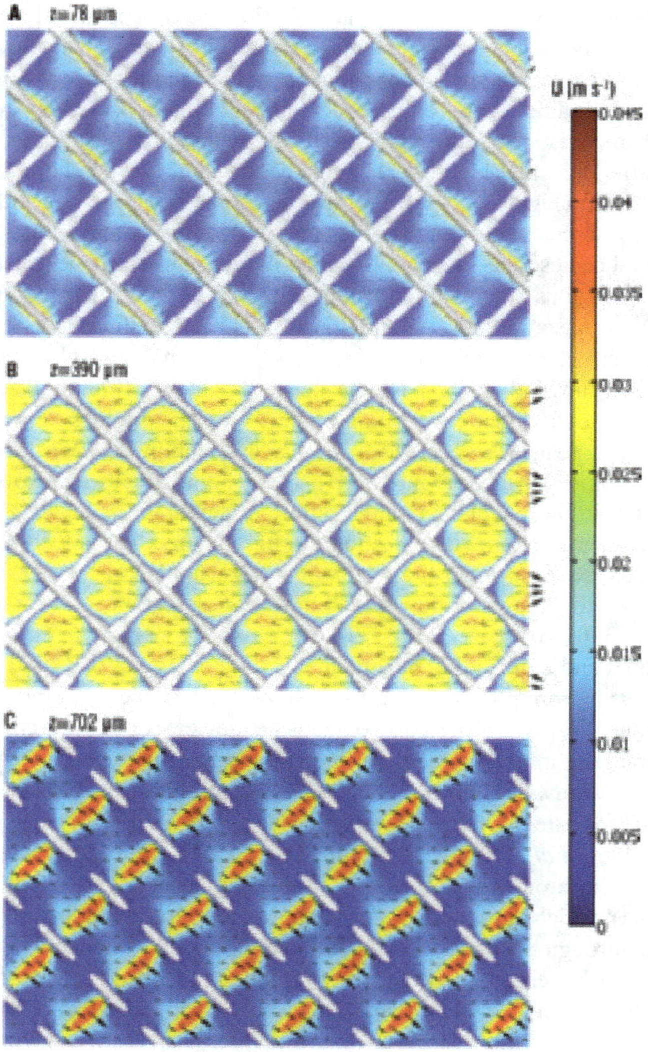

Figure 6.3 Two-dimensional flow velocity distributions for the **real spacer geometry**, calculated at **low flow velocity** $u_{in} = 0.0163\,\mathrm{m\,s^{-1}}$ without biofilm. The three *x-y* sections are at (**A**) $z = 78$, (**B**) $z = 390$ and (**C**) $z = 702\,\mu\mathrm{m}$. The colour scale shows the flow speed and the arrows show the flow velocity direction

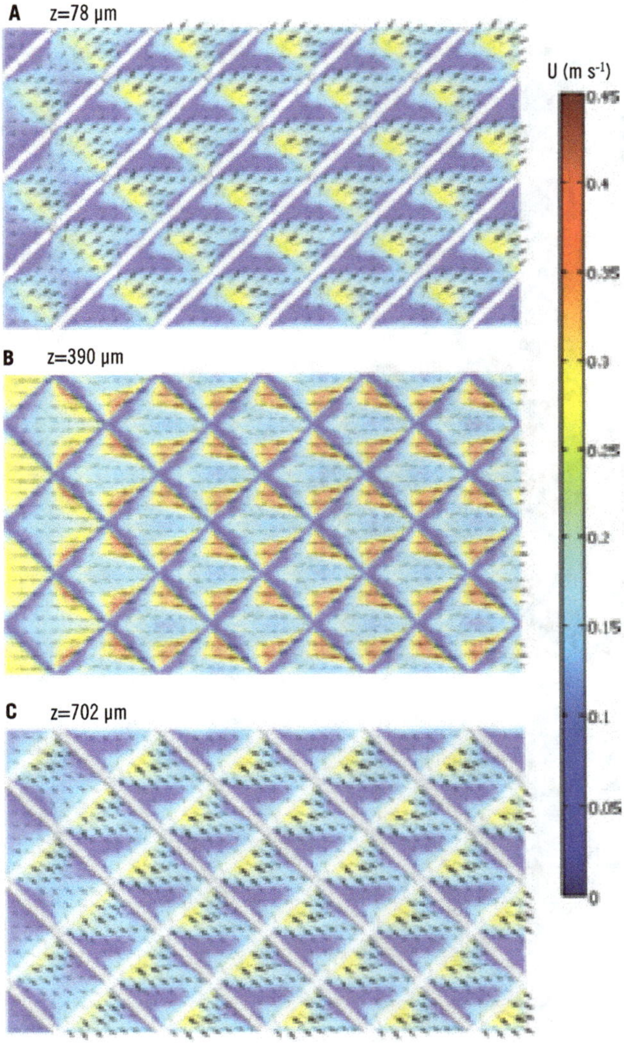

Figure 6.4 Two-dimensional flow velocity distributions for the **idealized spacer geometry**, calculated at **high flow velocity** u_{in} = 0.163 m s^{-1} without biofilm. The three *x*-*y* sections are at (**A**) z = 78, (**B**) z = 390 and (**C**) z = 702 µm. The colour scale shows the flow speed and the arrows show the flow velocity direction

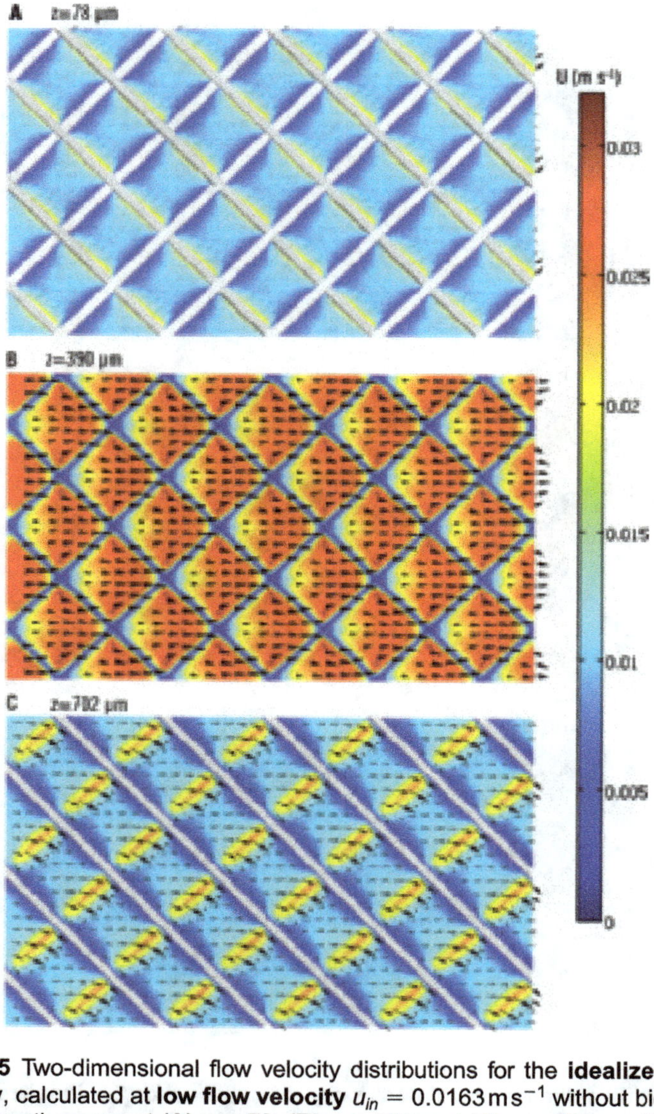

Figure 6.5 Two-dimensional flow velocity distributions for the **idealized spacer geometry**, calculated at **low flow velocity** $u_{in} = 0.0163\,\mathrm{m\,s^{-1}}$ without biofilm. The three *x-y* sections are at (**A**) $z = 78$, (**B**) $z = 390$ and (**C**) $z = 702\,\mu\mathrm{m}$. The colour scale shows the flow speed and the arrows show the flow velocity direction

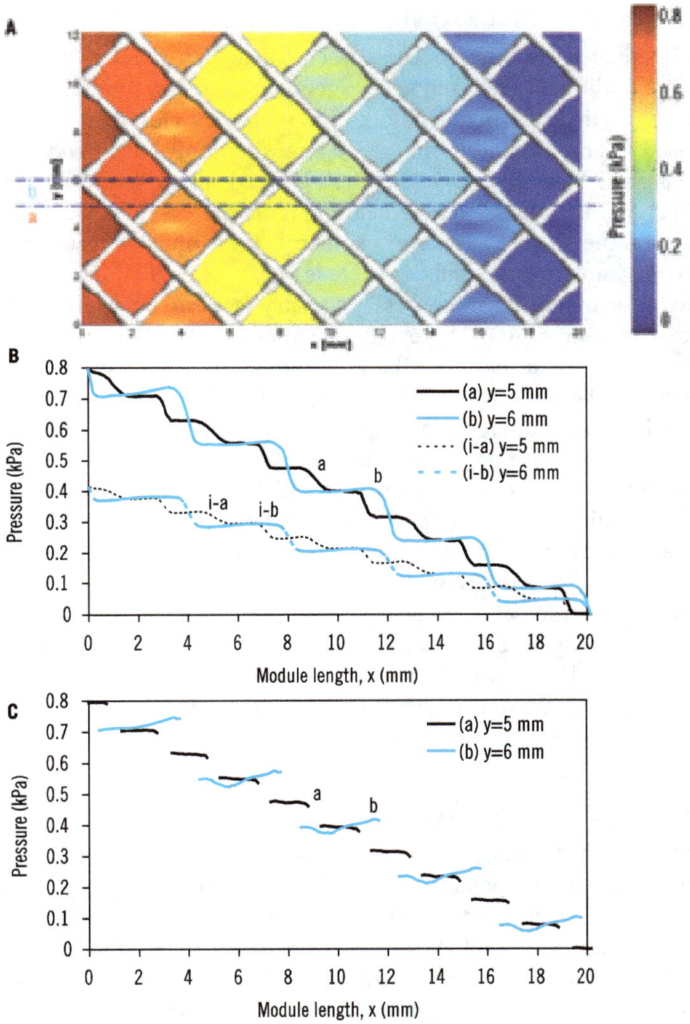

Figure 6.6 *Pressure distribution over the feed channel for the clean spacer.* (**A**) Two-dimensional pressure distribution in a section at $z = 390\,\mu m$ height in the feed spacer channel. Colour scale from high pressure (red, 0.8 kPa) to low pressure (blue, 0 kPa). Lines *a* and *b* indicate the sections for pressure profiles in panels (**B**) and (**C**). (B) Pressure along lines at the membrane surface, $z = 0$, at $y = 5$ (a) and $y = 6$ mm (b). Pressure drop for the idealized spacer geometry is smaller and indicated with lines (i-a) and (i-b). (C) Pressure along lines in the middle of the channel, $z = 390\,\mu m$, at $y = 5$ (a) and $y = 6$ mm (b)

(b) Importance of real vs. simplified spacer geometry

This case evaluates the differences between calculating pressure drop and flow fields with a traditional idealized spacer geometry (crossing arrays of spaghetti-like perfectly cylindrical filaments) and the more complex model geometry of spacers used in practice and in this study (crossing arrays of extruded filaments with necking and blended crossings). The idealized spacer geometry is shown in Figure 6.1C and the calculated flow field in Figure 6.4.

In case of the idealized spacer geometry, the principal result is that the calculated pressure drop along the module is only $0.2\,\mathrm{kPa\,cm^{-1}}$, which represents half of the value obtained for the 'real' geometry (see two pressure profiles along the length in Figure 6.6B). The 6% higher porosity of the idealized geometry (91%) can only explain partially the much lower resistance to flow. The main reason seems to be the lower mixing intensity induced by the idealized spacer, leading to less dissipation of momentum. This argument is supported by a numerical experiment we performed for the residence time distribution (at lower liquid velocity, for the reasons explained below). A comparison between the tracer concentrations in the outlet for the idealized and real spacer geometries shows less dispersion of the quasi-pulse tracer signal in the case of idealized spacer (Figure 6.7). Also visually, the tracer pulse looks more dispersed at the same time moments compared with the tracer in the real geometry (see Figure 6.8). Again, this is a consequence of the more spread out velocity directions for the flow past the real spacer (compare velocity distributions from Figure 6.4 with those from Figure 6.5). These simulations show that using an idealized geometry of the spacer is a too large simplification.

(c) Flow with biofilm growth

The second reason for choosing a 3-d model step is the characteristic way in which the biofilm development perturbs the flow field and creates preferential flow paths. For mass transport in a biofilm growing on planar support the 2-d and 3-d simulation results did not show significant qualitative differences (see Eberl *et al.*, 2000). However, in a porous medium, differences between the 2-d and 3-d simulation results are evident from the work of Graf von der Schulenburg *et al.* in granular packed beds (Graf von der Schulenburg *et al.*, 2009). In 2-d a closed pore neck (e.g. due to biofilm clogging or touching solids) will immediately prevent flow and substrate convection in that area. This leads to comparatively slower diffusive fluxes and only a low nutrient concentration. In 3-d the fluid in such a region will not be stagnant and the convective nutrient flux will penetrate via the third spatial dimension leading to higher nutrient concentrations or transport because of increased local flow rates.

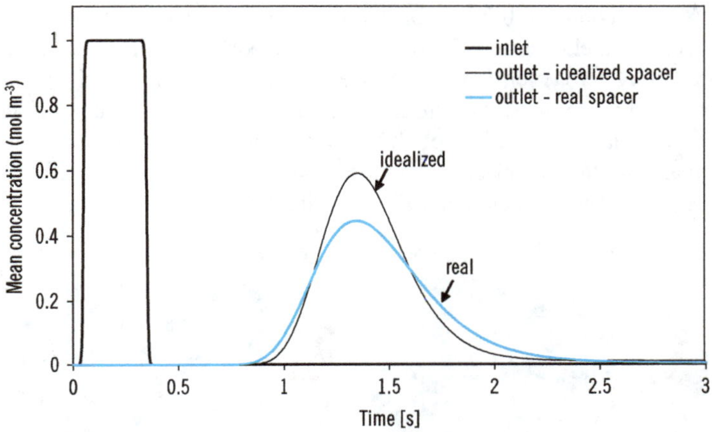

Figure 6.7 Solute residence time distribution in the feed spacer channels without biofilm compared for real and idealized spacer. The concentration of solute tracer in the inlet (red line) is presented together with the mean solute tracer concentration in the outlet recorded over time: idealized spacer geometry (black line) and the more realistic spacer geometry (blue line)

The simulations presented here calculate the flow pattern at the relatively high inlet flow velocity of $u_{in} = 0.163\,\mathrm{m\,s^{-1}}$, as used in practice. However, at this velocity (and moreover at the larger ones resulting when part of the channel is obstructed with biomass) the mass transport is clearly dominated by convection. The Péclet numbers ($Pe = u_{in}L_Z/D_S$, ratio between convective and diffusive transport rates) in this case are very high ($Pe > 50{,}000$), making the numerical solution of concentration fields very difficult. A good solution would avoid introduction of artificial diffusion to stabilize the numerical solution and require a very fine mesh that would generate a computational burden unrealistic with current computing power for such a large system size. In this case we chose to describe the biofilm biomass growth without limitations due to the mass transfer of substrate. Therefore, only the flow equations and the biomass balances are solved when $u_{in} = 0.163\,\mathrm{m\,s^{-1}}$. Three simulations with biofilm formation using different threshold shear values for biomass attachment were performed.

When the threshold shear for biomass attachment was 10 Pa the biofilm started growing only on the spacer filaments and not on the membrane. After 3.5 days (Figure 6.9C), two characteristics of the flow field are notable. First, the narrow flow channels resulted from biofouling lead to very high ($\sim 1\,\mathrm{m\,s^{-1}}$) fluid velocities. This artificial velocity enhancement was also studied in (Graf von der Schulenburg *et al.*, 2009). Second, the biofilm formation changes the regular flow pattern in the feed spacer channel (Figure 6.2) into a heterogeneous flow distribution, as shown in Figure 6.9. Heterogeneity means on one hand preferential flow channels, and on the other hand more dispersion in the cross-flow direction y. It can be seen from Figure 6.9C that in the middle of the flow channel ($z = 390\,\mu m$), the flow loses its dominant unidirectional characteristic (compared with clean spacer in Figure 6.2G), i.e., strong components in the lateral direction y can be noticed.

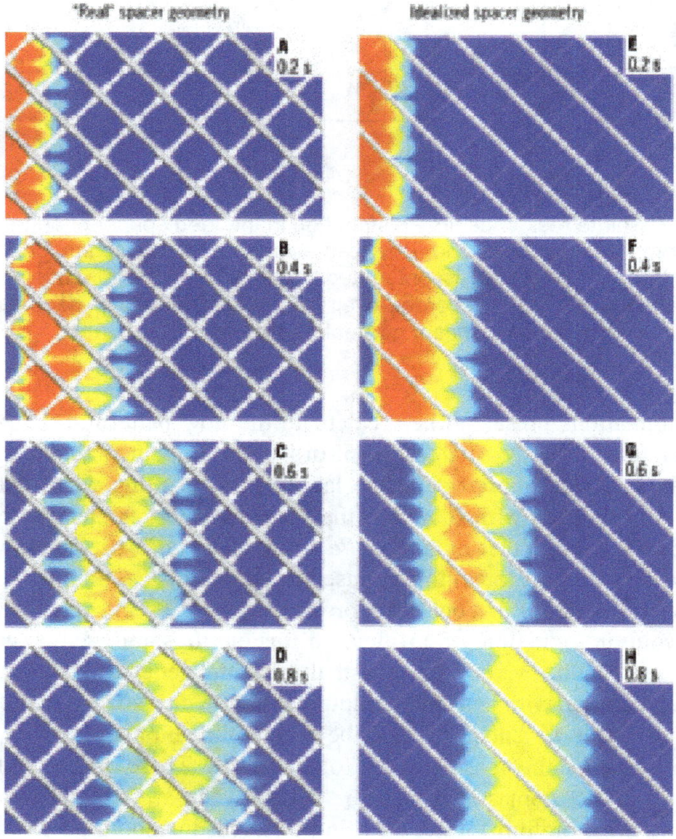

Figure 6.8 Two-dimensional tracer concentration distribution in middle sections of the feed spacer channel ($z = 390\,\mu m$) for the real spacer geometry (left: **A–D**) and idealized geometry (right: **E–H**), at 0.2, 0.4, 0.6 and 0.8 s. The tracer moves in the channels without biofilm. The average flow velocity in the inlet is $u_{in} = 0.0163\,m\,s^{-1}$. The colour scale for local tracer concentration is from 0 (blue) to 1 mol m^{-3} (red)

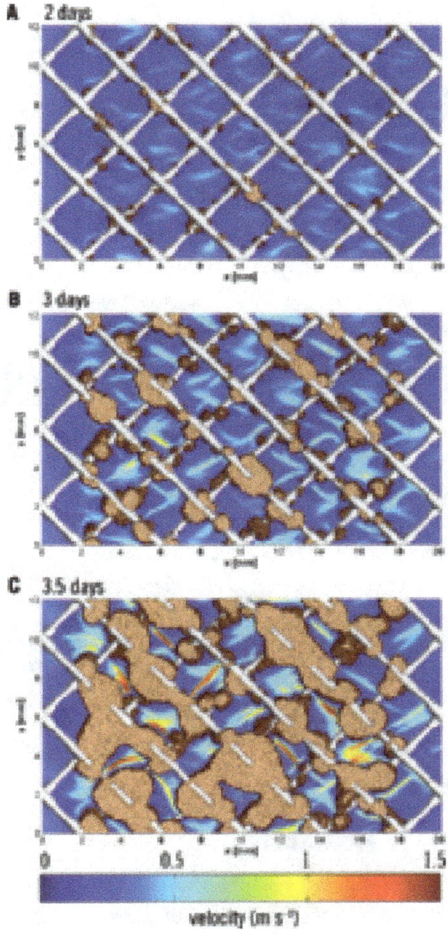

Figure 6.9 *Three-dimensional simulation of flow and biofilm formation in time.* Flow velocity profiles are shown in a two-dimensional section along the feed channel of a spiral-wound membrane device, at $z = 390\,\mu m$ and: (**A**) day 2, (**B**) day 3, (**C**) day 3.5. The development in time of the biofilm (brown volumes) on the spacer (white crossing filaments) changes the originally regular flow pattern into a heterogeneous flow with narrower preferential channels and higher velocities. The colour map shows the flow velocity from high values (red) to low values (blue). The flow is from left to right. New biomass attached where shear was less than 10 Pa

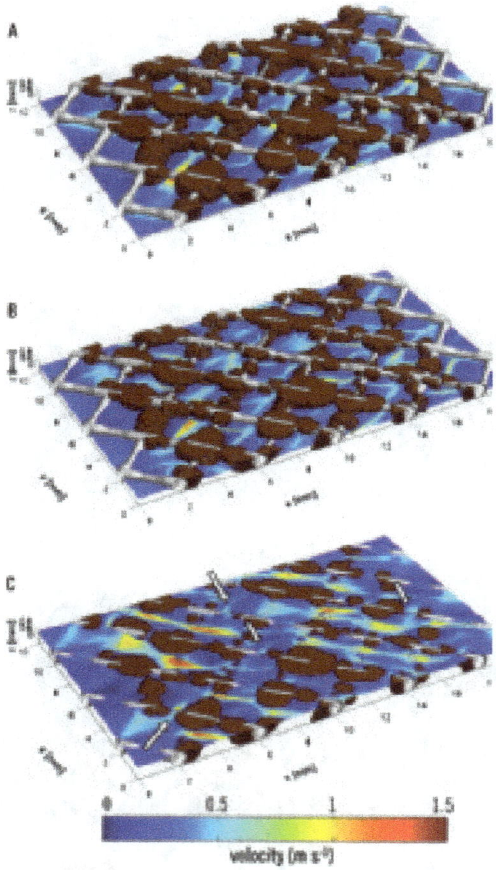

Figure 6.10 *Three-dimensional simulation of flow and biofilm formation.* Flow velocity profiles at day 3 are shown in two-dimensional sections at different heights in the feed channel of a spiral-wound membrane device, at (**A**) $z = 78\,\mu m$, (**B**) $z = 390\,\mu m$ and (**C**) $z = 702\,\mu m$. Colours show the biofilm (brown volumes), the spacer (white crossing filaments), and the flow velocity from high values (red) to low values (blue). The flow is from left to right. New biomass attached where shear was less than 10 Pa. Arrows in (C) point to channels with flow perpendicular to the main flow direction in that section

Flow field sections parallel with the membrane at different heights show first that, in general, there is a heterogeneous pattern of flow speed close to the membranes: there are regions of strongly accelerated flow alternating with quasi-stagnant zones (Figure 6.10AC). Secondly, although there is in general a clearly dominant flow direction at $z = 78$ and $702\,\mu m$, blocking of some channels forces the flow to change direction at 90 degrees in order to pass to the next spacer square element (see arrows on Figure 6.10C).

As a result of the overall biomass growth (Figure 6.11A) the overall pressure drop in the membrane module (per cm) quasi-exponentially increases in time (Figure 6.11B). The biomass growth leads also to a significant reduction of the channel porosity (fraction of liquid, e_L), from 84% initially to about 62% after 80 hours (~3.5 days). However, this overall decreased hydraulic section caused by the biofilm growth is not the only cause for the large increase in resistance to flow. The increased mixing and dispersion effects inherent to the more heterogeneous flow also contribute to this pressure drop.

The question remains, where are the places where biofilm growth affects most the pressure drop: on the spacer filaments or on the membrane surfaces? We performed therefore also simulations with biomass attached in places where the local shear is less than 100 Pa, i.e., leading to biomass growth both on spacer and membranes. Figure 6.12 presents profiles of pressure averaged over *y-z* planes, along the feed channel length at different moments. The biofilm growth on the spacer increases the pressure drop after 3 days up to 5 kPa ($2.5\,kPa\,cm^{-1}$) (see Figure 6.12A). Growth of the same biomass on the filaments only (at $\tau < 10\,Pa$) leads to more resistance to flow. Significantly, the pressure mostly drops in steps. Conversely, if the biomass is distributed also on the membrane, the pressure falls more uniformly (Figure 6.12B). This clearly demonstrates that the growth on spacers has the most effect on flow by further reducing an already diminished flow section when the fluid passes between filaments and membrane. An example of indiscriminate biofilm formation both on membrane and spacer, following the attachment rule of $\tau < 100\,Pa$, is shown in Figure 6.13A.

(d) Position of biofilm growth on the spacer

Experimental studies reveal that biofilm formation begins mainly in places behind the spacer filament crossings (Vrouwenvelder *et al.*, 2006b, 2009c, 2010a,b). An obvious reason for this observation may be that, by a lower shear level, the places behind the spacer filament crossings provide a shielded environment for better biomass attachment and/or less detachment. Figure 6.14A displays the viscous shear values on the spacer filaments. It is apparent that filament areas frontally exposed to flow experience most shear. Less shear is on the back on the filaments and the least behind the crossings. These are places where the biofilm starts to form when simulations were performed with the condition that attachment can occur only at $\tau < 1\,Pa$ (see Figure 6.14B). The simulated biofilm distribution in this case in the whole domain is shown in Figure 6.13B.

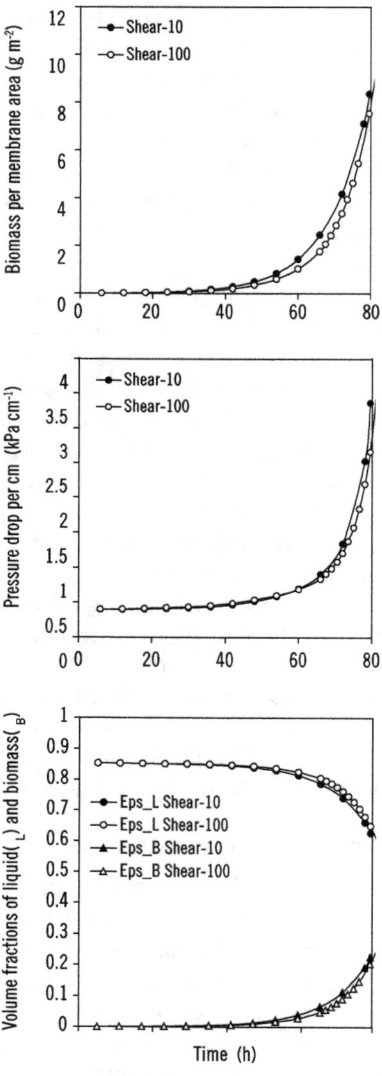

Figure 6.11 *Biomass, porosity and pressure drop in time.* (**A**) Biofilm biomass accumulated in the feed channel. The biomass is related to the surface area of one membrane side ($2.4\,cm^2$). (**B**) Evolution of pressure drop (per cm feed channel) in time. (**C**) Time profile of liquid (circles) and biomass (triangles) hold up in the feed channel. For all graphs there are two cases: when biomass attachment is mainly on the spacer filaments (places with shear <10 Pa, black symbols) and with attachment both on spacer and on membranes (places with shear <100 Pa, open symbols)

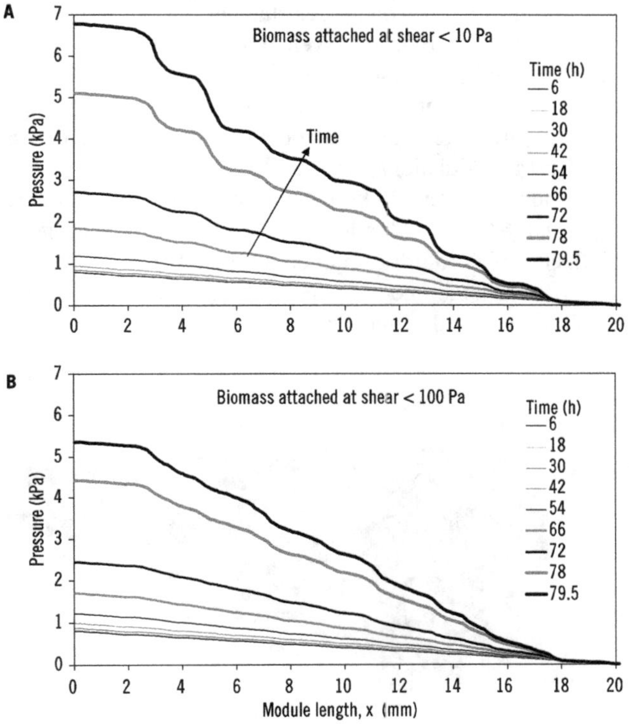

Figure 6.12 *Mean pressure over the feed channel length, in time.* (**A**) Case biomass attachment mainly on the spacer filaments (places with shear <10 Pa, black symbols) and (**B**) with attachment both on spacer and on membranes (places with shear <100 Pa, open symbols). Different lines on a graph show the mean pressure (average in *y-z* sections at each length *x* along the channel) at different times from 6 to 79.5 hours of biofilm formation

Effect of biofilm formation on the residence time distribution

The solutes residence time distribution (RTD) in the feed spacer channel of membrane devices is important for the process performance. Spreading of the RTD values indicates formation of preferential flow or substantial back mixing in the porous medium. In the case of water production in the RO membrane system this will lead to locally increased salt concentrations (concentration polarization) and therefore to decreased water production. We therefore performed 'numerical experiments' to determine the RTD in the membrane module at different levels of biofouling. RTD for a clean feed spacer channel (day 0) is then compared with RTDs after several days of biofilm development. Due to the very high Péclet numbers at the flow velocities used regularly in industrial devices however, as argued above, numerically stable calculations of solutes mass transport are very difficult for such a large computational domain. To illustrate the usefulness of tracer mass transport calculations, we simulated both the hydrodynamics and solute mass transport at a ten fold slower flow rate in the inlet, with a linear velocity of $u_{in} = 0.0163\,\mathrm{m\,s^{-1}}$. This lower flow rate still gives a good qualitative insight in the effect of the biofilm formation on the RTD. Note that also for NMR-imaging of the flow in membrane modules we had to decrease the flow rate (Graf von der Schulenburg *et al.*, 2008 and Vrouwenvelder *et al.*, 2009c, 2010a).

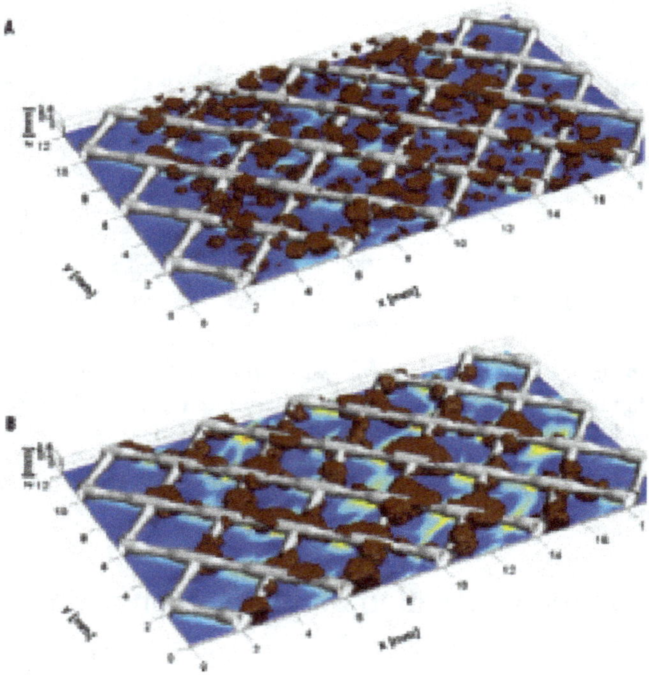

Figure 6.13 *Three-dimensional biofilm distributions at different attachment rules.* Colours show the biofilm (brown volumes), the spacer (white crossing filaments), and the flow velocity (colour scale in Figure 6.10) at day 3 in *x-y* two-dimensional sections at $z = 78\,\mu\mathrm{m}$ in the feed channel of a spiral-wound membrane device. (**A**) Case with attachment both on spacer and on membranes (places with shear $<100\,\mathrm{Pa}$) and (**B**) Case with biomass attachment on the spacer filaments (places with shear $<10\,\mathrm{Pa}$)

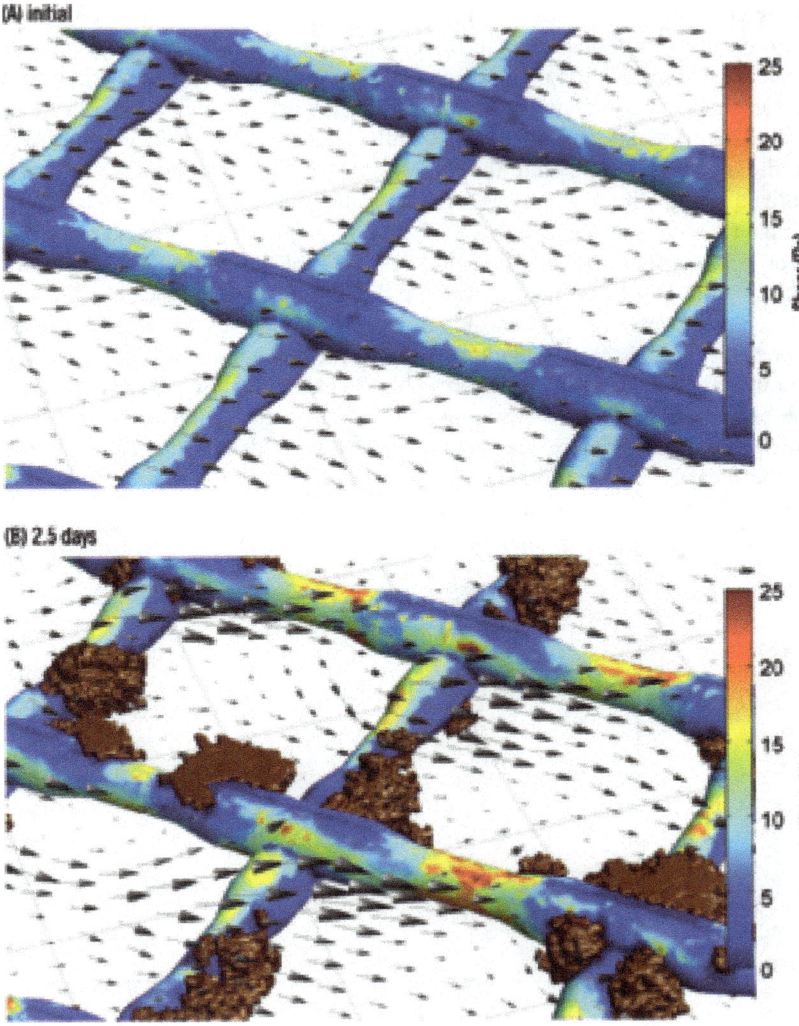

Figure 6.14 *Viscous shear and biofilm distributions on the spacer filaments.* Colours show the biofilm (brown volumes) and the viscous shear on the spacer filaments (red: high values, blue: low values). Arrows show the direction and magnitude of local flow velocity in an *x-y* two-dimensional section at $z = 650\,\mu m$ in the feed channel. (**A**) Day 0 – no biofilm; (**B**) Day 2.5 – biofilm formation

A signal of inert (non-reactive) tracer solute was sent for 0.3 s in the flow inlet and the response was measured in the outlet, after 2 cm of feed spacer channel. The average concentrations of tracer in the outlet are presented in Figure 6.15, for clean spacer and for biofilm developed on the spacer after 2.5 and 3.5 days. It is evident from the response in the outlet that the biofilm formation affects both the residence time and the dispersion of solutes. For a clean spacer the breakthrough time (when the solute reaches the outlet) is 0.75 s, which is significantly longer than for the case of 2.5-day-old biofilm in the channel (0.6 s) and for 3.5-day-old (0.45 s). This is a clear effect of the liquid flow acceleration induced by the biofilm growth narrowing the flow channels. However, the mean residence times are not shorter for the fouled channels (day 0: 1.27 s, day 3.5: 4.4 s) because the biofilm formation increases the dispersion both in the axial (x) and in the cross-flow (y) directions, and also because the tracer partially diffuses in the biofilm. Further numerical simulations will better evaluate the dispersion numbers from step signal experiments, because the pulse used here is rather far from the ideal Dirac function (i.e., very large concentrations introduced during very short time, in any case much shorter than the mean residence time).

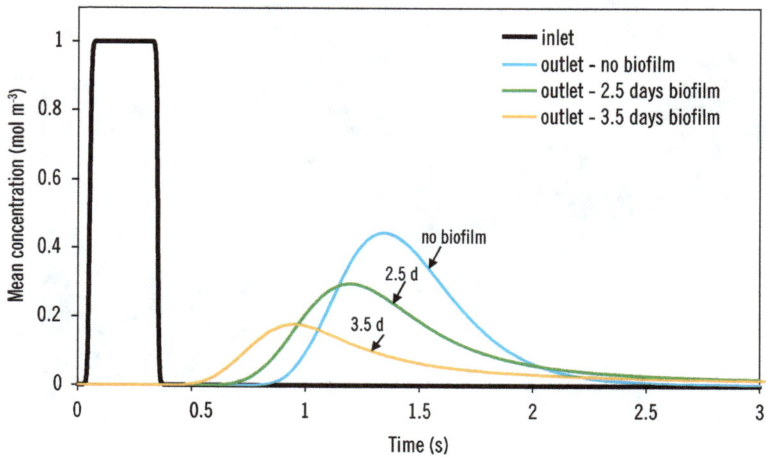

Figure 6.15 *Solute residence time distribution calculations.* The concentration of solute tracer in the inlet (red line) is presented together with the mean solute tracer concentration in the outlet recorded over time, for different stages of fouled spacer channel (at times 0 – blue line –, 2.5 days – green line – and 3.5 days – orange line –). The biomass distributions at 2.5 and 3.5 days were calculated in the conditions of the simulation presented in Figure 6.16

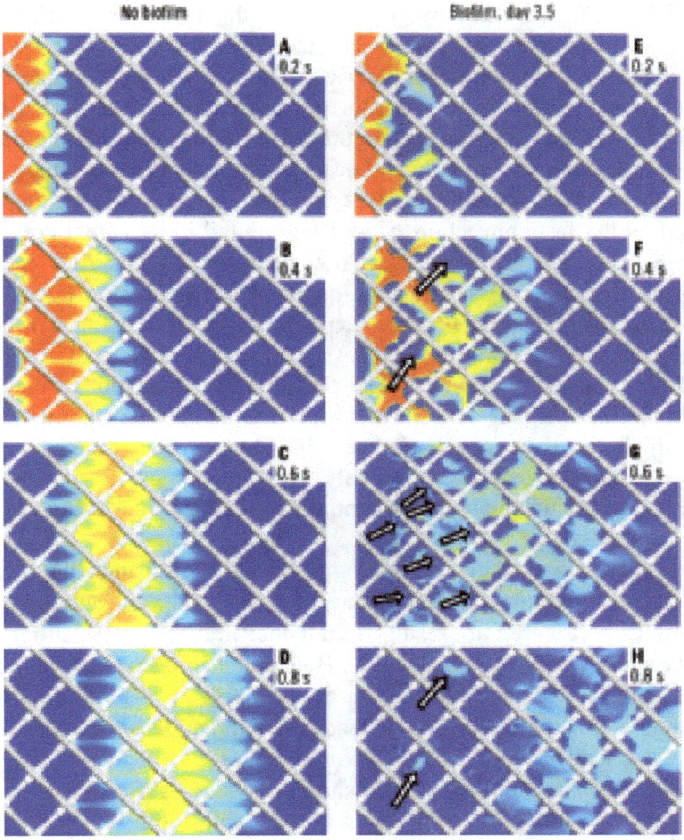

Figure 6.16 *Solute tracer movement in the feed channel.* Two-dimensional distributions of solute tracer concentration at 0.2, 0.4, 0.6 and 0.8 s are shown in a slice through the middle of the channel ($z = 390\,\mu m$). (**A**)–(**D**) Tracer movement in the channel without biofilm; (**E**)–(**H**) Tracer movement in the channel with a 3.5 day old biofilm, with biomass distribution calculated in the simulation presented in Figure 6.18. The colour scale for local tracer concentration is from 0 (blue) to $1\,mol\,m^{-3}$ (red). Arrows in (**F**) and (**H**) point to 'quasi-stagnant' zones. Arrows in (**G**) point to a few biofilm colonies

Dispersion of solute can be seen also on the 2-d sections of the concentration field through the middle of the flow channel, presented in Figure 6.16. For a clean spacer, a regular (symmetric) solute advection pattern can be observed, with a broadening of the pulse in time, as the tracer flows through the channel

(Figure 6.16A–D). For the spacer channel with very much biofilm formed (day 3.5) the flow pattern is irregular and heterogeneity leads not only to dispersion but also to significant flow acceleration. As an example, compare the lengths reached after 0.4 s: 8 mm for the clean condition (Figure 6.16B) and 14 mm for the 3.5-day-old biofilm (Figure 6.16F). Note also that the tracer only slowly penetrates the biofilm volume because the totally diffusive mass transport is here very slow compared with that in most of the liquid volume, clearly dominated by the fast convection rates. The long tail in the outlet concentration in Figure 6.15 for days 2.5 and 3.5 can now be clearly attributed to two kinds of delays: (i) formation of quasi-stagnant liquid zones, and (ii) slow tracer diffusion in the biofilm and then out. The tracer experiment reveals the zones of very slow liquid flow. Arrows in Figure 6.16F (0.4 s since tracer addition started) point to zones of low concentration situated in the liquid volume (not in the biofilm), which should have higher concentration like the neighboring liquid areas with areas in which now the concentration remained high although the pulse of tracer has passed. good convection. Conversely, arrows in Figure 6.16H (0.8 s) show the same areas in which now the concentration remained high although the pulse of tracer has passed.

This simulation proves that biofouling affects the liquid and solutes residence times also and not only the membrane area available. In general the effect of fouling is considered to result from concentration polarization near the membrane surface (Chiolle *et al.*, 1978). Both this theoretical study and previous experimental observations indicate that the real impact of biofouling is on the flow regime in the module leading to dead zones and an increase in the spread of the RTD. This leads to local accumulation of water with higher salinity thereby causing a lower permeate flux. Future simulation studies need therefore a more accurate description of the flow field as well as inclusion of the membrane flux and concentration polarization in order to have a more detailed analysis of the complex interaction between flow, mass transport and biofilm formation.

Effect of mass transport limitations on the biofilm development

In the last case presented in this paper we investigate the effects of substrate limitation on the biofilm growth and biofouling of the feed spacer channel. It has been observed experimentally that in certain cases most of the biofilm formation occurs in a region near the water inlet, probably due to substrate limitations (e.g., Gjaltema *et al.*, 1997). The questions which then arise are: (i) Is it important to consider substrates mass transfer or the biofilm develops anyway unrestricted by the substrate availability? (ii) Are there important substrate gradients over the module length? Are there important substrate gradients in the biofilm?

It is well known that the extent in which the substrate becomes limiting for the biomass growth depends on the balance between substrate consumption rate by biomass and substrate transport rates via diffusion and convection (equations (6.2) and (6.3)). These rates are further affected by the inlet substrate concentration, microbial yields and kinetics, diffusion coefficients, and liquid flow velocities. Here we did not vary these factors because we concentrated mainly on the effect of the biofilm on flow. However, under the applied biomass growth and substrate consumption parameters several useful observations can be made. The mean concentration of substrate collected in the outlet decreases in time (Figure 6.17A). Following the biomass accumulation seen in Figure 6.17B, the substrate conversion also increases and reaches 40% after 3.5 days (Figure 6.17A). The biomass formed under substrate limitation is, as expected, less than the amount formed without limitations. As time passes, the substrate limitation becomes stronger and the differences between the limited and unlimited biomass get larger (Figure 6.17B). It should be noted that compared to actual systems we used a relatively high substrate concentration, as often applied in the biofouling research in the laboratory (e.g., Vrouwenvelder *et al.*, 2009c). Using a lower substrate concentration will not lead to a significantly different result in terms of overall biomass production, only the time for the development of the biofilm will be much longer (Vrouwenvelder *et al.*, 2009a).

Important gradients form over the module length only after 80 hours, as seen in Figure 6.18 showing 2-d sections through the substrate concentration field at different moments of time. Moreover, slightly less substrate is available next to the membrane surface (Figure 6.18BC) compared to the middle of the channel (Figure 6.18EF) due to less convective transport. Several regions of almost zero substrate concentration can be observed at 84 hours (day 3.5 – Figure 6.18CF) as wakes downstream biofilm patches situated in zones with poor convection. This further illustrates the need to use a multi-dimensional biofilm model. The intrinsic assumption of a uniform mass transfer boundary thickness in the one-dimensional (1-d) models is clearly not valid here and will lead to erroneous results.

Substrate concentration is very low within the biofilm after 3.5 days, indicating strongly growth limitations (Figure 6.19). Even in places where high concentration is present in the adjacent liquid, as in the cross-section at $x = 4\,\text{mm}$ (Figure 6.19A), there is almost complete substrate depletion. This is of course accentuated further in the channel (at $x = 16\,\text{mm}$, Figure 6.19B) where the external substrate concentration is anyway lower.

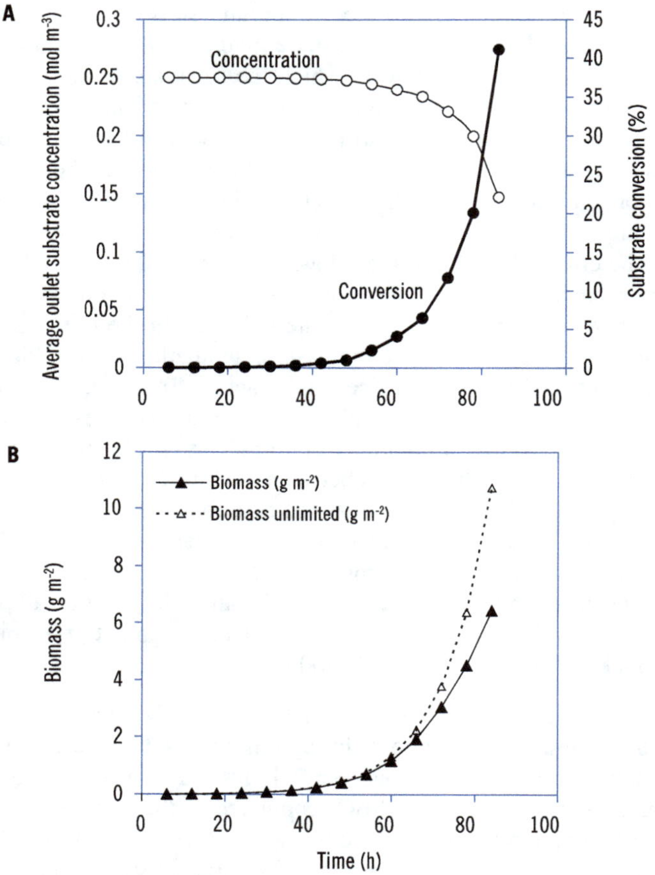

Figure 6.17 *Substrate conversion and biofilm biomass in time.* (**A**) Time profile of average substrate concentration in the outlet and of substrate conversion over the 2 cm feed spacer channel length. (**B**) Biofilm biomass accumulated in the substrate-limited case (black triangles) and in an unlimited case (open triangles). The biomass is related to the surface area of one membrane side (2.4 cm^2)

Model evaluation

The model presented in this article is a first successful attempt to describe numerically the complex process of biofouling in RO and NF membrane systems with feed channel spacers. The model is mainly based on first principles, including computational fluid dynamics, mass transport of solute matter (e.g., microbial substrates and products) and biofilm formation on the spacer and membranes.

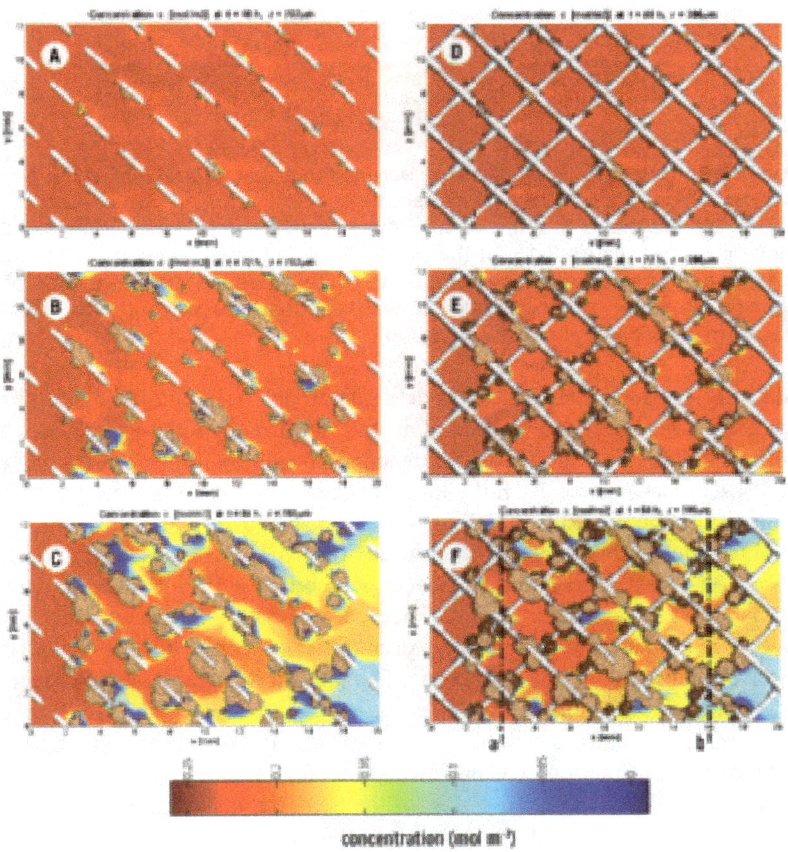

Figure 6.18 *Substrate concentrations along the feed channel at three times.* Two-dimensional distributions of substrate concentration are presented in two slices in the *x-y* plane: (**A**)–(**C**) slice at $z = 702\,\mu m$ and (**D**)–(**F**) slice at $z = 390\,\mu m$. The three times are: (A,D) day 2, (B,E) day 3 and (C,F) day 3.5. As the biofilm (brown volumes) grows on the spacer (white filaments), the available inlet substrate (red – $0.25\,mol\,m^{-3}$) is consumed (blue – zero concentration). Sections (a and b) across the channel width are shown in Figure 6.19

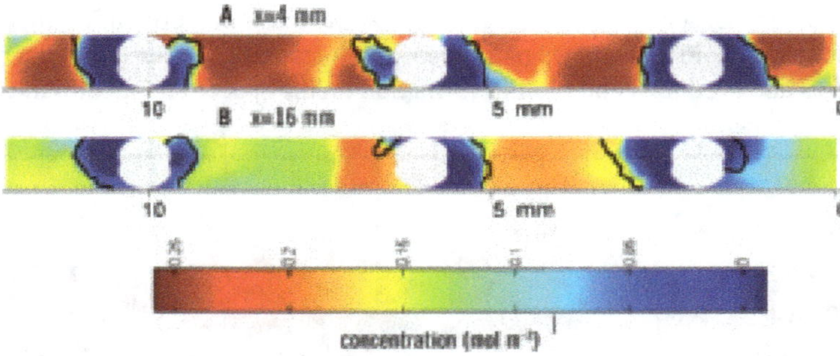

Figure 6.19 *Substrate concentrations across the feed channel and in the biofilm.* Two-dimensional distributions of substrate concentration are presented in two slices in the *y-z* plane at (A) $x = 4.03\,\text{mm}$ and (B) $x = 16.1\,\text{mm}$, after 3.5 days of biofouling (see Figure 6.18 for the positions of these two sections). The biofilm surface is indicated by a black line and the white areas are cross-sections of the spacer filaments. The colour scale shows the substrate concentration from red ($0.25\,\text{mol}\,\text{m}^{-3}$) to blue (zero concentration)

The merit of such a numerical modeling approach in biofouling of membrane modules is two-sided: on one hand it can confirm hypotheses and describe phenomena already observed experimentally, but, on the other hand it can also change established views from the literature. For example, it is clear that for the complex situation in a RO and NF membrane system a correct 3-d description of the solid material that supports the biofilm formation (i.e., spacer filaments) is needed. It has been shown that the spacer geometry significantly affects the flow patterns and the pressure drop in the membrane module. The main resistance to flow, thus the most important pressure drop occurs when the flow passes between the spacer filaments and the membrane. The biofilm formation on the spacer can only enhance this pressure drop.

The residence time distribution for water and solutes is also strongly affected by the biofilm formation. By narrowing the flow section, biofouling chiefly decreases the residence time for most of the fluid, but also creates zones of very low speed – regarded as 'dead zones' or quasi-stagnant. A significant effect of spreading the residence time distribution and increasing the solute dispersion in the module is clearly demonstrated here.

A series of aspects have been neglected here, but their relative importance will have to be evaluated by future model improvements. Biofilm accumulation in NF and RO systems seems to affect not only the feed channel pressure drop but also the biofilm-enhanced concentration polarization (CP), both effects having synergistic negative effect on membrane performance. Recent biofouling papers refer to a biofilm-enhanced osmotic pressure strongly contributing to a flux decline and salt passage increase (Chong *et al.*, 2006 and Herzberg and Elimelech, 2007). A flux decline due to the presence of extracellular polymeric substances (EPS) excreted by microorganisms was also described in a modeling study (Kim *et al.*, 2006). It is therefore evident that the flux of water through the membrane and the resulting diffusion of ions to and from the membrane system (concentration polarization, CP) must be represented in our model. Biofilm growth could also be affected by the higher substrate levels induced by concentration polarization effect caused by permeate production (Herzberg and Elimelich, 2008). Furthermore, the biomass detachment as an effect of high shear stress induced by the fast flow in narrow channels must be considered. Most importantly, model simulation results will be compared with experimental data (Vrouwenvelder *et al.*, 2010a).

It is therefore expected that this modelling approach will have significant implications for membrane system design and operation, as well as for the development of anti-biofouling strategies to have stable membrane installation performance at minimal costs.

SUMMARY

This study presents a new three-dimensional (3-d) computational model that couples fluid dynamics, solutes transport and biofouling by biofilm formation in nanofiltration and reverse osmosis membrane modules. A computational domain of 3 × 5 feed spacer frames with geometry as applied in practice was used in the model. Comparing the hydrodynamics computed with the realistic spacer geometry and with a spacer made from straight cylindrical filaments, like in previous modeling studies, showed that cylindrical filament feed spacers are too simplified for representative modeling studies. The 3-d numerical simulations showed that biomass accumulation, by attachment and biofilm growth in time, strongly affected the feed channel pressure drop, liquid velocity distribution and residence time distribution. The main pressure drop is encountered by the flow passing over the spacer filaments. Simulations showed the development of a heterogeneous flow pattern and formation of preferential flow channels. This study indicates that the real impact of biofouling is on the flow regime leading to quasi-stagnant zones and an increase in the dispersion of the residence time distribution. The presented 3-d mathematical modeling approach in (bio)fouling of membrane modules may have significant implications for membrane system design and operation to have stable membrane installation performance at minimal costs.

Additional data and animations of simulated biofilm development can be obtained from: http://www.biofilms.bt.tudelft.nl/Biofouling/index.html

Fundamental Studies

7

Biofouling, Pressure Drop and the Impact of flux

INTRODUCTION

Reverse osmosis and nanofiltration are high pressure membrane filtration processes that can produce water of high quality. A drawback in membrane filtration applications is membrane fouling (Shannon *et al.*, 2008). The corresponding increase of plant operation costs may be prohibitive in some situations for the application of membrane filtration in water treatment. A major type of fouling in reverse osmosis and nanofiltration membranes is biofouling, caused by biofilm formation in membrane elements (Paul, 1991; Flemming, 1993; Tasaka *et al.*, 1994; Ridgway and Flemming, 1996; Baker and Dudley, 1998; Schneider *et al.*, 2005 and Shannon *et al.*, 2008).

Membrane biofouling increases the pressure drop (frictional resistance) (Characklis and Marshall, 1990). In spiral wound membrane elements, two types of pressure drops can be discriminated: the trans-membrane pressure drop and the feed spacer channel pressure drop. The trans-membrane pressure drop is the differential pressure between feed and permeate lines, describing the frictional resistance over the membrane. The trans-membrane pressure drop is related to the membrane flux (permeation rate). When the trans-membrane pressure is increased by biofouling, the membrane flux is declined. The feed spacer channel pressure drop is the pressure drop between the feed and brine lines (Flemming *et al.*, 1994).

The definition of flux (permeation rate) is the water volume flowing through the membrane per unit area and time ($\mathrm{L\,m^{-2}\,h^{-1}}$). Many references mention a critical flux (Field *et al.*, 1995; Howell, 1995 and Bacchin *et al.*, 1995, 2006).

---

*This chapter is based, with permission from the copyright holder, on a paper previously published in Journal of Membrane Science Vol. 326 No. 1. pp. 36–44 doi: 10.1016/j.memsci.2008.09.029

It has been hypothesized that below this critical flux no fouling should occur. Originally, the critical flux concept is proposed for the fouling of microfiltration membranes by colloidal particles (Field *et al.*, 1995; Howell, 1995 and Bacchin *et al.*, 1995). Field *et al.* (1995) defined the critical flux as 'a flux below which a decline of flux with time does not occur; above it fouling is observed'. Howell (1995) defined 'the flux below which there is no deposition of colloids on the membrane'. Bacchin *et al.* (1995) defined 'the flux below which no fouling occurs'. Since the critical flux concept is introduced (1995), it has been studied extensively for the fouling of microfiltration and ultrafiltration by colloidal particles. A review paper of Bacchin *et al.* (2006) summarizes the literature on critical flux especially with regard to colloidal filtration.

In the water industry, extended pretreatment (microfiltration/ultrafiltration) is applied to remove colloidal material, particles and microorganisms from the feed water of nanofiltration and reverse osmosis. Therefore, it is expected that the critical flux concept is not valid for nanofiltration and reverse osmosis systems with microfiltration/ultrafiltration pretreatment. Nevertheless, literature often refers to critical flux applied for nanofiltration and reverse osmosis systems (Winters, 1997, 2001; Schwinge *et al.*, 2002a, 2004; Mänttäri *et al.*, 1997, 2000 and Lisitsin *et al.*, 2005). Several studies propose the critical flux concept for spiral wound nanofiltration and reverse osmosis membranes in relation to biofouling (Winters, 1997, 2001 and Schwinge *et al.*, 2004) as well because bacteria have about the size of colloidal particles. Biodegradable compounds are not rejected by microfiltration/ultrafiltration membranes whilst nanofiltration/ reverse osmosis membranes reject (part of the) nutrients. The rejection of nutrients in nanofiltration/reverse osmosis membranes may lead to elevated nutrient concentrations at the feed side of the membrane, resulting in more biofilm accumulation than in microfiltration/ultrafiltration membranes. However, systematic studies into the influence of flux on biofouling are lacking.

The main objective of this study was to determine the influence of flux on pressure drop and biofouling development in nanofiltration and reverse osmosis systems. When the flux does not influence pressure drop and biofouling development, the critical flux concept is not applicable for nanofiltration and reverse osmosis membrane processes after pretreatment with microfiltration/ultrafiltration.

MATERIALS AND METHODS

Experimental set-up

Laboratory study

The membrane fouling simulator (MFS) is a tool to study and monitor fouling in spiral wound membranes (Vrouwenvelder *et al.*, 2006b). With this monitor, fouling development can be monitored systematically by (i) pressure drop, (ii) in-situ and non-destructive (visual) observations using the sight glass and (iii) analysis of coupons sampled from the membrane sheet in the monitor (Vrouwenvelder *et al.*, 2006b, 2007a).

In parallel, three MFSs were operated in cross flow, without permeate production. New and unused spiral wound nanofiltration membrane elements (Trisep TS80, 4 inch diameter, polyamide membrane material) were used for preparation of membrane and spacer coupons for the MFS. The feed spacers in the MFS had the same spatial orientation as in spiral-wound membrane modules (45° rotation). The linear flow velocity in the monitors was $0.16\,m\,s^{-1}$. Feed water of the MFS installation was drinking water (Table 7.1) prepared from ground water by aeration and rapid sand filtration at treatment plant Tull en 't Waal (Water Supply Company Vitens). The water had a constant temperature of 13°C. Acetate, a biodegradable compound, was added to the water of one MFS to promote biofouling (Table 7.1). From a sterile vessel, containing a 5 L solution of concentrated sodium acetate, $25\,mL\,h^{-1}$ substrate was dosed to the feed water prior to the MFS by a peristaltic pump (Masterflex) achieving a concentration of 0.100 mg acetate carbon per litre feed water. The dosage of substrate was checked periodically by measuring the weight of the stock solution.

Table 7.1 Acetate dosage and DOC and conductivity average values of feed water

Study	Substrate dosage (acetate) to feed water	Dissolved Organic Carbon (DOC) $mg\,C\,L^{-1}$	Conductivity $mS\,cm^{-1}$
Laboratory	Yes	1.9	27
Monitors, test rigs and full-scale RO	No	3.2	45
NF pilot plant	No	7.5	50

Monitors, test rigs and full-scale plant

A production site with severe reverse osmosis fouling problems (elevated feed channel pressure drop) was selected for a comparative study. At this site, Baanhoek (Evides Industriewater) process water is produced from surface water using ultrafiltration and reverse osmosis. The full-scale installation contained RO membrane elements (DOW Filmtec BWROLE, 8 inch diameter). New and unused spiral wound reverse osmosis (RO) and nanofiltration (NF) membrane elements (DOW Filmtec BWROLE and Trisep TS80 TSF, 4 inch diameter) were used for preparation of membrane and spacer coupons for the MFS and for test rig studies. Two MFSs (NF, RO), two test rigs (NF, RO) and the full-scale installation (RO) were fed with surface water after pretreatment with ultrafiltration (Table 7.1) and dosage of a scale-inhibitor Hypersperse AF 200 (3 mg L^{-1}) with a relatively high growth potential; class III (Vrouwenvelder *et al.*, 2000). As a control, a blank MFS (feed: drinking water prepared from ground water without antiscalant) was operated. During these comparison studies, the test rigs and the full-scale installation were operated with the same flux (20 L m^{-2} h^{-1}). The MFSs were operated without permeate production. The MFSs and the test rigs were operated with the same linear flow velocity. No substrate (acetate) was dosed to the feed water (Table 7.1).

NF pilot plant: membrane elements with/without flux

Water Supply Company Vitens produces drinking water (25,000,000 m^3 $year^{-1}$) from anaerobic ground water at water treatment plant Spannenburg by aeration, rapid sand filtration, deacidification, softening and rapid sand filtration. Drinking water from this treatment plant was the feed for a pilot plant consisting of ultrafiltration and nanofiltration. No substrate (acetate) was dosed to the feed water (Table 7.1). The two stage nanofiltration installation (Figure 7.1), contained a total of 18 spiral wound Trisep TS80 TSF membrane elements (4 inch diameter), with a molecular weight cut-off of 200 Da. Nanofiltration membranes were used in the study because of potential use in practice to improve removal of dissolved (in)organic compounds and biodegradable compounds. For this research project, one third stage single element (3.1 in Figure 7.1) was added to the installation. An antiscalant (7 mg L^{-1} 4Aqua OSM 92) with a low growth potential (class I, Vrouwenvelder *et al.*, 2000) was dosed to the UF permeate (Table 7.1) to prevent NF scaling. For 146 days, the NF installation was operated continuously at a feed flow of 3.2 $m^3 h^{-1}$ with a recovery of 75%. At the start of the study, the feed pressure of the NF installation was about 600 kPa. The process parameters monitored were pressure drop, flow, normalized flux and conductivity. The pressure drop and flux were normalized (Verdouw and Folmer, 1997 and Huiting *et al.*, 1999). Usually, all membrane elements produce permeate. In the pilot plant study at treatment plant Spannenburg, the permeate pipes of element 1-2-1 and 3-1 were isolated and closed, resulting in membrane elements without flux. The other elements including 1-1-1 and 2-6 were operated with permeate production (Figure 7.1). Periodically, the permeate side of elements 1-2-1 and 3-1 was shortly (about 20 minutes) opened to determine permeate production, normalized flux, pressure drop and flow velocity. In all MFS, test rig and membrane installation studies, no disinfectants were dosed to the feed water and no cleanings were applied. For assessment of the static pressure and the pressure drop over individual membrane elements, quick connectors were installed on the NF pressure vessels.

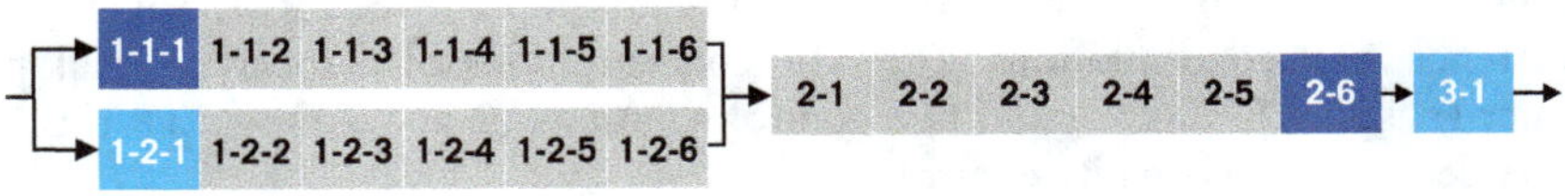

Figure 7.1 Scheme of the NF installation with coding and order of membrane elements. The membrane modules 1-1-1 and 2-6 were operated with permeate production and the modules 1-2-1 and 3-1 were operated without permeate production

Pressure drop

A differential pressure transmitter (Labom) was used to determine the pressure difference over the feed spacer channel of the MFS and membrane elements in the test rig. The differential pressure transmitter was connected to the MFS and test rig with quick connectors. For the pilot nanofiltration installation, continuous pressure drop measurements were performed of the feed pressure, concentrate pressure and permeate pressure for stage 1, stage 2 and the total nanofiltration installation with different transmitter types (Vrouwenvelder *et al.*, 2009d). A differential pressure transmitter (Labom) was used for periodical (manual) measurements applying quick connectors on the pressure vessels to determine the feed spacer channel pressure drop over individual membrane elements in the nanofiltration installation.

Membrane autopsy

To determine the accumulated fouling, sections of membrane, feed and product spacers were taken from the MFS and membrane modules. The sections (4 cm^2) were placed in 10 mL of autoclaved tap water in a capped tube. To determine the amount of biomass, the tubes with the membrane sections were placed in an ultrasonic cleaning bath (Bransonic, model 5510E-DTH, output 135 Watts, 42 kHz). The low energy sonic treatment (2 minutes) followed by mixing on a Vortex (few seconds) was repeated four times. Next, volumes of water collected from the tubes were used to determine biomass parameters (ATP and TDC).

Active biomass was determined in duplicate by measuring the adenosinetriphosphate (ATP) concentration from 100 μL samples (Holm-Hansen and Booth, 1966). The luminometer (Celcis, model Advance) added 100 μL of

LuminEX-B reagent (Celcis) to a sample to release ATP from the bacterial cells. Subsequently, 100 μL of LumATE-PM (Celcis) was added for light production. The amount of light produced was measured with the luminometer (relative light units, RLU). The concentration of ATP was derived from the RLU values using the conversion factors of the linear relationship between RLU values and reference ATP concentrations. The ATP concentrations of autoclaved drinking water (<1 pg ATP mL^{-1}) and two control solutions (2 and 100 pg ATP mL^{-1}) were determined as quality control.

Total Direct Cell counts (TDC values, cells cm^{-2}) were performed with epi-fluorescence microscopy using acridine orange as fluorochrome by applying a slightly adapted method to eliminate fading (Hobbie *et al.*, 1977). The slides were examined with a Leica DM RXA microscope equipped with a 100 W power supply and phase and filter sets (I2/3) appropriate for acridine orange. Microorganisms were enumerated through a 100x/1.30 PL Fluotar oil immersion objective. Non-fluorescence immersion oil (Olympus, code 35505) was used because of low background fluorescence. A total of 10 random fields were counted.

The maximum values for biomass parameters in membrane elements and membrane and spacer sheets from the MFS were used (Vrouwenvelder *et al.*, 2008b).

Calculation of the ratio of diffusive and convective flux

In this paper, studies are described to determine the influence of permeate production on development of membrane biofouling. Experiments were performed operating membranes with and without permeate production. Biofilm accumulation on membranes is related to the availability of biodegradable nutrients near the membrane surface. Transport of nutrients to the membrane surface is controlled by two processes:

- Diffusive flux, J_{dif}, the transport of nutrients to the membrane material caused by a concentration difference;
- Convective flux, J_{con}, the transport of nutrients to the membrane material due to a pressure difference (generating permeate production).

To predict the influence of permeate production on biofilm formation, the diffusive and convective flux were calculated. Membrane specifications from Van der Meer (2003a) were used (Table 7.2). The feed water was assumed to contain one biodegradable nutrient (acetate). Acetate was selected for the calculations because (i) it is an easily biodegradable compound and (ii) the compound has been used as a model substrate in many biofouling studies (Vrouwenvelder *et al.*, 2006b, 2008b) providing a database for result comparison. Therefore, acetate was used also in the laboratory experiments of this study. The concentration of acetate in the bulk liquid, C_{bulk}, is 1.0×10^{-4} kg C m^{-3}. The concentration of nutrient at the biofilm surface, $C_{membrane}$, is 0 kg C m^{-3}, assuming that the biofilm on the membrane utilizes all nutrients and that the transport of nutrient to the biofilm is entirely governed by the external mass transfer resistance. Under this condition, the biofilm is growing completely mass transfer limited.

Table 7.2 Properties of membrane module, feed water, nutrient and operational aspects

	Unit	Description
Module diameter	m	10.16×10^{-2} (4 inch)
Membrane surface area of module	m^2	7.5
Membrane envelope width	m	0.75
Number of envelopes	–	5
Module membrane envelope width (w)	m	3.50
Feed channel height (h)	m	0.00068
Porosity of feed spacer (ε)	–	0.89
Specific surface of feed spacer (S)	$m^2 m^{-3}$	11,600
Volumetric feed flow rate (Q)	$m^3 s^{-1}$	1.25×10^{-3} ($4.5 m^3 h^{-1}$)
Diffusivity coefficient of acetate (D)	$m^2 s^{-1}$	1.29×10^{-9}
Specific density (ρ)	$kg m^{-3}$	997
Dynamic viscosity (η)	$kg m^{-1} s^{-1}$	8.9×10^{-4}
Permeation rate (flux, Jv)	$m^3 m^{-2} s^{-1}$	5.56×10^{-6} ($20 L m^{-2} h^{-1}$)

Diffusive flux, J_{dif}

The diffusive mass flux is given by:

$$J_{dif} = k\,(C_{bulk} - C_{membrane}) \tag{7.1}$$

The mass transport coefficient k in spacer filled channels is described by the Sherwood correlation (Sh):

$$Sh = \frac{k\,d_h}{D_x} = a\,\mathrm{Re}^b\,\mathrm{Sc}^c \tag{7.2}$$

In this correlation, d_h is the hydraulic diameter of the flow channel, D the diffusion coefficient of the nutrient and a, b and c are coefficients, Re is the Reynolds number and Sc the Schmidt number:

$$\mathrm{Re} = \frac{\rho\,v\,d_h}{\eta} \tag{7.3}$$

$$\mathrm{Sc} = \frac{\eta}{\rho D} \tag{7.4}$$

In which ρ is the specific density of the liquid, ν the flow velocity and η the dynamic viscosity of the solute.

Shock and Miquel (1987) have determined the coefficients a, b and c (eq. (7.2)) in experiments using spiral wound membrane modules:

$$\mathrm{Sh} = 0.065\,\mathrm{Re}^{0.875}\,\mathrm{Sc}^{0.25} \tag{7.5}$$

The hydraulic diameter of the channel (d_h) can be derived, according to Shock and Miquel (1987), using:

$$d_h = \frac{4\,\varepsilon}{2\dfrac{(w+h)}{w\,h} + (1-\varepsilon)\,S} \tag{7.6}$$

Where ε is the porosity, w the total membrane leaf width in the membrane module, h the channel height and S the specific surface of feed spacer.

The flow velocity (v) is determined using:

$$v = \frac{Q}{A_{eff}} = \frac{Q}{h\,w\,\varepsilon} \tag{7.7}$$

Where Q is the volumetric feed flow rate of the membrane module and A_{eff} the effective hydraulic surface area, which is determined using the flow channel height (h), width (w) and porosity (ε).

Data from Table 7.2 were used to determine the mass transport coefficient k (equation 7.2) and the diffusive flux (equation 7.1): k is $1.2 \times 10^{-4}\,\mathrm{m\,s^{-1}}$. J_{dif} is $1.2 \times 10^{-8}\,\mathrm{kg\,m^{-2}\,s^{-1}}$.

Convective flux, J_{con}

The convective mass flux (J_{con}) is determined using:

$$J_{\mathrm{con}} = J_v\,C_{bulk} \tag{7.8}$$

Where J_v is the permeation rate or volume flux perpendicular to the membrane surface. The convective flux is $5.6 \times 10^{-10}\,\mathrm{kg\,m^{-2}\,s^{-1}}$.

The ratio between diffusive and convective flux is

$$\text{ratio} = \frac{J_{\text{dif}}}{J_{con}} = 22 \tag{7.9}$$

Obviously, the ratio between diffusive and convective flux is influenced by parameters used in the calculation (such as feed channel height, see Table 7.2). But variation of the parameters within practical limits showed that the diffusive flux remained always much larger than the convective flux. In addition, the calculation is based on a worse case scenario, in which the biofilm on the membrane utilizes all nutrients and transport of nutrient to the biofilm is entirely governed by the external mass transfer resistance only.

Evaluation

Biofilm development on membrane surfaces of spiral wound nanofiltration and reverse osmosis membrane modules is predominantly caused by utilisation of biodegradable nutrients supplied by *diffusion*: the diffusive flux will be considerably larger (factor 22) than the convective flux. The calculations underline that the supply of biodegradable nutrients by a convective flux caused by permeate production is playing an insignificant role in biofilm formation in spiral wound membranes.

RESULTS

The development of feed spacer channel pressure drop, flux and biofouling was investigated with (i) a membrane fouling simulator (MFS, Figure 7.2), (ii) a MFS, single membrane element test rigs and a RO installation (Figure 7.3) and (iii) a nanofiltration installation (Figures 7.4, 7.5 and 7.6).

Fouling in monitor without flux

Three MFSs were operated without permeate production in a parallel study (Vrouwenvelder *et al.*, 2006b). A biodegradable compound (acetate) was dosed to the feed water of one MFS to enhance biofouling. A concentration of 0.100 mg acetate C L^{-1} in the feed water of the MFS resulted in a strong increase of the pressure drop over the MFS, while the pressure drop over the two MFSs without dosage remained constant (Figure 7.2A). At the end of the research period the MFS units were opened for analyses of the accumulated material. On the membrane and feed spacer taken from the MFS with acetate dosage, higher concentrations of active biomass (ATP) were found compared to the biomass concentrations on samples from the MFSs without dosage (Figure 7.2B). Also, the number of cells (TDC) was higher in the MFS with acetate dosage.

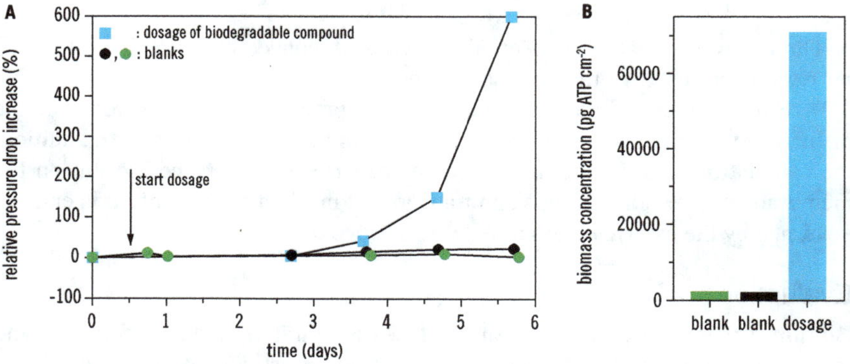

Figure 7.2 Pressure drop (**A**) in time over MFSs operated without permeate production, with and without dosage of a biodegradable compound (0.100 mg acetate-C L^{-1}) to the feed water of the MFS. Biomass concentration (**B**) on the membrane and spacer in the MFS after 6 days of operation without permeate production, with and without acetate dosage. Adapted from reference Vrouwenvelder *et al.* (2006b)

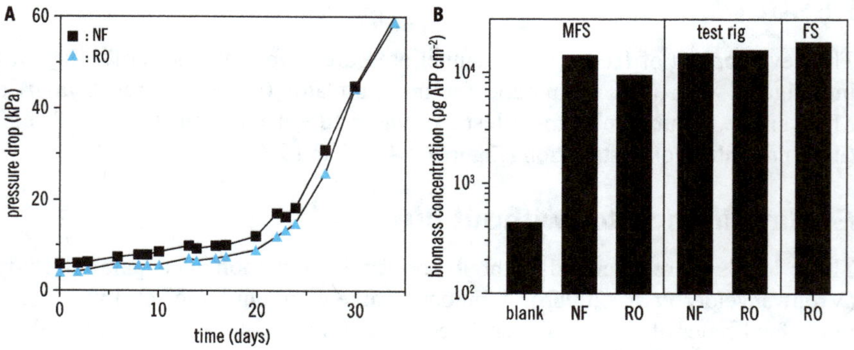

Figure 7.3 Development of pressure drop (**A**) over the MFS without permeate production, containing NF and RO membranes and spacers. Biomass concentration (**B**) in the non-permeate producing MFS with NF and RO membranes and spacers, in permeate producing spiral wound membranes in a test rig with respectively a NF and RO membrane element and in a RO membrane element from the full-scale installation (FS). Adapted from reference Vrouwenvelder *et al.* (2007b)

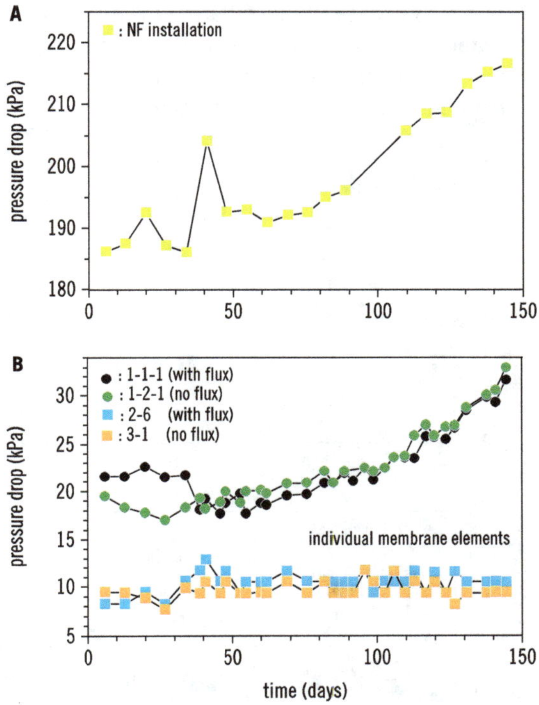

Figure 7.4 Pressure drop in time over the nanofiltration installation (**A**) and over membrane elements with and without permeate production (**B**) on the feed side (1-1-1, 1-2-1) and concentrate side (2-6, 3-1) of the nanofiltration installation

The background levels of the feed water for ATP ($\leq$2 ng ATP L^{-1}), DOC ($\leq$2 mg L^{-1}) and AOC ($\leq$2 μg AOC-C equivalents L^{-1}) were low and did not contribute significantly to fouling accumulation in the monitors (Figure 7.2, blanks). This experiment shows that in a membrane fouling simulator membrane fouling occurred without permeate production. The pressure drop increased and biomass developed due to biodegradable compounds in the feed water.

Fouling in monitors, test rigs and full-scale plant

In the MFS without flux, biofouling developed by dosage of biodegradable nutrients to the feed water (Figure 7.2). To evaluate the influence of permeate production on biofouling development, comparative studies – without nutrient dosage – were carried out. Two MFSs (NF, RO) were operated in parallel to two test rigs (NF, RO) and a full-scale RO installation (Vrouwenvelder *et al.*, 2006b, 2007b). The feed water was produced from surface water by ultrafiltration. The MFSs were operated without permeate production; the test rigs and full-scale installation were operated with permeate production.

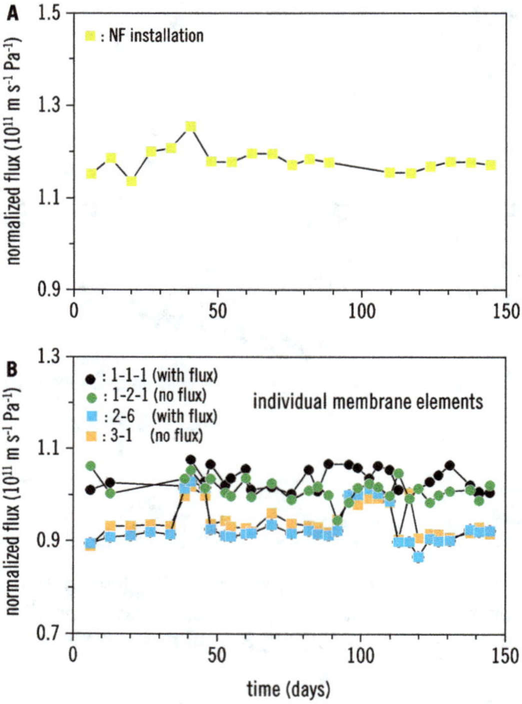

Figure 7.5 Normalized flux in time over the nanofiltration installation (**A**) and over membrane elements with and without permeate production (**B**) on the feed side (1-1-1, 1-2-1) and concentrate side (2-6, 3-1) of the nanofiltration installation

Over the two MFSs and two test rigs, the pressure drop increased strongly within 35 days of operation. The results of the MFS are shown (Figure 7.3A). Relative high and similar concentrations of active biomass were found on the membranes from the MFSs and test rigs (Figure 7.3B). A lead membrane module taken from a full-scale installation – supplied with the same water as the MFSs and test rigs – contained about the same concentration active biomass (Figure 7.3B). Since the pressure vessel of the full-scale installation contained six elements in series, no data was available on the pressure drop over the lead membrane module. The full-scale installation suffered from an elevated pressure drop. Lower biomass concentrations were observed in the blank MFS after 41 days of operation fed with drinking water prepared from groundwater (Figure 7.3B). Apparently, the feed water at the site of the full-scale installation had a much higher biofouling potential than the blank.

The same concentrations of biomass were observed in the two MFSs (without permeate production), in the test rigs and in the full-scale installation (with permeate production, Figure 7.3B), indicating that permeate production was not influencing biofouling.

Fouling in membrane elements with/without flux in NF pilot plant

In a pilot installation, lead membrane elements from the first stage were operated with (1-1-1) and without (1-2-1) permeate production (Figure 7.1). The membrane elements on the last positions of the second stage were operated with (2-6) and without (3-1) permeate production as well. No substrate was dosed to the feed water (Table 7.1).

The nanofiltration installation feed water had a neutral pH (7.6 ± 0.1). Average retention values of the nanofiltration installation operated with a recovery of 75% were DOC (98%), UV extinction (98%), colour (89%), and conductivity (85%). After the research period dosing experiments with organic acids were performed with the nanofiltration installation loaded with new membrane modules (from the same supplier) at recovery 87% and pH 7.8. In these dosing experiments the retention of five organic acids was higher than 86% (the retention of acetate was higher than 90%). The operational aspects of the NF installation ensured a significant rejection of (biodegradable) compounds.

After a run of 146 days, the feed channel pressure drop over the total installation was increased by about 15% (Figure 7.4A). Operation with and without permeate production showed that (i) the elements on the same position in the installation had the same development of pressure drop and that (ii) the lead elements had a stronger pressure drop increase than the membranes on the concentrate side of the nanofiltration installation (Figure 7.4B). The pressure drop increase over the total installation (Figure 7.4A) is caused almost completely by the pressure drop increase over the lead membrane module (Figure 7.4B).

The normalized flux of the nanofiltration installation was not declining during the 146 days research period (Figure 7.5A). Operation with and without permeate production showed that (i) the elements on the same position in the installation had the same normalized flux which was constant in time and that

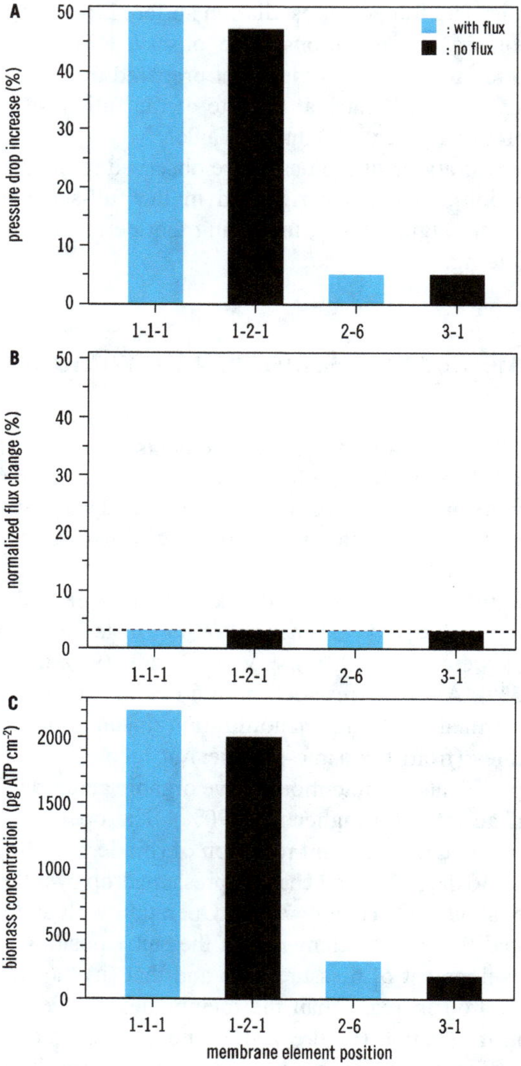

Figure 7.6 Pressure drop increase (**A**), normalized flux change (**B**) and biomass concentrations (**C**) of individual membrane modules on the feed side (1-1-1, 1-2-1) and concentrate side (2-6, 3-1) of the NF installation after 146 days of continuous operation. Membrane modules were operated with (1-1-1, 2-6) and without permeate production (1-1-2, 3-1)

(ii) the lead elements had a slightly higher flux than the elements on the concentrate side of the nanofiltration installation (Figure 7.5B). Irrespective whether or not a flux was applied, the same pressure drop (Figure 7.6A) and normalized flux (Figure 7.6B) were observed for membrane elements in the same position of the nanofiltration installation.

For fouling diagnosis, autopsies were performed of membrane elements operated with (1-1-1, 2-6) and without (1-2-1, 3-1) permeate production after 146 days continuous operation. The total concentration of inorganic compounds in all membrane elements was below the detection limit of the analysis performed ($\leq 0.2\,\mathrm{mg\,m^{-2}}$ membrane surface area). In practice, these low values ($\leq 0.2\,\mathrm{mg\,m^{-2}}$) were observed in installations without pressure drop problems, suggesting that the performance of the installation in this study was not influenced by accumulated inorganic compounds (Vrouwenvelder *et al.*, 2008b).

The membrane elements operated with and without permeate production showed that (i) the elements on the same position in the installation had the same biomass concentration and that (ii) the lead elements had a higher biomass concentration than the elements on the concentrate side of the nanofiltration installation (Figure 7.6C). The biomass concentrations (ATP) related well with the pressure drop increase. High biomass concentrations were found in the membrane elements with a high pressure drop increase and low biomass concentrations were found in elements with a low pressure drop increase (Vrouwenvelder *et al.*, 2009d). The relationship between pressure drop increase and biomass concentration is not linear (Figure 7.6). Pressure drop and biofilm development were not affected by the (absence of) flux in membrane elements.

The high biomass concentrations in the lead modules were not affecting the normalized flux (Figure 7.6B). So, based on flux data no indications were found for reduced membrane performance. The biofouling resulting in a pressure drop increase over the feed spacer channel of the membrane modules had no detectable influence on the normalized flux (Figure 7.6).

DISCUSSION

Flux and critical flux

Bacteria accumulate on material surfaces by two processes: attachment and growth (by cell multiplication). Attachment is a prerequisite for biofilm formation (Van Loosdrecht *et al.*, 1990). The presence of microorganisms on membranes was already observed after a short contact time, in the order of minutes, with water containing micro-organisms (Schaule, 1992 and Flemming, 1993). Microorganisms are commonly observed on water exposed surfaces, even in pure water systems (Mittelman, 1991). Initial adhesion of microorganisms to certain materials have shown not to be predictive for biofilm formation (Gjaltema *et al.*, 1997). The presence of attached microorganisms is not indicative for biofouling. Once attached, the cell may grow, reproduce and produce extracellular polymers, utilizing nutrients transported by the bulk fluid and nutrients originating from the materials (Characklis and Marshall, 1990). Obviously, biofouling of nanofiltration and reverse osmosis systems is a problem of biofilm growth. Thus, nutrients can be considered as potential biofouling. In water treatment, nutrient removal can be obtained by biological treatment such as slow sand filtration and activated carbon filtration. The dominance of biofouling in nanofiltration and reverse osmosis membrane installations with microfiltration/ultrafiltration pretreatment indicates that nutrients are not rejected by microfiltration and ultrafiltration membranes (Vrouwenvelder *et al.*, 2008b).

The critical flux concept 'the flux below which no fouling should occur' has been studied extensively for the fouling of microfiltration and ultrafiltration membranes by colloidal particles. In fouling of microfiltration and ultrafiltration membranes, colloidal particles accumulation is a major factor of importance (Bacchin *et al.*, 2006).

In relation to biofouling of membranes, the critical flux concept was proposed as well (Winters, 1997, 2001; Schwinge *et al.*, 2004; Kang *et al.*, 2004 and Goosen *et al.*, 2004) because bacteria have about the size of colloidal particles. The biofouling process is based on the transformation of dissolved into non-dissolved compounds (nutrients into biofilm), clearly differing from fouling by filtration of colloidal particles. In other words, biofouling involves a completely different fouling process than particulate fouling. Based on theoretical considerations, the critical flux concept can not be valid for biofouling of membranes.

For reverse osmosis and nanofiltration membranes, the critical flux concept was proposed in relation to biofouling (Winters, 1997, 2001 and Schwinge *et al.*, 2004). In the water industry, the feed water of nanofiltration and reverse osmosis installations is generally pretreated with microfiltration/ultrafiltration. Microfiltration/ultrafiltration completely retains colloidal material, particles and microorganisms. Therefore, fouling processes in nanofiltration and reverse osmosis systems differ from fouling processes in microfiltration/ultrafiltration systems. In our study on nanofiltration and reverse osmosis membranes, it is shown that the flux is not influencing biofouling and biofouling is not influencing the flux (Figures 7.3, 7.4, 7.5 and 7.6), demonstrating that the critical flux concept is not valid for biofouling in reverse osmosis and nanofiltration membranes. Calculations also support the observations that the flux is playing an insignificant role in biofouling. Monitoring of the flux is not appropriate for biofouling detection. Evidently, the critical flux concept 'below a critical flux no fouling occurs' is incorrect to apply for the control of biofouling of spiral wound reverse osmosis and nanofiltration membranes with microfiltration/ultrafiltration pretreatment.

Nutrient rejection

Biodegradable compounds are (partly) rejected by nanofiltration/reverse osmosis membranes, whilst microfiltration/ultrafiltration membranes do not reject nutrients. In nanofiltration/reverse osmosis membranes, the rejection of nutrients may lead to elevated nutrient concentrations near the membrane feed side, possibly resulting in more biofilm accumulation than in microfiltration/ultrafiltration membranes. Since the flux is not influencing biofilm formation (Figure 7.6), the rejection of nutrients is not clearly contributing to biofouling in nanofiltration and reverse osmosis elements, illustrating that the influence of nutrient rejection on biofilm formation is negligible at conditions applied in nanofiltration/reverse osmosis systems. This is also underlined in the theoretical calculations which indicate that diffusive flux will be a more important nutrient supply to the biofilm than convective flux.

Biofouling is a feed spacer problem

Membrane biofouling increases the pressure drop (Characklis and Marshall, 1990). In spiral wound membrane elements, two types of pressure drops can be discriminated: the feed spacer channel pressure drop and the trans membrane pressure drop, which is related to the membrane flux, Figure 7.7 (Flemming *et al.*, 1994). In our study, only the feed spacer channel pressure drop was affected by biofouling. The feed channel pressure drop is mainly caused by the feed spacer. Biofouling is therefore a feed spacer problem.

Irrespective whether nanofiltration or reverse osmosis membranes were used, the same feed spacer channel pressure drop increase and biomass concentrations were observed (Figure 7.3), suggesting that the type of membrane may be of minor importance compared to the feed spacer. The feed spacers from the nanofiltration and reverse osmosis membrane elements seemed to be identical (thickness, structure, orientation) during visual inspection. For biofouling control, research should be focused on reducing the feed spacer channel pressure drop. Therefore, adaptation of the feed spacer and fluid flow hydrodynamics in the feed spacer channel may be suitable approaches for biofouling control. Other configurations of (smaller) pressure vessels in the membrane installations or alternative operation of the installation may reduce pressure drop increase problems (Vrouwenvelder *et al.*, 2009d).

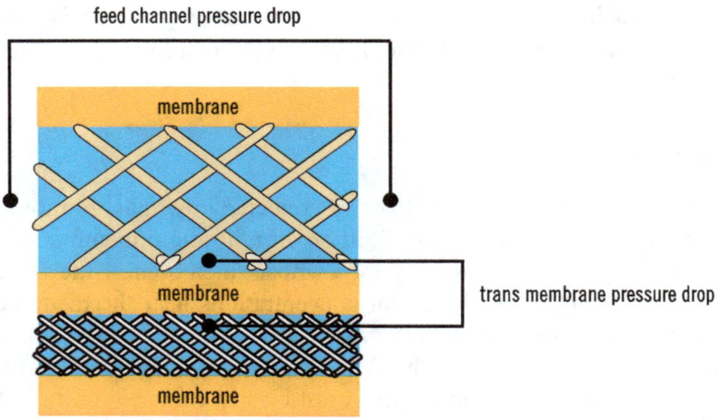

Figure 7.7 Scheme of pressure drops in membrane sheets showing the feed spacer channel pressure drop and trans membrane pressure drop. The feed channel pressure drop is between the feed and brine lines. The trans membrane pressure drop is the differential pressure between feed and permeate lines

SUMMARY

The relation between biofouling and membrane flux in spiral wound nanofiltration and reverse osmosis membranes in drinking water stations extensive pretreatment such as ultrafiltration pretreatment has been studied. The flux – water volume flowing through the membrane per unit area and time – is not influencing the development of membrane biofouling. Irrespective whether a flux was applied or not, the feed spacer channel pressure drop and biofilm concentration increased in reverse osmosis and nanofiltration membranes in a monitor, test rigs, a pilot scale and a full scale installation. Identical behavior with respect to biofouling and feed channel pressure drop development was observed in membrane elements in the same position in a nanofiltration installation operated with and without flux. Calculation of the ratio of diffusive and convective flux showed that the diffusive flux is considerably larger than the convective flux, supporting the observations that the convective flux due to permeate production is playing an insignificant role in biofouling. Since fouling occurred irrespective of the actual flux, the critical flux concept stating that 'below a critical flux no fouling occurs' is not a suitable approach to control biofouling of spiral wound reverse osmosis and nanofiltration membranes.

8

Biomass and Pressure Drop Development: Role of Feed Spacer

INTRODUCTION

Biofouling – excessive growth of biomass, i.e. biofilms – is the major fouling type in spiral-wound nanofiltration (NF) and reverse osmosis (RO) systems, resulting in a pressure drop increase (Characklis and Marshall, 1990; Ridgway and Flemming, 1996 and Patching and Fleming, 2003). In NF/RO membranes, pressure drop occurs over the feed spacer channel and membrane (Flemming *et al.*, 1994 and Patching and Fleming, 2003). Autopsies on membrane modules from full-scale and pilot installations show biofouling of the spacer, located in the feed channel (Figure 8.1).

In the late 1990s, two strategies were strongly proposed to prevent and control membrane biofouling: (i) physical removal of bacteria from the feed water of membrane systems (for example by microfiltration or ultrafiltration pretreatment), and (ii) metabolic inactivation of bacteria by applying biocide dosage or UV irradiation (Ridgway, 1997). At present, the focus is on nutrient removal by biological pretreatment (e.g. sand filtration) and modification of membranes (disinfectant resistant and low fouling). Already in 1997, Ridgway was the first author to point out that adaptation of hydrodynamics and feed channel spacers may be an approach to control membrane biofouling. Nevertheless, since then research to control biofouling was predominantly focused on development of low fouling membranes and not on feed channel and spacer modification. The number of publications in journals satisfying the search criteria 'biofouling' and 'modified and/or adapted membrane' in the article title, abstract and keywords in the scopus database of March 2008 amounted 59. The search criteria 'biofouling'

---

*This chapter is based, with permission from the copyright holder, on a paper previously published in Water Research Vol. 43 No. 3 pp. 583–594 doi: 10.1016/j.watres.2008.11.019

and 'modified and/or adapted spacer' yielded no references at all. At the North American Membrane Society (NAMS) conference 2007, to control biofouling membrane modification was addressed in 6 presentations while feed spacer modification was addressed in 1 presentation only. Evidently, until now biofouling control was considered a membrane problem and not a feed channel problem.

Figure 8.1 Feed spacer taken during autopsy of spiral-wound membrane module from a full-scale installation suffering from a prolonged elevated feed channel pressure drop

The main items of this study were as follows:

1. an overview of biofouling problems in practice,
2. the effect of biofouling on the feed channel pressure drop and trans-membrane pressure drop and
3. the role of the feed spacer on the development of feed channel pressure drop caused by biofouling.

MATERIALS AND METHODS

Terminology

Biofouling – excessive growth of biomass, i.e. biofilms – is the major fouling type in nanofiltration and reverse osmosis systems after extended pretreatment with for example ultrafiltration. Biofouling increases the pressure drop (Characklis and Marshall, 1990), thereby increasing the process costs (Ridgway, 2003).

In spiral-wound membrane modules, two types of pressure drop can be discriminated: the trans-membrane pressure drop (TMP) and the feed channel pressure drop (FCP, Figure 8.2; Flemming *et al.*, 1994). The TMP is the differential pressure between feed and permeate lines, caused by the frictional resistance over the membrane. The TMP is related to the membrane flux. When the TMP is increased by biofouling, the membrane flux is declined. The definition of flux is the water volume passing a membrane per unit area and time (L m^{-2} h^{-1}). The flux normalized for temperature and pressure, the Mass Transfer Coefficient (MTC), is expressed in m s^{-1} Pa^{-1}. The FCP is the pressure drop between feed and concentrate lines.

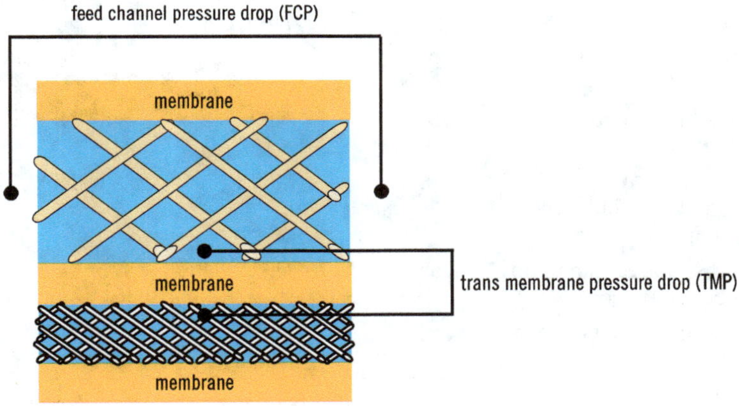

Figure 8.2 Scheme of pressure drops in membrane sheets showing feed channel pressure drop (FCP) and trans-membrane pressure drop (TMP)

Experimental set-up

In this study, the development of biofouling in the feed channel was investigated using (i) full-scale installations (ii) a nanofiltration pilot plant, (iii) test-rigs and (iv) MFSs with and without feed spacer. In-situ visual observations on fouling accumulation using the Membrane Fouling Simulator (MFS) window and in-situ observations on fouling accumulation and velocity distribution profiles using Magnetic Resonance Imaging (MRI) were performed.

An overview of the experimental conditions is shown in Table 8.1. In all studies with modules and MFSs nanofiltration membranes and spacers from the same supplier (TriSep TS80) were used. RO membranes used in the full-scale investigations and the comparison study (both module and MFS) were from supplier Filmtec (BW30LE). The NF membrane is a polyamide membrane material. For the MFS studies, sheets of membrane and spacers taken from an new and unused nanofiltration TriSep TS 80 membrane module were used. The feed spacer taken from the module was a 31 mil thick diamond-shaped polypropylene spacer with porosity 0.85. The feed spacers in the MFS had the same spatial orientation as in spiral-wound membrane modules (45° rotation).

Table 8.1 Scheme experimental conditions of the studies (substrate dosage, type of membrane and feed spacer).

Study	Substrate dosage	Membrane type	Feed Spacer
Full-scale and test-rig investigations	no[a]	several RO and NF types	RO and NF
Comparison full-scale, test-rig and MFS	no	RO and NF[b]	RO and NF[b]
Influence of permeate production	no	NF[b]	NF[b]
In-situ visual direct observations on fouling accumulation	yes[c]	NF[b]	NF[b]
In-situ MRI observations on fouling accumulation	yes	NF[b]	NF[b]
Feed spacer impact on biofouling	yes[c]	NF[b]	NF[b]

[a]*Substrate was dosed in some of the test-rig studies performed during the full-scale investigations only.*
[b]*All NF membranes and feed spacers were from the same type from the same manufacturer (TriSep TS80 module). Sheets of membrane and spacers taken from unused TriSep TS80 membrane modules were used for all indicated studies.*
[c]*Blanks where operated without substrate dosage.*

Full-scale and test-rig investigations with different feed water types

In order to quantify biofouling, selected biomass parameters in membrane elements were investigated by membrane autopsies and compared with the FCP-increase in test-rigs and 15 full-scale nanofiltration and reverse osmosis installations supplied with different types of feed water (Vrouwenvelder *et al.*, 2008b). The 15 membrane filtration installations differed in staging of pressure vessels and number of membrane elements per pressure vessel. The study included polyamide and polysulfone thin film composite membranes from different suppliers. No disinfectant residual was present in the feed water.

Membrane elements (spiral wound 4' or 8' NF or RO membrane elements) were taken from different plants varying in performance to determine biomass parameters by autopsy.

Comparison full-scale, test-rig and MFS studies

In a comparison study, the development of biofouling was studied in a full-scale RO installation and two single element test-rigs and two MFSs. The test-rigs were operated with a RO (Filmtec BW30LE) and NF membrane module (TriSep TS 80). The MFSs were operated with membrane and spacer sheets from an unused RO (Filmtec BW30LE) and NF element (TriSep TS80). The RO installation was selected because of severe biofouling (Vrouwenvelder *et al.*, 2006b, 2007a).

NF pilot plant: membrane elements with/without permeate production

The development of FCP and flux, normalized for temperature and pressure (Verdouw and Folmer, 1997 and Huiting *et al.*, 1999), was studied in a pilot plant consisting of a 2 staged nanofiltration installation (Figure 8.3) with ultrafiltration pretreatment. An antiscalant ($7\,mg\,L^{-1}$ 4Aqua OSM 92) with a low growth potential (class I, Vrouwenvelder *et al.*, 2000) was dosed to the ultrafiltration permeate to prevent NF-scaling. Details of the installation, pretreatment and water quality are provided by Vrouwenvelder *et al.* (2009b). For 146 days, the NF installation was operated with a continuous feed flow of $3.2\,m^3\,h^{-1}$ at a recovery of 75%. In practice, all NF elements produce permeate. In this pilot plant study, the permeate pipes of selected NF elements were isolated and closed, resulting in no permeate production from those membrane elements. Parallel staged elements were selected to operate with and without permeate production. Periodically, the permeate side of elements operated without flux was opened for about 20 minutes to determine permeate production, normalized flux, pressure drop and flow velocity. For assessment of the pressure and the pressure drop over individual membrane elements, quick connectors were installed on the NF pressure vessels.

Laboratory study

The membrane fouling simulator (MFS) is a tool developed to study and monitor fouling in spiral-wound membranes (Vrouwenvelder *et al.*, 2006b, 2007a). Using the MFS, fouling development can be monitored by (i) pressure drop measurements, (ii) in-situ and non-destructive (visual) observations using the sight window and (iii) analysis of coupons sampled from the membrane and spacer sheet in the monitor. The feed spacer channel dimensions of the MFS are $0.00077\,m \times 0.040\,m \times 0.200\,m$. In the MFS, water flows through the spacer attached between a membrane sheet and the window.

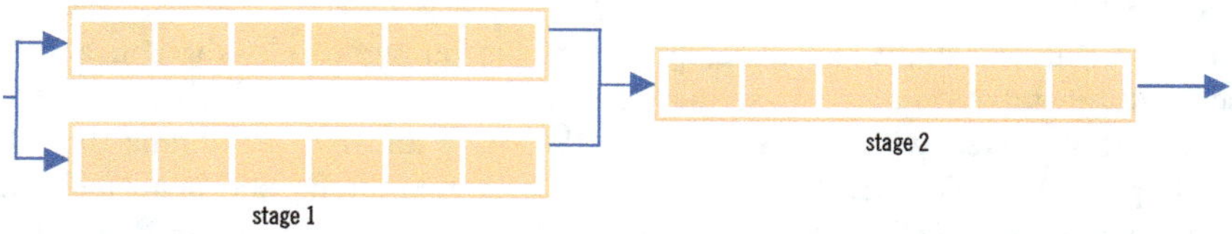

Figure 8.3 Scheme of NF installation. Stage 1 consisted of two parallel pressure vessels and stage 2 consisted of one pressure vessel. Each pressure vessel contained six spiral-wound membrane modules

Membrane and spacer coupons of unused spiral-wound nanofiltration membrane elements (Trisep TS80) were placed in the MFS. For the visual observations using the window, two MFSs were operated in parallel in cross-flow, without permeate production. Feed water of the MFSs was drinking water, distributed by the drinking water supply company without any disinfectant dosage or residual. One monitor was operated as a reference – without substrate dosage – to determine the possible influence of the drinking water on biofilm formation. Acetate, a biodegradable compound, was added to the water of the second MFS to promote biofouling. From a sterile vessel, containing a 5-L solution of concentrated sodium acetate, $28\,mL\,h^{-1}$ substrate was dosed to the feed water prior to the MFS by a peristaltic pump (Masterflex) achieving a concentration of 0.100 mg acetate carbon per litre feed water. The substrate dosage was checked periodically by measuring the weight of the stock solution and/or pump calibration.

For determining the influence of spacer presence on biofouling development, four MFSs were operated in parallel with and without spacer. Drinking water was used as feed without and with substrate dosing resulting in a concentration of 0.200 mg acetate carbon per litre feed water. The linear flow velocity in all MFSs was $0.16\,m\,s^{-1}$.

MRI study

Nuclear Magnetic Resonance (NMR) is a powerful tool to study membrane fouling, since it is a quantitative non-invasive measurement technique. Magnetic materials can not be used since magnetic material influence the NMR signal.

Because of the interference of the metal MFS with the NMR signal, a PVC version of the MFS had to be developed (named S-MFS, Figure 8.4). In order to fit in a 200 MHz super wide-bore magnet (Figure 8.5), the S-MFS dimensions had to be scaled down. The S-MFS has feed channel dimensions of 0.00077 m × 0.016 m × 0.040 m. In the S-MFS, the water flows through the spacer placed between two membrane sheets. The S-MFS was produced with a manufacturing accuracy of 20 μm. The feed spacer and membranes sheets used in the NMR studies were taken from an unused TriSep TS80 nanofiltration membrane module. The PVC flow cell was operated with the same linear flow velocity and the same substrate dosage method was applied, as described for the laboratory study. A biodegradable compound was dosed to the feed water (1.00 mg acetate C L^{-1}) of the S-MFS to accelerate the biofouling process. Using ^{1}H NMR, the evolution of spatial biofilm distribution and velocity field was determined in time. The flow velocity in the S-MFS was 0.16 m s^{-1}. For the NMR studies after 2, 3 and 4 days, the flow was interrupted for about 5 minutes for imaging and the flow was temporarily lowered for about 45 minutes, enabling detection of the range of local flow velocities.

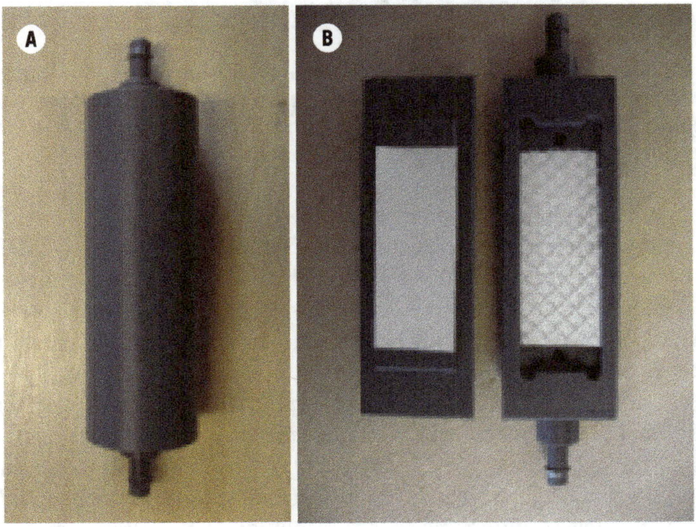

Figure 8.4 Spacer Membrane Fouling Simulator (S-MFS) made of plastic, closed (**A**) and open (**B**). The feed channel dimensions are 0.00077 m × 0.016 m × 0.040 m, designed to contain a feed spacer of the same dimensions placed between two membrane sheets. The S-MFS fits in a NMR radio frequency coil with a 2.5 cm inner diameter

Figure 8.5 Wide-bore magnet used for the NMR study on the background, containing the S-MFS. Water containing substrate was fed to and from the S-MFS using tubing. The differential pressure transmitter (PMD70) used to monitor the pressure drop over the S-MFS is shown in the bottom right corner

Structural two-dimensional images of 512×256 pixel^2 of the flow cell where acquired employing a Rapid Acquisition Relaxation Enhancement (RARE) pulse sequence (Henning *et al.*, 1986) with a slice thickness of 2 mm. The field of view was $50 \times 25\,\text{mm}^2$ resulting in an isotropic in-plane image resolution of ~98 μm/pixel. A RARE factor of 64 (effective echo time 650 ms) was employed to obtain the T_2 weighting enabling biofilm imaging, as shown previously (Hoskins *et al.*, 1999 and Graf von der Schulenburg *et al.*, 2007). Biofilm will appear in the NMR images as regions of low signal intensity. Two-dimensional velocity images of the superficial flow component (axial direction) were acquired employing a standard phase shift velocity encoding pulse sequence with a size of 256×128 pixels resulting in a spatial resolution of ~210 μm/pixel. For phase shift velocity encoding, gradient increments, ΔG, and pulse gradient durations, δ, were decreased with increasing biofilm accumulation to increase the velocity imaging range and avoid fold-over artefacts. More details on NMR microscopy can be found elsewhere in the literature (Callaghan, 1991).

Pressure drop

A differential pressure transmitter (Endress+Hauser Deltabar S: PMD70-AAA7FKYAAA) was used to determine the pressure difference over the feed spacer channel of the (S-)MFS. The differential pressure transmitter was connected to the (S-)MFS with quick connectors. For the pilot nanofiltration installation, continuous pressure drop measurements were performed of the feed pressure, concentrate pressure and permeate pressure for stage 1, stage 2 and the total nanofiltration installation with different transmitter types. The different transmitter types differed in accuracy and long term stability (Vrouwenvelder *et al.*, 2009d). Pressure transmitters (Endress + Hauser, PMC131) were used for continuous monitoring of the feed pressure, concentrate pressure and permeate pressure for stage 1, stage 2 and the total NF installation (Figure 8.3). A digital precision manometer (Endress + Hauser, type PMC41) and a differential pressure transmitter (Labom) were used for periodical (manual) measurements applying quick connectors on the pressure vessels to determine the FCP over individual membrane elements in the nanofiltration installation.

Membrane autopsy

To characterize the accumulated fouling, sections of membrane, feed and product spacers were taken from the (S-)MFS and membrane modules. The sections (4 cm^2) were placed in capped tubes filled with 30 mL of autoclaved tap water. To determine the amount of biomass, the tubes with the membrane sections were placed in an ultrasonic cleaning bath (Bransonic, model 5510E-DTH, output 135 W, 42 kHz). The low energy sonic treatment (2 minutes) followed by mixing on a Vortex (few seconds) was repeated three times. Water out of the tubes was used to determine active biomass, which was determined in duplicate by measuring the adenosinetriphosphate (ATP) concentration (Holm-Hansen and Booth, 1966). The luminometer (Celcis, model Advance) added 100 μL of LuminEX-B reagent (Celcis) to a 100 μL sample to release ATP from the bacterial cells. Subsequently, 100 μL of LumATE-PM (Celcis) was added for light production. The amount of light produced was measured with the luminometer (relative light units, RLU). The concentration of ATP was derived from the RLU values using the conversion factors of the linear relationship between RLU values and reference ATP concentrations. The ATP concentrations of autoclaved drinking water (<1 pg ATP mL^{-1}) and two control solutions (2 and 100 pg ATP mL^{-1}) were determined as quality control.

RESULTS

Full-scale and test-rig investigations with different feed water types

Full-scale and test-rig investigations showed that measurements of biological parameters in the water were not appropriate to quantify biofouling. ATP determined during destructive membrane studies proved to be a suitable parameter to diagnose biofouling (Vrouwenvelder *et al.*, 2008b). The studies performed at 15 full-scale installations showed that the FCP-increase is a practical concern (Figure 8.6). Some plants operated with FCP-increase values exceeding 300% (Figure 8.6).

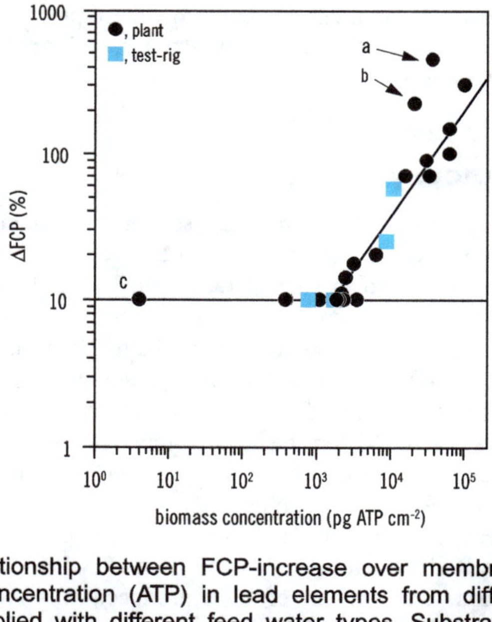

Figure 8.6 Relationship between FCP-increase over membrane installations and biomass concentration (ATP) in lead elements from different membrane installations supplied with different feed water types. Substrate was dosed to the feed water in selected test-rig studies. The line represents the calculated relationship between FCP-increase and ATP. Case studies a,b,c deviate from the calculated relationship since other types of fouling than biofouling were affecting membrane performance as well. Adapted from Vrouwenvelder *et al.*, 2008b

Comparison full-scale, test-rig and MFS studies

A comparison study with spiral-wound membrane elements in a full-scale installation, test-rigs, and MFSs showed a similar development of FCP and biomass accumulation (Figure 8.7, Vrouwenvelder *et al.*, 2006b, 2007a). The full-scale installation and the test-rigs were operated with permeate production ($20\,L\,m^{-2}\,h^{-1}$), whilst the MFSs were operated without permeate production. The development of FCP and biomass accumulation were not affected by permeate production and membrane type (NF or RO, Figure 8.7).

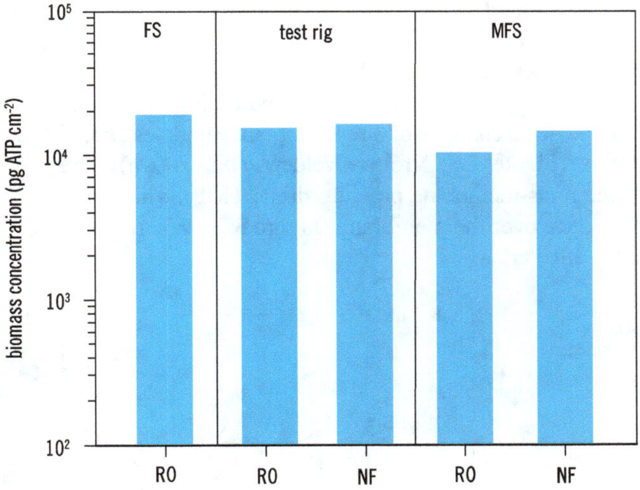

Figure 8.7 Biomass concentrations in the spiral-wound RO membrane module from the full-scale installation (FS), spiral-wound RO and NF membrane modules from the test-rigs and on the RO and NF membrane and spacer sheets from the MFSs. The blank operated with tap water had a factor 50 lower biomass concentration (not shown), indicating that the feed water of the full-scale installation had a high biofouling potential. The modules were operated with permeate production and the monitor were operated with cross-flow only. Adapted from Vrouwenvelder *et al.*, 2007b

Influence of permeate production on biofouling

The FCP-increase was the same for membrane elements operated with and without permeate production at the same position in the nanofiltration pilot plant

(Figure 8.8A). The normalized flux of the nanofiltration installation was not declining during 146 days of operation. With and without permeate production, elements at the same position in the installation had the same normalized flux, which was constant in time (Figure 8.8B). For fouling diagnosis, after 146 days of continuous operation autopsies of membrane elements operated with and without flux were performed. In all membrane elements, the total concentration of inorganic compounds was below the detection limit of the analysis performed ($\leq 0.2\,mg\,m^{-2}$ membrane surface area), illustrating that the performance of the installation in this study was not hampered by accumulation of inorganic compounds. In membrane elements operated with and without permeate production at the same position in the installation the same biomass concentration was found (Figure 8.8C, lead membrane modules from stage 1). The pressure drop and biofilm development were not affected by the (absence of) flux in membrane elements. The high biomass concentrations in the lead modules were not affecting the normalized flux. The biofouling resulted in a pressure drop increase over the feed channel of the membrane modules, but did not affect the normalized flux (Figure 8.8, Vrouwenvelder *et al.*, 2009b). The membrane flux is related to the trans-membrane pressure drop (TMP). The TMP is caused by the frictional resistance over the membrane (Figure 8.2). When the flux is not affected, the TMP is not affected.

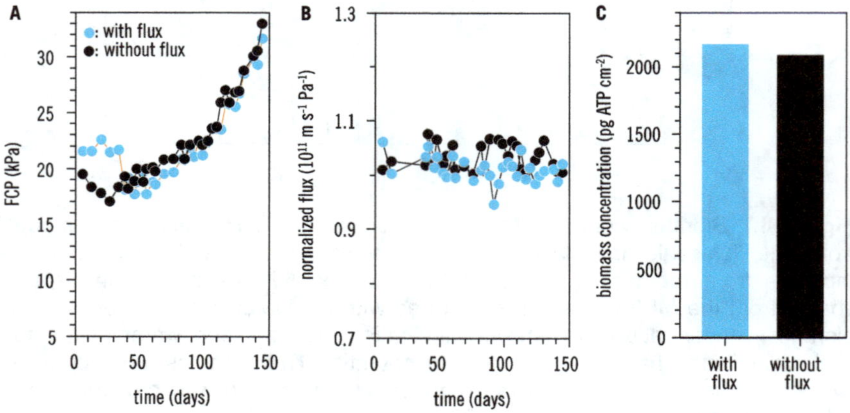

Figure 8.8 FCP (**A**) and normalized flux (**B**) over nanofiltration lead membrane modules in time (pilot plant data). Biomass concentrations (**C**) in the membrane modules determined after 146 days continuous operation. The elements were operated with and without flux for 146 days. Periodically, the permeate side of the element operated without flux was shortly (about 20 minutes) opened to determine the flux. The flux was normalized for pressure and temperature. Adapted from Vrouwenvelder *et al.*, 2009b

This experiment confirmed the earlier results (Figure 8.7) and showed that biomass accumulation increased the FCP. The FCP and biomass concentration increased irrespective whether (i) a NF or RO membrane was used (Figure 8.7), and whether or not (ii) the membranes produced permeate (Figures 8.7 and 8.8).

In-situ visual observations on fouling accumulation

Biodegradable compounds in feed water of membrane installations may cause biofouling (Vrouwenvelder *et al.*, 1998 and Van der Hoek *et al.*, 2000). Two MFS's operated without permeate production were supplied with feed water with and without dosage of a substrate (acetate). A concentration of 0.100 mg acetate-C L^{-1} in the MFS feed water resulted in a strong FCP-increase over the MFS, while the FCP over the MFS without substrate dosage remained constant (Vrouwenvelder *et al.*, 2006b). After 6 days, the MFS units were opened for analyses of accumulated material on membrane and spacer sheets. High-biomass concentrations were found in the MFS with substrate dosage while low-biomass concentrations were found in the MFS without dosage.

In-situ microscopic visual observations through the MFS window showed biomass accumulation at the feed spacer junctions and to a lesser extent on the spacer branches and the membrane in the MFS supplied with water containing substrate (Figure 8.9). The biomass structure at the spacer junctions looked like filamentous biofilm streamers (Figure 8.9B). In the MFS supplied with water containing substrate (Figure 8.9B), more fouling was observed visually and a higher pressure drop and a higher biomass concentration were found compared to the MFS supplied with water without substrate (Figure 8.9A). In other studies we also have observed streamer forming in cases of biofouling.

In-situ MRI observations of fouling accumulation and velocity distribution profiles

MRI studies were performed with an S-MFS flow cell completely produced out of plastic designed for NMR studies (Figure 8.4). The water flowed through the spacer channel sandwiched between two membrane sheets. A low dosage of substrate was fed to the water. The spatial biofilm distribution and velocity field were determined in time.

The feed spacer and the flow channel were clearly visible, applying a 2-d-MRI structural image of the S-MFS (Figure 8.10, day 0). The intensity of the MRI image signal corresponded to the water density at each spatial location. The spacer junctions were clearly visible because of the low water intensity.

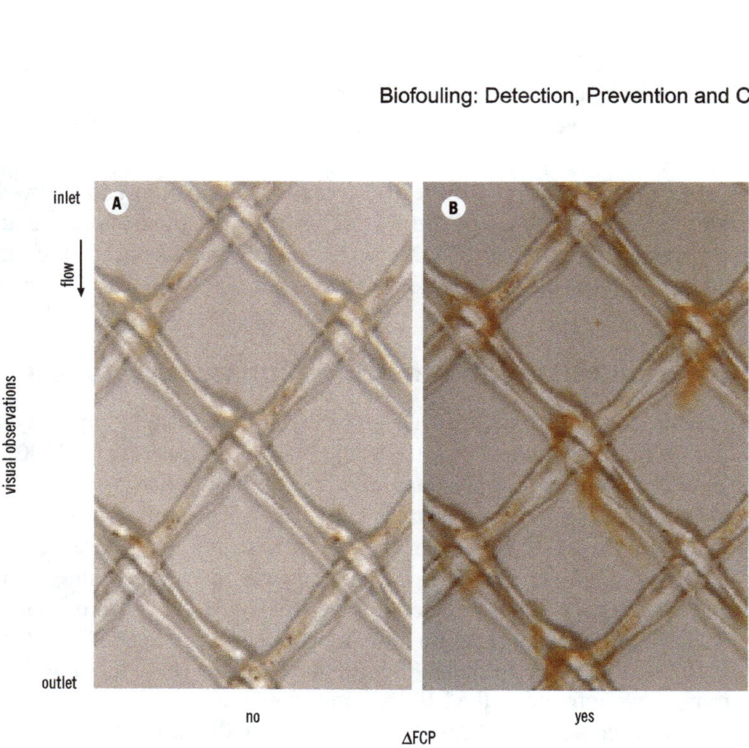

Figure 8.9 In-situ visual observations of the feed spacer and membrane in the MFS – using the sight window during operation – without (**A**) and with substrate (**B**) dosage after a run time of 6 days. Substrate dosage resulted in a FCP-increase (ΔFCP) over the MFS and biomass accumulation on the spacers (B). Without substrate dosage no FCP-increase and limited biomass accumulation was observed (A). Adapted from Vrouwenvelder *et al.*, 2006b

In time, biomass accumulated on the spacer junctions and branches (Figure 8.10). Biofilm growth preferentially started at the spacer junctions close to the inlet of the S-MFS. Filament biofilm streamers were observed. These observations of biofilm localization and morphology were in good agreement with visual examinations of biomass accumulation on the feed spacer using the MFS window (Figure 8.9). In time, the streamers reached lengths of up to ~6 mm, exceeding the distance between adjacent spacer junctions and parallel spacer branches.

The corresponding NMR velocity images after 0, 2, 3 and 4 days of substrate dosage showed that the biomass accumulation had a severe impact on the flow field (Figure 8.11). Before starting substrate dosage, the flow field was homogeneous with parabolic flow profiles in the regions between spacer branches (Figure 8.11, day 0). There was no flow at the spacer junctions and regions under the spacer branches tended to flow faster due to the liquid volume reduction at that position. After 2 days, the biofilm growth at the spacer junctions (Figure 8.10) caused a distortion of the flow field and regions of low velocities close to the spacer junctions (Figure 8.11, day 2). An increase in flow field heterogeneity with large stagnant regions and fast flowing channels (Figure 8.11, day 3 and 4) are caused by biomass blocking of entire regions of the feed spacer channel. In time, the velocity distribution range broadened and lowered, and even negative flow velocities were observed (Figure 8.12). Negative flow velocities or back-flow by eddies should be avoided in membrane units because of the involved energy loss.

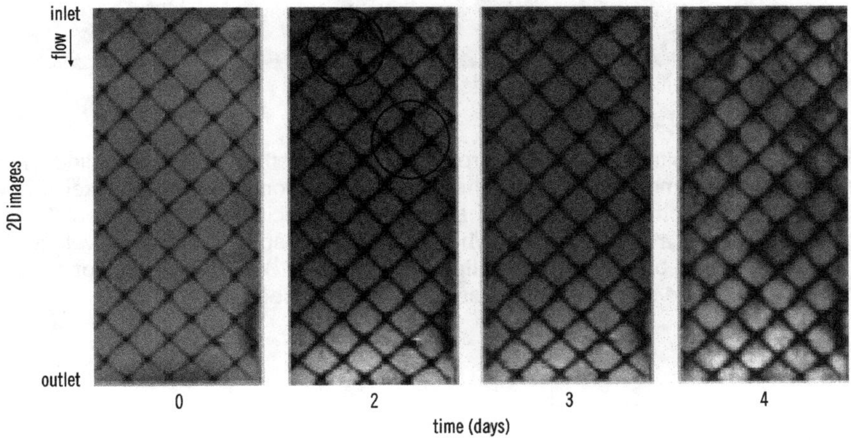

Figure 8.10 In-situ NMR 2-d images in time of the S-MFS supplied with substrate to promote biofouling. The image resolution is ~98 μm/pixel. The flow direction is from top to bottom. At day 2, the highlighted circles mark some regions of biomass accumulation

After 2, 3 and 4 days of the NMR study the FCP over the S-MFS was increased by respectively 95%, 155% and 680% compared to the initial pressure drop at day 0 (Figure 8.13). The FCP-data related well with the results of MRI imaging and velocity profiles measurements. A low-pressure drop corresponded with low-biomass and a homogeneous flow field while a high-pressure drop corresponded with strong-biomass accumulation and a distorted flow field in the S-MFS.

Notably already limited biomass accumulation on the feed spacer had strong impact on the flow field heterogeneity and thus on the FCP. Both visual and MRI studies showed the same fouling accumulation in the MFS (Figures 8.9 and 8.10), emphasizing the strong and elegant potential of MRI studies to provide a quantitative insight in the biofouling process.

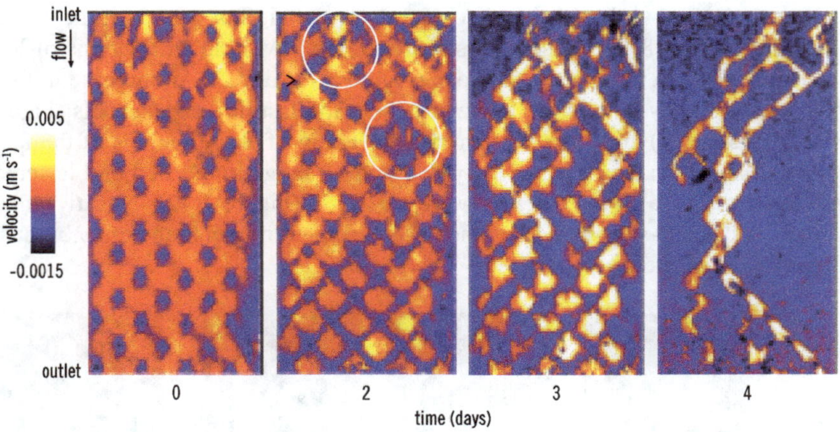

Figure 8.11 In-situ 2-d velocity profiles in time of the S-MFS supplied with substrate to promote biofouling. The image resolution is ~210 μm/pixel. Flow is from top to bottom. The images show the superficial flow component (axial component) on a colour scale where white corresponds to high velocities ($0.005\,m\,s^{-1}$) and black to low velocities ($-0.0015\,m\,s^{-1}$). At day 2, the highlighted circles mark some regions of biomass accumulation (see Figure 8.10)

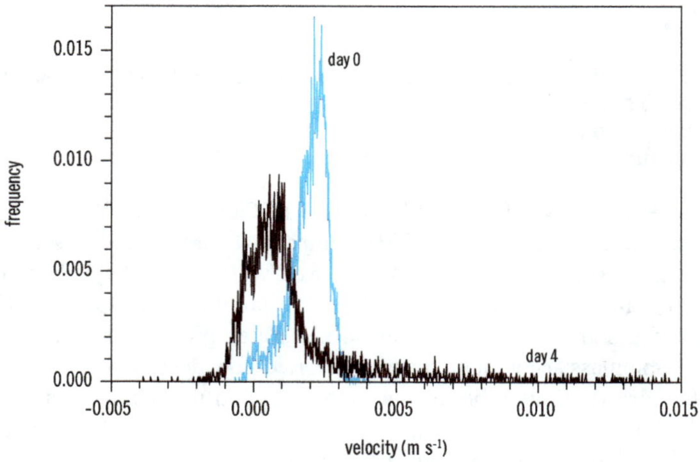

Figure 8.12 Velocity distribution in the S-MFS after 0 and 4 days of substrate supply. The velocity profiles were derived from the velocity images (Figure 8.11)

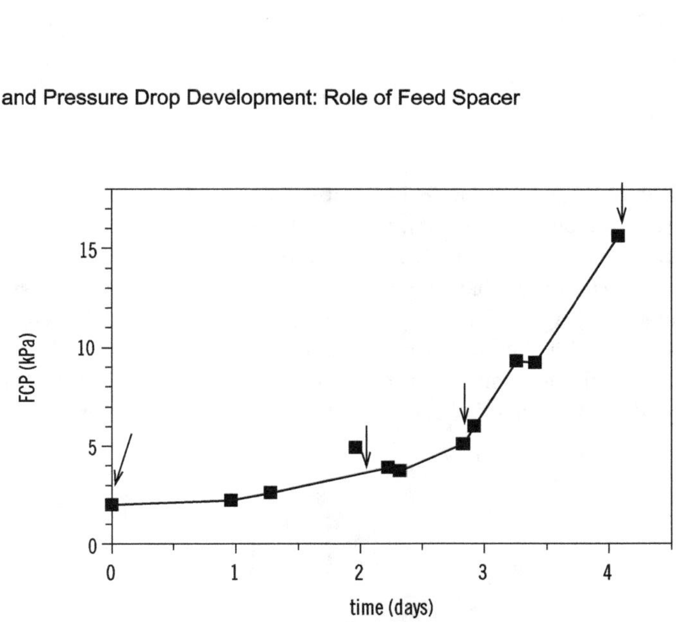

Figure 8.13 FCP development over S-MFS in time. The arrows indicate the moments of NMR imaging and velocity field imaging (Figures 8.10 and 8.11). The biomass concentration found on the spacer and membrane sheets in the S-MFS at the end of the study confirmed biomass accumulation (3×10^4 pg ATP cm^{-2})

Feed spacer impact on biofouling

When biofouling occurs in spiral-wound membranes, biofilm accumulation causes an FCP-increase. (Vrouwenvelder *et al.*, 2006b, 2008b, 2009d, Figures 8.6 and 8.8). The influence of feed spacer presence on FCP-development in time was studied using MFSs with and without feed spacer. Both MFSs were fed with water containing a biodegradable substrate (0.200 mg acetate-C L^{-1}). The monitors were operated with the same linear flow velocity. As control, monitors with and without feed spacer were operated in parallel with the same linear flow velocity without substrate addition.

The FCP was affected by the feed spacer. Over the MFS with feed spacer, a significantly higher FCP was observed compared to a MFS without feed spacer operated at the same linear flow velocity and channel height (Figure 8.14: blanks). During a 9 day research period, the FCP increased over both monitors supplied with substrate dosage while the FCP was constant over the control monitors without substrate dosing. The relative FCP-increase was the same for the MFS with and without feed spacer (factor 3.6). However, the absolute FCP – sum of FCP and FCP-increase – over the feed channel will determine the operational costs. The absolute FCP in the monitor with feed spacer was significantly higher compared to the monitor without spacer (factor 6) after

9 days substrate dosage (Figure 8.14). Impact on the FCP-increase of membrane fouling itself is insignificant compared with the FCP-increase caused by feed spacer fouling. The absolute FCP over the monitor without feed spacer supplied with substrate was lower than the FCP over the blank monitor with spacer and without substrate (Figure 8.14: comparison of 'with spacer blank' with 'without spacer +S'). Biofilm accumulation in the feed channel increased the FCP. With feed spacer, biofilm accumulation resulted in a stronger absolute FCP-increase. Biofouling of the feed channel for NF/RO is clearly a feed spacer problem.

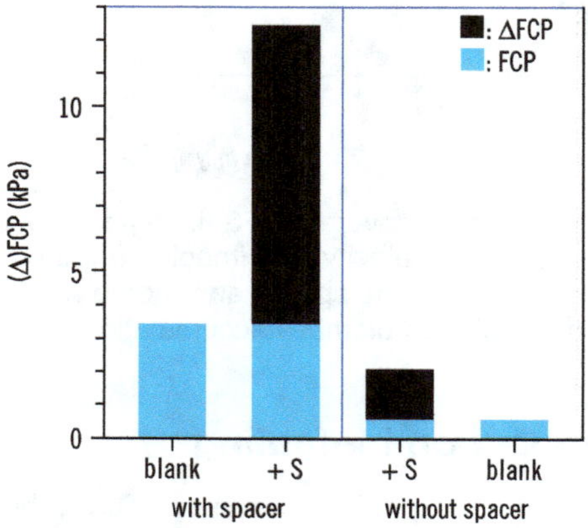

Figure 8.14 FCP and FCP-increase in time in a MFS with and without feed spacer after nine days operation for feed water with and without substrate dosing

DISCUSSION

Biofouling of NF and RO membranes increased the feed channel pressure drop (FCP, Figure 8.6). The same FCP-increase and biomass accumulation were observed irrespective of the membrane type (Figure 8.7) and permeate production (Figures 8.7 and 8.8). Biofouling was not influencing the transmembrane pressure (Figure 8.8B). Biomass accumulation (Figures 8.9 and 8.10) in the feed channel had strong impact on the flow field (Figure 8.11) resulting in a strong FCP-increase over the feed channel (Figures 8.13 and 8.14). Preventive actions to control the FCP-increase are most likely more efficient than waiting until the FCP-increase becomes unmanageable. Studies at different scales (from full-scale to miniature flow cells: S-MFS) provided a coherent database, supporting the presented conclusions.

Biomass accumulates on the location with highest impact on feed channel pressure drop

In pressure vessels, membrane modules are placed in series of up to eight modules (Figure 8.3). A strong decline of flow velocity occurs over the membrane modules in the pressure vessels with increasing distance to the feed side as a consequence of the reduction of water flow in the feed channel by permeate production. The lead module in the pressure vessel contributes most to the total FCP over the membrane installation because of the highest flow rate.

In practice, in most cases biofouling is observed at the feed side of the installation, especially in the first half of the lead module (Carnahan *et al.*, 1995 and Vrouwenvelder *et al.*, 1998, 2006b, 2008b, 2009d). The accumulated biomass volume increases the effective water flow velocity, resulting in a more than proportional FCP-increase (Figure 8.11). Biofilm accumulation in membrane systems therefore occurs on locations where the influence on the FCP is highest.

Biofouling is a feed spacer problem

Recent biofouling papers (Herzberg and Elimelech, 2007 and Chong *et al.*, 2008) refer to a biofilm enhanced osmotic pressure strongly contributing to a flux decline. For a flux decline caused by biofouling, two mechanisms were identified: (i) an increase of the hydraulic resistance over the membrane and (ii) hindering the back diffusion of salts. In their studies, flow cells were used containing membranes but no feed spacers. It is likely that their findings would have been different when feed spacers had been present. Research on pretreatment to reduce biofouling also shows a flux decline in membrane test cells, but spacer use and FCP were not described (Griebe and Flemming, 1998 and Wend *et al.*, 2003). In practice, increased FCP over the membrane module is a dominant reason for operational problems caused by biofouling (Figure 8.6, Vrouwenvelder *et al.*, 2008b). Schneider *et al.* (2005) mentioned that 'membrane fouling was so severe that feed spacers were forced out of the feed channels'. An increased FCP over the membrane unit at constant inlet pressure will also lead to less permeate production. To unravel the causes of reduced membrane performance, research is needed into the effect of both spacers and membranes, including monitoring of membrane performance (FCP, flux and salt passage) under representative conditions. Use of recently developed tools – such as the MFS (Vrouwenvelder *et al.*, 2006b) and MRI (Graf von der Schulenburg *et al.*, 2008; this paper) – to study fouling and membrane performance under practical conditions is recommended.

Biomass may accumulate on all surfaces in spiral-wound membrane modules: both on membranes and on spacers. Figure 8.7 showed that irrespective whether nanofiltration or reverse osmosis membranes were used, the same FCP-increase and biomass concentrations were observed, suggesting that the type of membrane may be irrelevant for biofouling of the feed channel. Irrespective of permeate production, FCP and biomass concentration increased similarly in a nanofiltration test installation (Figure 8.8). So, the FCP-increase in membrane elements caused by biomass accumulation was not influenced by both permeate production or membrane type.

Biomass accumulation in a small part of the feed channel increases the FCP over the total membrane installation. When flow channels develop (Figure 8.11), the FCP increases. The FCP-increase results in a lower driving force for permeation over the total installation, reducing permeate production. The influence of flow channels on the reduction of permeate production may dominate the effect of concentration polarization. Biomass accumulation in only a small part of the membrane installation may therefore be responsible for both FCP-increase and flux decline.

In literature, the effect of biofouling on membrane performance is not univocal. In membrane (test) installations biofouling is observed together with either a FCP-increase (Carnahan *et al.*, 1995; Ghedr, 1998; Vrouwenvelder *et al.*, 1998, 2000, 2008b; Schneider *et al.*, 2005 and Brouwer *et al.*, 2006) or flux decline (Ridgway *et al.*, 1983; Speth *et al.*, 2000; Van der Hoek *et al.*, 2000 and Ivnitsky *et al.*, 2005) or the combination thereof (Tasaka *et al.*, 1994 and Vrouwenvelder *et al.*, 1998). A literature overview on symptoms of membrane biofouling is given by Ridgway and Flemming (1996). Usually, the effect of biofouling on one operational parameter is described. The effect of two parameters, i.e. data on the FCP and flux are rare. A possible explanation may be that installations are operated at either constant feed pressure or at constant permeation rate. Data reported from practice are usually 'too anecdotal' to relate membrane biofouling to flux reduction and/or FCP-increase. In our research effort, increased FCPs caused by biofouling have been observed in the absence of a flux decline (Figure 8.8). Therefore, the flux reduction may have different causes than feed spacer biofouling such as the concentration and composition of natural organic matter (NOM) of the feed water, other fouling types and/or combinations thereof.

This study showed that spacer biofouling is much more important than membrane biofouling (Figures 8.7, 8.8 and 8.14): biofouling is primarily a feed spacer problem. Until now research to control biofouling has been focused predominantly on modification of membranes rather than on feed channel modification, as already mentioned in the introduction of this paper. Prevention of channeling is vital (Paul, 2002, Figure 8.11). Based on the results of this study feed spacer channel and hydrodynamic modifications should be pursued to restrict feed channel biofouling.

Reduction of biofouling by adaptation of spacer geometry and hydrodynamics

The use of feed spacers in spiral-wound membrane modules is a paradox: economic reasons require a low pressure drop while feed spacers are used to cause a high pressure drop in order to create turbulence along the membrane to prevent concentration polarization. Spiral-wound membrane modules without spacers are not expected to be feasible since spacers are needed for separating the membrane envelopes and creating turbulence. However, spacer modification may be a suitable approach to develop membrane systems with sufficient turbulence but with less susceptibility to a feed channel pressure drop increase caused by biomass accumulation.

Over a period of 13 years about 85 membrane modules from different pilot and full-scale nanofiltration and reverse osmosis installations were autopsied to determine and quantify the accumulated material. In all these membrane modules from different membrane manufacturers the feed spacers seemed to be of similar thickness, structure and orientation (unpublished results). Until now, research to control biofouling was predominantly focused on membrane adaptation and not on feed channel/spacer modification (see introduction). Clearly, feed spacer adaptation (e.g. coating, thickness and structure/geometry) in membrane modules deserves much more attention. In addition, spacer modification should cause more efficient cleaning.

An additional approach to reduce the pressure drop increase is to include hydrodynamics in the spacer studies. The pressure drop increase is mainly caused at the feed side of the installation where the flow velocity is highest. A lower flow velocity will result in a lower pressure drop increase. Without influencing the accumulated biomass, the impact of biomass on the pressure drop may be reduced by lowering the flow velocity. Reducing the flow velocity in lead membrane modules in combination with advanced feed spacers may be an elegant approach to control biofouling.

The presented flow cells enable to study membrane fouling processes in relation to membrane performance. MRI is an ideal tool to non-invasively study biofouling development in spiral-wound membranes and enables in-situ real-time non-invasive quantitative measurements for (combinations of) 2-d/3-d imaging, velocity imaging, and propagators (Graf von der Schulenburg *et al.*, 2008). The presented use of flow cells in combination of NMR measurements is a highly recommended approach for developing and testing advanced spacers and hydrodynamics.

SUMMARY

Biofouling was studied in full-scale and pilot-scale installations, test-rigs and membrane fouling monitors by conventional methods as well as Magnetic Resonance Imaging (MRI). Independent of permeate production, the feed spacer channel pressure drop and biomass concentration increased similarly in a nanofiltration pilot installation. In the presence of a feed spacer the absolute feed channel pressure drop increase caused by biomass accumulation was much higher than when a feed spacer was absent: in both spiral-wound nanofiltration and reverse osmosis systems biofouling is dominantly a feed spacer problem. This conclusion is based on (i) in-situ visual observations of the fouling accumulation (ii) in-situ non destructive observations of the fouling accumulation and velocity distribution profiles using MRI, and (iii) differences in pressure drop and biomass development in monitors with and without feed spacer. MRI studies showed that even a restricted biofilm accumulation on the feed channel spacer influenced the velocity distribution profile strongly. Biofouling control should be focused on the development of low fouling feed spacers and hydrodynamic conditions to restrict the impact of biomass accumulation on the feed channel pressure drop increase.

9

Evaluation of Experimental Data through the 3D Numerical Model

INTRODUCTION

Worldwide, the drinking water demand is increasing and regulations on drinking water quality become stricter. High pressure membrane filtration processes like reverse osmosis (RO) and nanofiltration (NF) produce high quality drinking water. Declining costs of membrane processes stimulate the application of membrane filtration processes in advanced water treatment practice (Mallevialle *et al.*, 1996 and Shannon *et al.*, 2008). Fouling affects the performance of the RO/NF systems due to e.g. (i) increase in pressure drop across membrane elements (feed-concentrate channel), (ii) decrease in membrane permeability, (iii) increase in salt passage. These phenomena result in the need to increase the feed pressure to maintain constant production and to clean the membrane elements chemically. The major fouling mechanisms of RO and NF membranes are scaling, particulate and organic fouling and biofouling. Scaling by inorganic compounds is usually controlled using a scale inhibitor, such as a polymer or an acid. Particulate fouling can be controlled by extensive pretreatment, including ultrafiltration. Thus, all types of fouling except biofouling and organic fouling – related types of fouling – are controllable. At the moment, biofouling is considered the major problem for RO and NF (Tasaka *et al.*, 1994; Ridgway and Flemming, 1996; Baker and Dudley, 1998; Vrouwenvelder *et al.*, 1998, 2008b; Van Hoof *et al.*, 2002; Patching and Fleming, 2003; Huisman and Feng Kong, 2004; Schneider *et al.*, 2005; Karime *et al.*, 2008 and Shannon *et al.*, 2008).

---

*This chapter is based, with permission from the copyright holder, on a paper previously published in Journal of Membrane Science Vol. 346 No. 1 pp. 71–85 doi: 10.1016/j.memsci.2009.09.025

In spiral wound membrane modules, two types of pressure drop can be discriminated: the trans-membrane pressure drop (TMP) and the feed spacer channel pressure drop (FCP), also named longitudinal pressure drop. The trans-membrane pressure drop, the differential pressure between feed and permeate lines, is related to the membrane flux (permeation rate). The definition of flux (permeation rate) is the water volume passing the membrane per unit area and time (L m^{-2} hr^{-1}). The FCP is the pressure drop between the feed and concentrate lines. In practice, the FCP is critical for operation.

Many authors mention a critical flux (Field *et al.*, 1995; Howell, 1995 and Bacchin *et al.*, 1995, 2006). It has been hypothesized that 'below a critical flux no fouling should occur'. The critical flux has been studied extensively for the fouling of microfiltration and ultrafiltration by colloidal particles (Bacchin *et al.*, 2006). Several studies propose the critical flux concept for spiral wound nanofiltration and reverse osmosis membranes in relation to biofouling (Winters, 1997, 2001 and Schwinge, 2004) as well. The influence of biofouling on flux decline and FCP increase has been studied on monitor, test rig, pilot and full-scale for extensively pretreated water (Vrouwenvelder *et al.*, 2008b, 2009b,c). Irrespective whether a flux was applied or not, the FCP and biofilm concentration increased. A previous study showed that the critical flux concept, stating that 'below a critical flux no fouling occurs', is not a suitable approach to control biofouling of NF and RO systems with extensive pretreatment (Vrouwenvelder *et al.*, 2009b). It appears that in extensively pretreated water of NF and RO, biofouling is a feed channel pressure drop problem (Vrouwenvelder *et al.*, 2009c).

Insight in biofilm formation and fluid dynamics is indispensable. Such an integrated model requires at least two essential components: a model for fluid flow (hydrodynamics or computational fluid dynamics, CFD) and a model for biofilm development. These two models are coupled because on one hand the fluid flow pattern is influenced by the biofilm position. On the other hand, the biofilm growth is determined by the water flow which supports the convective transport of nutrient and cell attachment and biofilm detachment are affected by the local shear stress. Currently available two-dimensional (2-d) and three-dimensional (3-d) computational fluid dynamics models on feed spacers in spiral wound membrane modules focus predominantly on improvement of mass transfer and flux (Cao *et al.*, 2001; Karode and Kumar, 2001; Schwinge *et al.*, 2002c,d; Li *et al.*, 2002, 2009; Geraldes *et al.*, 2002a; Koutsou *et al.*, 2004, 2007, 2009; Ranade and Kumar, 2006a; Shakaib *et al.*, 2007, 2009; Santos *et al.*, 2007; Fimbres-Weihs and Wiley, 2007 and Shrivastava *et al.*, 2008). While giving a detailed insight in local flow patterns around the spacer fibres, these CFD studies do not focus on the biofilm development and its influence on biofouling. 2-d and 3-d biofilm models, on the other hand, have evolved in the recent years to include the effects of flow and mass transfer on biofilm formation not only on simple planar surfaces (Picioreanu *et al.*, 2000a, 2001 and Eberl *et al.*, 2000), but also in complex geometries such as porous media (Kapellos *et al.*, 2007; Graf von der Schulenburg *et al.*, 2009 and Picioreanu *et al.*, 2010). Because no models combining CFD and biofilm growth for spiral wound systems with spacers have been found in literature, recently a 3-d numerical model is developed to determine hydrodynamics and biofouling in the feed spacer channel (Picioreanu *et al.*, 2009).

The main objective of this study was to qualitatively evaluate the developed mathematical 3-d model with experimental data and previously derived conceptual insights. When the model matches with the experimental data, then 3-d numerical simulations of biofilm formation and fluid dynamics will be a suitable technique to understand membrane fouling processes.

MATERIALS AND METHODS

Feed spacer characterization

An inventory of feed spacers used in practice, produced by four major global manufacturers of spiral-wound NF and RO membrane modules, was made to determine spacer geometry and material (Figure 9.1, manufacturers are coded I to IV). The feed spacers were made of polypropylene. Spacer geometry was characterized with a stereo microscope and a calibrated digital camera. The spacer thickness was determined with a sensitive accurate digital calliper (Mitutoyo absolute, ID-C112BS).

The spacers commonly used in practice had a diamond-shaped structure (Figure 9.1) and thicknesses varying between 26 and 34 mil (1 mil equals 25.4 μm), according to manufacturer specifications. In some cases thicker spacer (45 to 50 mil) or spacers with a parallel-shaped structure were used. Membrane modules from manufacturer II (Figure 9.1) with feed spacer thickness 31 mil (780 μm) were used in several membrane installations in the Netherlands. As being the most common, this spacer (Figure 9.2A) was selected for our studies. Similar spacer dimensions were used in the model (Figure 9.2B). An overview of spacer dimensions is presented in Table 9.1 and Figure 9.2.

Feed spacer porosity was determined with (i) a defined spacer sheet surface area and volume measurements in a volumetric flask and (ii) calculations based on average spacer strand diameter and strand woven density. The determined porosity of the feed spacer was ~0.85 (Table 9.1), matching the porosity in the model geometry (~0.84).

The specific surface area of feed spacers could not be determined accurately, even when using a confocal laser scanning microscope and coated spacer. However, from the model the feed spacer specific surface area can be derived as 7700 m^2 spacer m^{-3} spacer or 1230 m^2 spacer m^{-3} total volume.

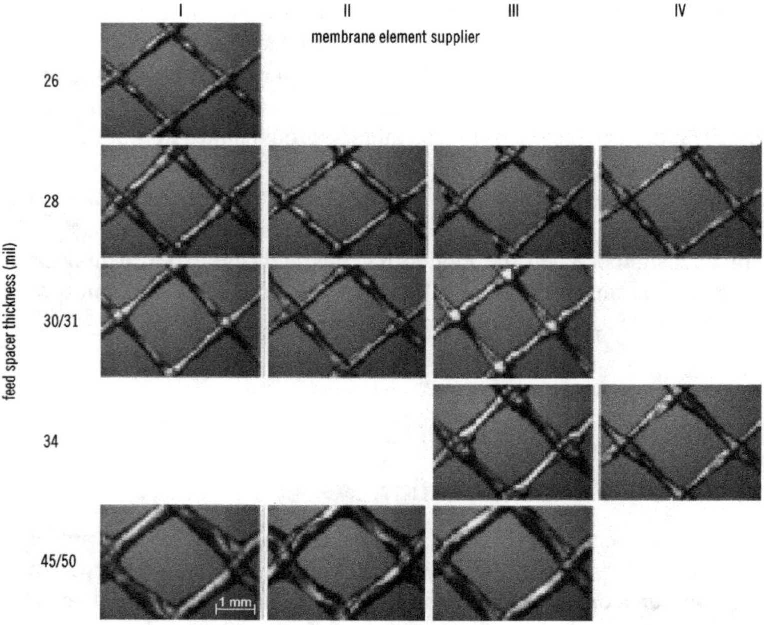

Figure 9.1 Inventory of feed spacers commonly used in practice in commercially available spiral wound NF and RO membrane modules by four global membrane manufacturers. The membrane module manufacturers are coded I to IV. The feed spacer thickness is commonly expressed in mil (1 mil equals 25.4 μm). The bar length indicates the scale of all spacers

Model description

The numerical model used in this study and water quality parameters have been described (Picioreanu *et al.*, 2009). Therefore, only the basic assumptions and features are summarized here.

The computational domain is a small area from the membrane separation device of 2000 μm length (along main flow direction) by 1200 μm width, with the membranes kept apart at 780 μm by the spacer (Figure 9.2C). The feed spacer geometry was derived from feed spacers used in practice in spiral wound membrane modules, using a stereo microscope and calibrated digital camera to determine spacer spatial dimensions. Spacer orientation in the model was the same as applied in practice: 45° spacer rotation towards feed flow direction ('diamond configuration'). The computational domain contained five by three diamond spacer frames (Figure 9.2C). Periodic boundary conditions were applied for the two sides of the domain (perpendicular to the main flow direction), both for fluid flow and for biofilm formation. Using periodic boundary in a smaller model, means that two domain sides are in open connection with each other, improving model representativeness (since wall effects of the model domain are excluded). The use of periodic boundary conditions reduces the domain size needed to obtain quantitative information about a larger system (Li and Tung, 2008a and Darcovich *et al.*, 2009), and therefore the computational requirements. An industrial spiral wound membrane module (8 inch diameter and 1 m length) can have up to 41 m^2 membrane surface area and approximately 2×10^6 diamond spacer frames distributed over several membrane envelopes, requiring massive computational efforts.

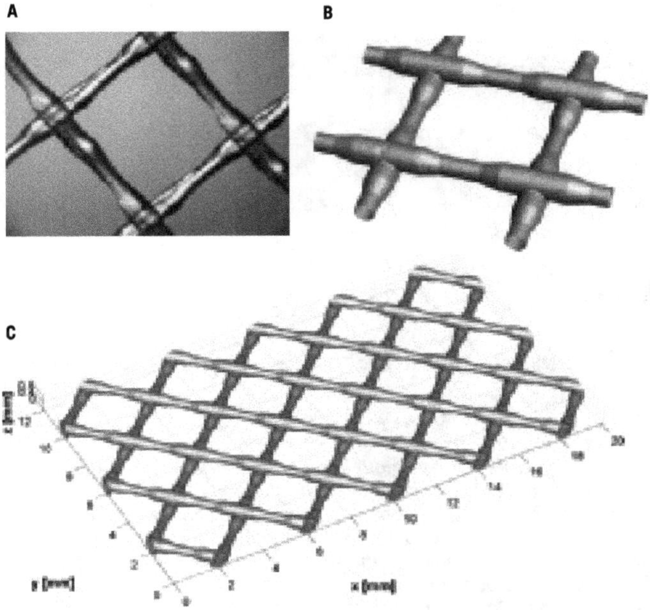

Figure 9.2 Feed spacer geometry from practice (**A**) and model (**B**) with the same geometry. Feed spacer with thickness 31 mil from manufacturer II was selected for the studies. (**C**) The computational domain of 3 × 5 spacer frames: membranes on the top and bottom sides (z = 0 and 0.78 mm), periodic lateral boundaries (y = 0 and 12 mm), liquid inlet (x = 0) and liquid outlet (x = 20 mm). Adapted from (Picioreanu *et al.*, 2009)

The hydrodynamic model includes a feed flow of water, entering the domain with an already established laminar flow profile (linear velocity $0.16\,m\,s^{-1}$). Navier-Stokes equations for steady-state laminar flow are solved at several moments in time, for different stages of biofilm growth, thus corresponding to different biofouling states. The membrane and the spacer are no-slip (zero-velocity) walls for the flow. The biofilm surface is also a no-slip wall, but its spatial position changes in time as new cells attach and as the biofilm grows.

Table 9.1 System geometry and operation parameters in experimental conditions compared with the model input

Characteristic	Unit	Practice	Model	Monitor MFS/(S-MFS)
Feed spacer porosity	$m^3\ m^{-3}$	0.85	0.84	0.85
Specific surface area feed spacer	$m^2\ m^{-3}$	n.a.	7700	n.a.
Specific surface area feed spacer per total volume	$m^2\ m^{-3}$	n.a.	1230	n.a.
Thickness feed spacer	μm	780	780	780
Thickness flow channel	μm	780 [a]	780	780
Structure feed spacer (non woven)	–	similar	similar	similar
Linear flow velocity	$m\ s^{-1}$	0.16	0.16	0.16
Membrane and spacer length	m	0.96[b]	0.02	0.20 / (0.07)
Pressure drop over flow channel[c]	kPa	18–~40[d]	0.80	3–8
Normalized pressure drop[e]	$kPa\ cm^{-1}$	0.2–~0.4[d]	0.40	0.15–0.40

[a] = estimated value; [b] = membrane length in membrane module; [c] = under non-fouled conditions at 0.16 $m\,s^{-1}$; [d] = indicative data for lead membrane modules from (Vrouwenvelder et al., *2009d) and a consulting company. The maximum pressure drop over the module flow channel can be up to 140 kPa according membrane module manufacturers' specifications. In membrane modules at the end of a pressure vessel the pressure drop and normalized pressure drop values are clearly lower (like 7 kPa and 0.07 $kPa\,cm^{-1}$ (Vrouwenvelder* et al., *2009d). [e] = normalized for system length; n.a. = not available.*

The biofilm model includes microbial (biomass) attachment in random positions on the membrane or spacer, but with an increased attachment chance for decreasing values of shear stress on the walls. Further, the attached microbial cells grow and form colonies (cellular automata algorithm from Picioreanu *et al.* (1998a,b). Microbial growth and attachment are calculated in time increments of 3 hours, in between which the flow pattern is re-calculated. For purpose of simplicity, it was assumed here that substrate and oxygen mass transport limitations are negligible and therefore the biomass can grow unlimited. However, the effect of mass transport on biofilm formation was presented in (Picioreanu *et al.*, 2009).

The model implementation (Picioreanu *et al.*, 2009) consists of a combination of MATLAB code (MATLAB 2007b, MathWorks, Natick, MA, www.mathworks.com) as the main algorithm script, COMSOL Multiphysics (COMSOL 3.5a, Comsol Inc., Burlington, MA, www.comsol.com) finite element methods for solving the partial differential equations governing the flow field and the solute mass balances, and own Java routines for the biofilm model (Picioreanu *et al.*, 1998a,b). This hybrid approach allows highly efficient model solution on commercially parallelized code and with the large memory requirements inherent for 3-d models. The 3-d mesh for finite element solution of the hydrodynamics and mass balances contained ~430,000 tetrahedral elements (quadratic accuracy) with a maximum size of 200 μm both in the volume and near the spacer boundaries. This means ~2,100,000 equations were simultaneously solved for hydrodynamics and ~663,000 for the solute mass balance. The 3-d mesh for biomass and pseudo-viscosity distribution contained 250 × 150 × 10 rectangular elements (cube size ~80 μm). The computational code was run for all simulations on a workstation with two parallel quad-core Intel Xeon E5430 processors, 32 GB DDR2 memory and Quadro 3700 FX graphics card.

Experimental set-up

The development of biofouling in the feed channel was investigated at two scales: (i) using a spiral wound membrane module (Graf von der Schulenburg *et al.*, 2008) and (ii) Membrane Fouling Simulator (MFS, Vrouwenvelder *et al.*, 2006b) with and without feed spacer. Visual microscopic observations on fouling accumulation using the MFS window and in-situ observations on fouling accumulation and velocity distribution profiles using Magnetic Resonance Imaging (MRI) were performed. In all modules and MFS studies nanofiltration membranes and spacers from the same supplier were used. The NF membrane is a polyamide membrane. The feed spacer taken from the module was a 31 mil thick diamond-shaped polypropylene spacer with porosity 0.85 (Table 9.1). The feed spacers in the MFS had the same spatial orientation as in spiral-wound membrane modules (45° rotation).

The membrane fouling simulator (MFS) is a tool developed to study and monitor fouling in spiral-wound membranes (Vrouwenvelder *et al.*, 2006b, 2007a). Using the MFS, fouling development can be monitored by (i) pressure drop measurements, (ii) in-situ and non-destructive (visual) observations using the sight window and (iii) analysis of coupons sampled from the membrane and spacer sheet in the monitor. The feed spacer channel dimensions of the MFS are 0.00078 m × 0.040 m × 0.200 m. In the MFS, water flows through the spacer attached between a membrane sheet and the window. Membrane and spacer coupons of new and unused spiral-wound nanofiltration membrane elements were placed in the MFS.

To quantify linear flow velocities in lead modules, an inventory has been made at several full-scale installations containing 8 inch diameter spiral wound NF or RO membrane modules (Vrouwenvelder *et al.*, 2009a). Linear flow velocities in lead modules ranged between 0.07 and $0.20\,m\,s^{-1}$. The same range of velocities was found for installations containing 4 inch diameter lead modules. Biofouling was predominantly observed in lead modules with a high linear velocity. Therefore, a high linear flow velocity ($0.16\,m\,s^{-1}$) was selected for the studies described in the following sections.

Feed water for the MFSs was drinking water, distributed by a drinking water supply company without any disinfectant dosage or residual.

Substrate was dosed to the feed water of MFSs to stimulate biomass growth. From a sterile vessel, containing a 5 L solution of concentrated substrate, the substrate was dosed into the feed water prior to the MFS by a peristaltic pump (Masterflex) at a flow of $0.03\,L\,h^{-1}$. The dosage of substrate was checked periodically by measuring the weight of the dosing bottle. The chemicals $NaCH_3COO$, $NaNO_3$, and NaH_2PO_4 and were used with a ratio C:N:P of 100:20:10 for the dosage solution. C was the growth limiting compound. N and P were dosed to eliminate growth limiting conditions for N and P. The substrates were dissolved in ultrapure water. To restrict bacterial growth in the substrate dosage bottle, the solution pH was set to 10.5 using NaOH. Dosage bottles were replaced every 5 days. The substrate dosage ($0.03\,L\,h^{-1}$) was low compared to the feed water flow rate ($16\,L\,h^{-1}$). Thus, the pH of the feed water was not measurably influenced by substrate dosage.

MRI study

Based on the principles of proton Nuclear Magnetic Resonance (^1H NMR), the evolution of spatial biofilm distribution and velocity field can both be visualized by a technique called Magnetic Resonance Imaging (MRI). MRI has already been applied to study fouling with biofilms in other systems such as circular pipes (Lens *et al.*, 1999; Van As and Lens, 2001 and Manz *et al.*, 2003, 2005) and recently in membranes (Graf von der Schulenburg *et al.*, 2008 and Vrouwenvelder *et al.*, 2009c). Magnetic materials can not be used since they interfere with the NMR signal. Therefore, a PVC version of the MFS had to be developed, named S-MFS (Vrouwenvelder *et al.*, 2009c). In order to fit in a 200 MHz super wide-bore magnet, the S-MFS had to be scaled down to channel dimensions of $0.00078\,m \times 0.016\,m \times 0.040\,m$. In the S-MFS, the water flows through the spacer placed between two membrane sheets. The S-MFS was produced with a manufacturing accuracy of $20\,\mu m$. The feed spacer and membrane sheets used in the NMR studies were taken from a new nanofiltration membrane module. The PVC flow cell was operated with the same linear flow velocity and the same substrate dosage method was applied, as described for the MFS studies. For the determination of the spatial flow velocity distribution, the flow was temporarily lowered to $0.09\,m\,s^{-1}$ (56% of the original linear flow velocity) for about 45 minutes.

Two-dimensional velocity images of the superficial flow component (axial direction) were acquired employing a standard phase shift velocity encoding pulse sequence with a size of 256×128 pixels resulting in a spatial resolution of ~210 μm/pixel (Graf von der Schulenburg *et al.*, 2008). More details on MRI velocimetry can be found elsewhere in the literature (Callaghan, 1991).

Pressure drop

A differential pressure transmitter (Endress + Hauser Deltabar S: PMD70-AAA7FKYAAA) was used to determine the pressure difference over the feed spacer channel of the (S-)MFS (Vrouwenvelder *et al.*, 2009d).

Membrane autopsy

To characterize the accumulated fouling, $4\,cm^2$ sections of membrane and feed spacers were taken from the (S-)MFS. The sections of spacer and membrane were together placed in capped tubes filled with 30 mL of autoclaved tap water. To determine the amount of biomass, the tubes with the membrane sections were placed in an ultrasonic cleaning bath (Bransonic, model 5510E-DTH, output 135 W, 42 kHz). The low energy sonic treatment (2 minutes) followed by mixing on a Vortex (few seconds) was repeated three times. The biomass-water suspension from the tubes was used to determine active biomass in duplicate by measuring the adenosinetriphosphate (ATP) concentration by luminescence (Holm-Hansen and Booth, 1966). The luminometer (Celcis, model Advance) added 100 μL of LuminEX-B reagent (Celcis) to a 100 μL sample to release ATP from the bacterial cells. Subsequently, 100 μL of LumATE-PM (Celcis) was added for light production. The amount of light produced was measured with the luminometer (relative light units, RLU). The concentration of ATP was derived from the RLU values using the conversion factors of the linear relationship between RLU values and reference ATP concentrations. The ATP concentrations of autoclaved drinking water (<1 pg ATP mL^{-1}) and two control solutions (2 and 100 pg ATP $\times$ mL^{-1}) were determined as quality control.

Total organic carbon (TOC) of the accumulated biomass amount was determined as non-purgeable organic carbon by infra-red gas analysis. Sample pretreatment was identical as described for ATP analysis except for the water type in the capped tubes, which was demineralized water for the TOC analysis.

A spiral wound membrane module was fed with tap water supplemented with substrate (1.00 mg acetate-C L^{-1}). The center tube of the membrane was blocked during both fouling and MRI experiments to ensure that no permeate was produced (Graf von der Schulenburg *et al.*, 2008). A membrane module from a full-scale installation suffering from an elevated pressure drop caused by biofouling was cut open lengthwise and unwrapped to visualize flow patterns and determine and quantify accumulated material.

The plot of the natural logarithm of biomass concentration against time gave a straight line; the slope of the line is the specific biomass growth rate.

RESULTS

Inventory of feed spacers used in practice

Feed spacers used in practice in spiral wound nanofiltration and reverse osmosis membrane modules were characterized on a number of aspects such as material type, structure and thickness. The feed spacers provided by 4 membrane manufacturers were made of polypropylene. The spacers commonly used in practice had a diamond-shaped structure (Figure 9.1) and thicknesses varying between 26 and 34 mil (1 mil equals 25.4 μm), according to manufacturer specifications. In some cases thicker spacer (45 to 50 mil) or spacers with a parallel-shaped structure were used.

Membrane modules from manufacturer II (Figure 9.1) with feed spacer thickness 31 mil (780 μm) were used in several membrane installations in the Netherlands. Therefore, this spacer (Figure 9.2) was selected for our studies. The determined porosity of the feed spacer was ~0.85 (Table 9.1). The spatial dimensions of the feed spacer were determined using a stereo microscope with a calibrated digital camera (Figure 9.2A). Spatially identical spacer dimensions were used in the model (Figure 9.2B). The calculated porosity of the feed spacer in the model was 0.84, matching the determined feed spacer porosity (~0.85). The specific surface area of the feed spacer was not determined experimentally. However, the feed spacer specific surface area (7700 $m^2 m^{-3}$, Table 9.1) can be derived from the model geometry. The model feed spacer had very similar geometry to the feed spacer used in practice and experimental studies (Table 9.1, Figure 9.2).

Biomass growth parameters and pressure drop increase

The specific biomass growth rate (μ_{max} in day^{-1}) is a key parameter in the numerical model. To determine the biomass growth rate, studies with five MFSs in parallel were performed with the same feed water substrate concentration (0.60 mg acetate-C L^{-1}) and same linear flow velocity in the MFS (0.16 $m s^{-1}$). The monitors were sampled after different operating periods to obtain information about the biomass development after different running times. Determined were the pressure drop and biomass accumulation.

The biomass concentration increased exponentially (Figure 9.3C). The biomass was equally distributed over the monitor width and length (Figure 9.4). Therefore, biomass concentrations averaged over the module length were used for data evaluation. The average biomass concentration increased exponentially (Figure 9.3C) at least until day 3, indicating that biomass growth was not limited by substrate within this period. The derived net biomass growth rate $\mu_{max} = 2.8\,d^{-1}$ was used in the model. This is actually a lumped biomass accumulation rate, as the difference between the growth and the detachment rates. The model does not explicitly include biomass detachment; therefore the experimental value was used as a net growth rate.

Comparison model with experimental data

The three-dimensional computational model (Picioreanu *et al.*, 2009) was based on spacer geometry, biomass growth rates and operating conditions (feed water quality, linear flow velocity) from practice (Table 9.1). The linear flow velocity was within the range as applied in practice (Vrouwenvelder *et al.*, 2009a). The effect of substrate in the feed water on biomass accumulation, caused by substrate utilization, was determined with the model and experiments. In all model simulations we used default values for the model parameters as mentioned by Picioreanu *et al.* (2009). No parameter fitting or model calibration was done.

The pressure drop remained constant during the experiments with monitors fed with water without substrate (Figure 9.5). The pressure drop increased strongly in the presence of substrate in the feed water, in relatively good agreement between model and experimental data. After 4 days, the model calculated a pressure drop that deviated from the measured pressure drop (Figure 9.5). There are two reasons for this overestimation of the pressure drop by the model. First, substrate limitations appear when thick biofilms develop in the modules, leading to a linear biomass growth rate vs. the unlimited exponential growth rate. Although the biomass distribution along the module remained approximately the same (Figure 9.4) indicating no growth limitations in the flow direction, substrate limitations can still occur in the biofilm depth. Herzberg and Elimelech (2008) reported a higher distribution of active cells in a *Pseudomonas aeruginosa* biofilm close a RO membrane surface; likely due to higher substrate levels induced by concentration polarization effect caused by permeate production. In other words, the substrate concentration may be higher near the membrane. Biomass detachment was not included in the model. The biofilm detachment certainly becomes important when the local flow velocities increase due to fouling. This leads to high shear stress and biomass removal, affecting the pressure drop.

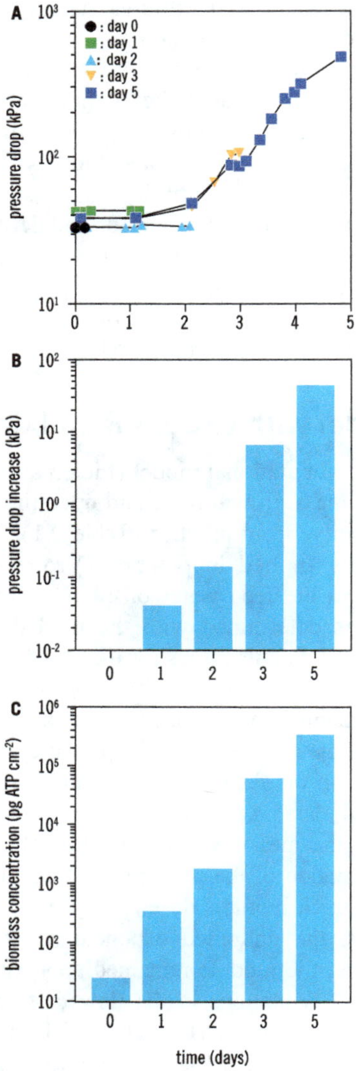

Figure 9.3 Measured pressure drop in time (**A**), pressure drop increase (**B**) and biomass accumulation (**C**) in monitors operated in parallel under identical conditions after 0, 1, 2, 3, and 5 days. Note the logarithmic scale in (B) and (C). The nutrient concentration in the feed water was 0.60 mg acetate-C L^{-1} and linear flow velocity in the monitors was 0.16 $m s^{-1}$

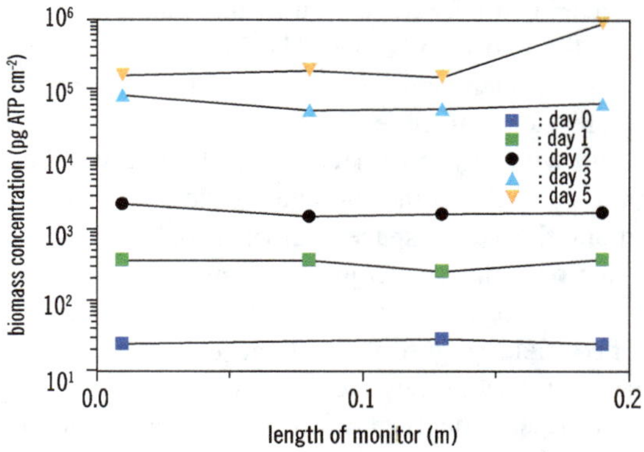

Figure 9.4 Measured biomass distribution over length of spacer and membrane in the monitors operated in parallel under the same conditions and analyzed at different running times

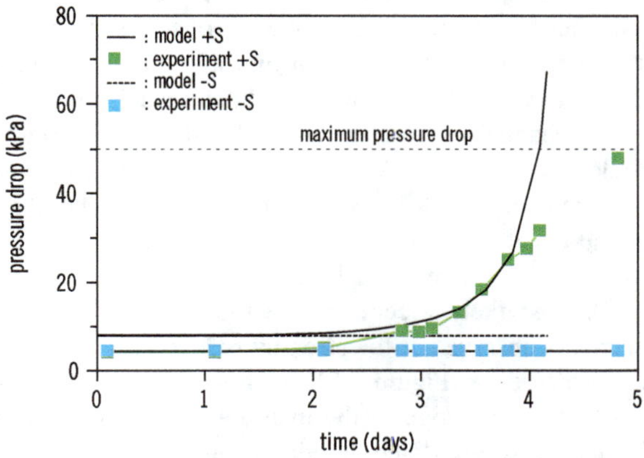

Figure 9.5 Compared pressure drop over feed spacer channel in time in computational model with (model +S) and without substrate (model -S) and experimental data with (experiment +S) and without substrate (experiment -S) dosage. The horizontal line represents the maximum pressure drop reading of the differential pressure drop transmitter

Velocity distribution profiles were (i) calculated using the mathematical model and (ii) determined experimentally using in-situ MRI observations with an S-MFS flow cell designed for MRI studies (Vrouwenvelder *et al.*, 2009c) and with a spiral wound membrane module (Graf von der Schulenburg *et al.*, 2008). The calculated velocity field was averaged over the height of the computational domain, i.e., the direction *z* between the two membranes. The 2-d average velocity was relatively the same in all square elements formed by spacer filaments before biomass accumulation (Figure 9.6A). In time, the flow field profiles gradually changed to a distorted flow field with flow channels (Figure 9.6D). MRI measurements with the S-MFS monitor showed that the average 2-d flow velocity field was relatively homogeneous (among the square spacer elements) before substrate dosage (Figure 9.6E). There was no flow at the spacer junctions and in the regions under the spacer filaments to flow was faster due to the reduction in the hydraulic area. In time, a distortion of the flow field (Figure 9.6FG) and eventually flow channels (Figure 9.6H) were observed, indicating flow heterogeneity across the main flow direction (Figure 9.6C). MRI measurements of a full-scale spiral wound membrane module showed that biofouling led to a heterogeneous flow distribution also along the main flow direction (Figure 9.7). This also led to stagnant zones with low velocities and zones with high velocities to compensate for the low velocity regions (Figure 9.7).

Flow channels can be observed during membrane autopsy, however these are not always easily detected or often overlooked due to the opaque nature of biofilms. The effect of flow channel formation has been clearly observed in cases where a high feed channel pressure drop caused by major biofouling occurred (Figure 9.8). The biofilm entrapping inorganic compounds facilitated visual observation of channel formation. The flow channels were mainly found on the first 0.20 to 0.50 m of lead membrane modules from the installation. The flow channels – caused by biofilm accumulation – were evenly distributed over the individual membrane leaves and over the membrane envelope width (visual observations). The effective feed side membrane surface area were reduced by approximately 50% (Figure 9.8A) and 90% (Figure 9.8B), most likely affecting the membrane flux strongly. Displacement of the feed spacer was observed at the strongly reduced (by ~90%) effective feed side membrane surface area (Figure 9.8B). Figure 9.8A clearly shows the same fouling amount on the glued edge of the membrane and on the membrane where permeate production occurred. Since the glued edge was not producing permeate, this observation supports previous findings that biofouling is a predominantly a feed spacer channel related problem (Vrouwenvelder *et al.*, 2008b, 2009c).

The localization of biomass growth is predominantly at the feed side of membrane installations, in the first half of lead modules (Carnahan *et al.*, 1995 and Vrouwenvelder *et al.*, 1998, 2008b, 2009c,d). Recently it has been clearly shown that biofouling is a problem of biofilm growth and not a filtration effect leading to accumulation of biomass on the membrane and spacer (Chen *et al.*, 2004a; Pang *et al.*, 2005, 2007 and Bereschenko *et al.*, 2008). In the model, biomass attached predominantly on the feed spacer (Figure 9.9B) at linear flow velocities as applied in practice. Biomass attachment occurred at positions with low shear. Multiplication of microorganisms resulted in a biomass volume increase, reducing the water volume fraction. Experiments with the MFS showed that biomass accumulated mainly on the feed spacer, at the downstream side of the spacer (Figure 9.9D). Filamentous biofilm structures, called streamers, were observed and also reported earlier (Vrouwenvelder *et al.*, 2006b, 2009c). The longest streamers were attached to the spacer crosses. The model showed the same localization of biomass growth as in experimental studies using the MFS sight window (Figure 9.9, Vrouwenvelder *et al.*, 2006b, 2009c) and MRI studies (Graf von der Schulenburg *et al.*, 2008).

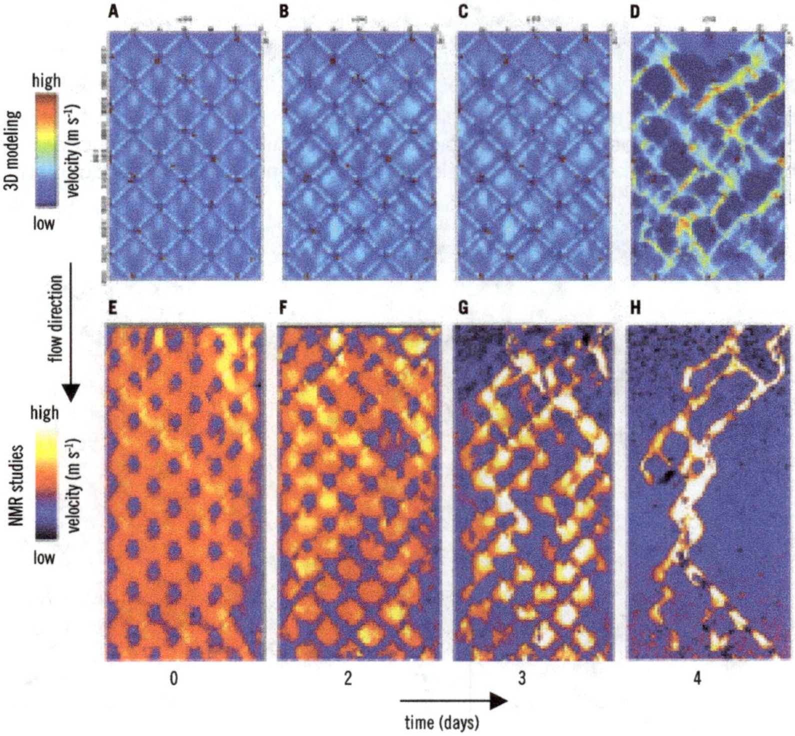

Figure 9.6 Compared time series of velocity profiles averaged over the channel height obtained with 3-d-modeling (**A–D**) and NMR experiments (**E–H**). The flow channel contained spacers and the liquid was supplied with substrate to promote faster biofouling. Flow is from top to bottom. The images show the velocity values on colour scales different between the model and NMR studies. Panels (E)–(H) adapted from Vrouwenvelder *et al.*, 2009c

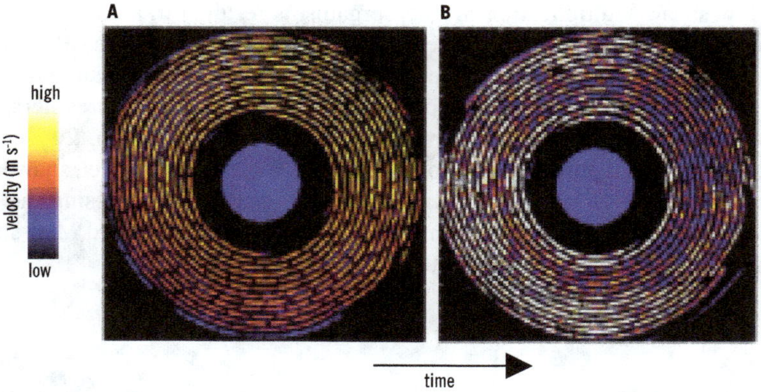

Figure 9.7 Measured 2-d radial velocity images of a spiral wound membrane module measured before and after biofouling. The images show the z component of the flow velocity on a colour scale (adapted from Graf von der Schulenburg *et al.*, 2008)

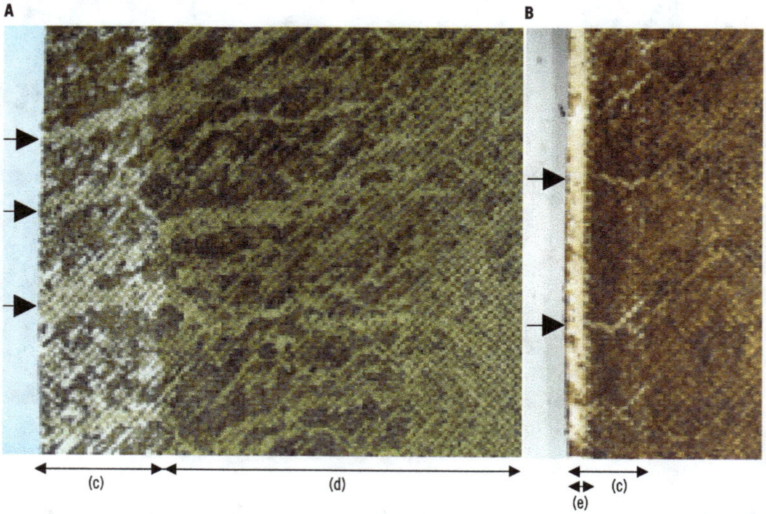

Figure 9.8 Unwound spiral-wound membrane modules from full-scale installations suffering from severe biofouling showing flow channels caused by biofouling. The flow direction is from left to right. (**A**) Flow channels covering ~50% of the feed inlet. In region (c) the membrane is glued and has a lighter colour. In region (d) water was produced and the membrane has a darker colour. (**B**) Flow channels covering ~10% of the feed inlet, resulting in displacement of the feed spacer (e). The arrows in the figures indicate a few examples of flow channel locations. The biofilm entrapping inorganic compounds facilitated visual observation of channel formation

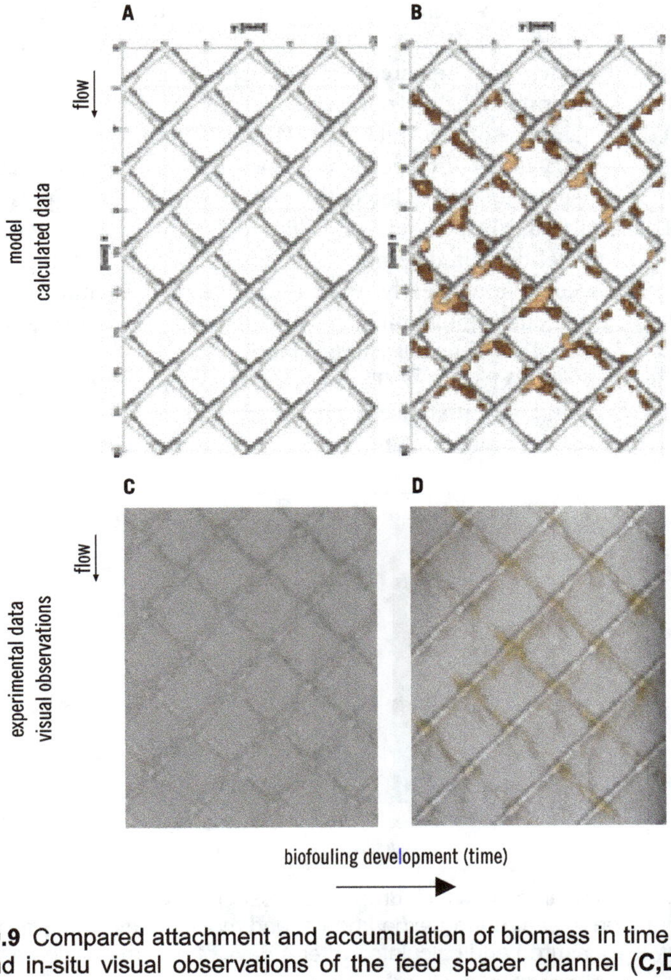

Figure 9.9 Compared attachment and accumulation of biomass in time in model (**A,B**) and in-situ visual observations of the feed spacer channel (**C,D**). Visual observations were made using the sight window of the MFS during operation with dosage of a compound containing substrate. Analysis of the accumulated material confirmed biomass accumulation

Influence feed spacer: model and experimental data

Comparative studies were performed with and without feed spacer at the same cross flow velocity and substrate concentration in the feed water to evaluate the influence of feed spacer on biofouling development. As control, studies with and without feed spacer at the same cross flow velocity were done with and without substrate.

As expected, the feed channel pressure drop is strongly affected by the presence of a feed spacer (Figure 9.10). The feed spacer presence increased 10 times the pressure drop compared with the experiment without spacer, at the corresponding experimental flow rates and without biofilm (clean device flow channels). In the absence of substrate there was no biofilm growth and obviously no increase in the pressure drop relative to the clean device flow channels. In the presence of substrate a biofilm developed and the relative pressure drop increased. The absolute FCP – sum of FCP and FCP-increase – over the feed channel of the model domain determines the costs caused by fouling. With feed spacer, biofilm accumulation clearly resulted in a stronger absolute FCP-increase (Figure 9.10). The model simulation matched rather well the experimental results in the case of clean flow channels without biofilm, with (0.35–0.4 kPa cm^{-1}) or without spacer (~0.03 kPa cm^{-1}). The current model predicted more pressure drop after 4 days, but the difference is attributed to more biofilm formation due to both absences of substrate limitation and biomass detachment in the model.

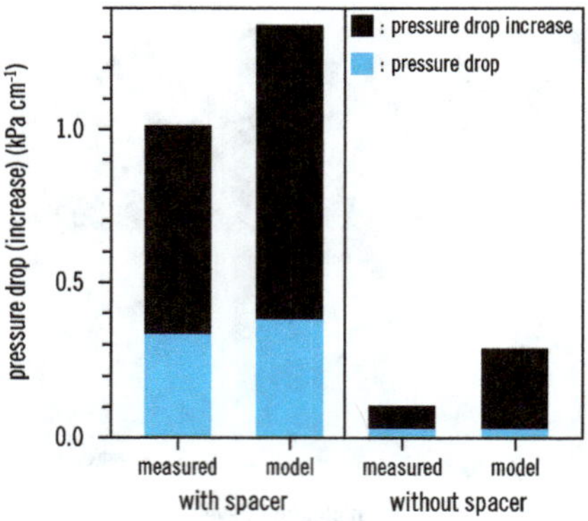

Figure 9.10 Compared pressure drop and pressure drop increase in time with and without feed spacer during experimental and model study. The pressure drop increase was measured and calculated after 4 days of biofilm development

Modeling the overall pressure drop, biomass accumulation and volume fractions in time showed that the same biomass concentration in the system with and without spacer had different impact on the pressure drop (Figure 9.11). Although the same amount of biomass was formed with/without spacer (Figure 9.11B), the overall pressure drop increased more rapidly with biofilm formation when the spacer was present (Figure 9.11A). With spacer the initial pressure drop was 10 times higher, which cannot be explained by the 15% less volume liquid (Figure 9.11C), but rather by the severely tortuous flow path imposed by the spacer (Picioreanu *et al.*, 2009).

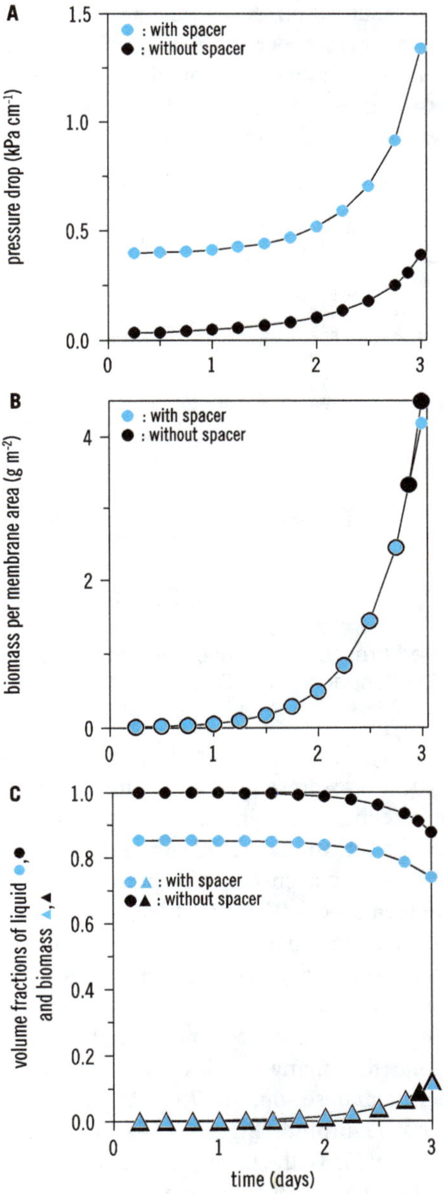

Figure 9.11 Calculated overall pressure drop FCP (**A**), biomass (**B**) and volume fractions of liquid and biomass (**C**) in time with and without feed spacer

With spacer, the pressure drop development over the model domain length showed that the high biomass concentration after 3 days resulted in a stepwise decline of the pressure drop along the domain length (Figure 9.12A). Without feed spacer, the pressure drop was lower and a continuous (no stepwise) decline of pressure drop was observed even at high biomass concentrations (Figure 9.12B).

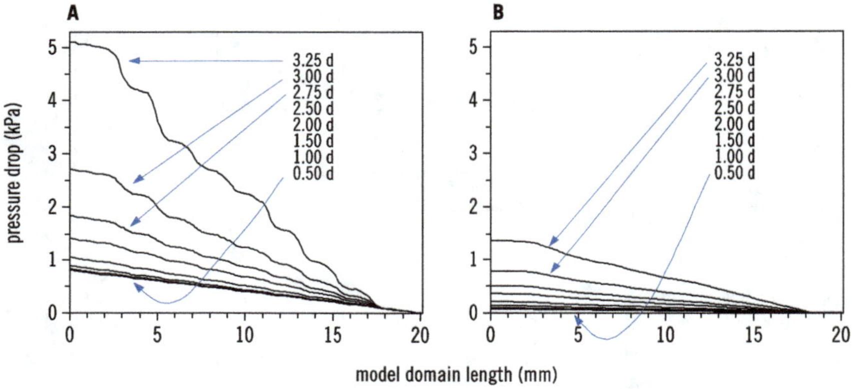

Figure 9.12 Calculated pressure drop over the model domain length, at different moments in time: (**A**) with spacer, and (**B**) without feed spacer

The simulated velocity fields in the feed spacer channel (height is 780 μm) with and without spacer are presented in Figures 9.13 and 9.14. The model results are shown in a section through the middle of the flow channel (390 μm, Figure 9.13) and near the membrane (at 585 μm, i.e., 195 μm from the membrane surface, Figure 9.14). When a spacer was present the biomass accumulated predominantly on the feed spacer (Figure 9.13BC). Without feed spacer, biomass attachment could occur at the membrane only (Figure 9.13DE). The biomass growth clearly affected the velocity field. The change in velocity field distribution was much stronger with spacer (Figure 9.13C) than without feed spacer (Figure 9.13F). Clearly the effect of the spacer on biofouling in membrane processes is predominant, despite that many studies do not consider the presence of a spacer in their experimental set-up. At 75% of the channel height (585 μm) the velocity field shows a profile (Figure 9.14) that differs strongly from the channel centre (Figure 9.13). With spacer, nearer to the membrane the velocity flow is mostly parallel to the spacer filaments present in this planar field (Figure 9.14) which is not observed in the channel centre (Figure 9.13). The fact that the velocity field shows great variations over the channel height illustrates that 3-d modeling is essential for fluid dynamics and biofouling studies in membrane systems.

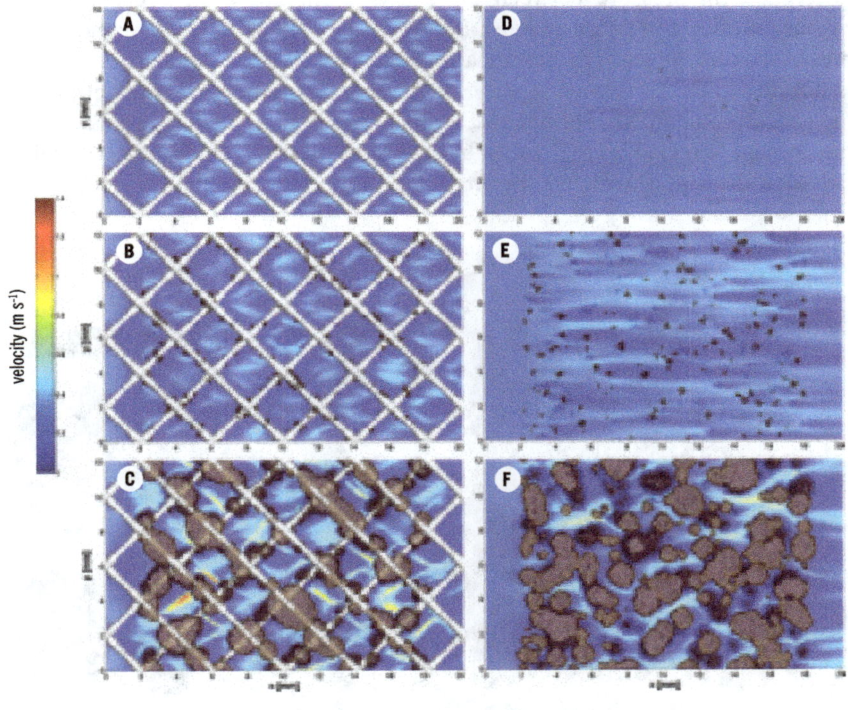

Figure 9.13 Calculated flow velocity field at different times with (**A, B, C**) and without (**D, E, F**) feed spacer, presented in sections at z = 390 μm (middle of the feed channel). The times are: (A,D) 6, (B,E) 42 and (C,F) 80 hours

In Figure 9.6 a comparison is made between the computed velocity fields averaged over the channel height (Figure 9.15) with the average velocity field results from NMR studies with the S-MFS. The calculated average velocity fields visualize the change in flow channels caused by biomass accumulation. With feed spacer, more regions of higher flow velocities were observed (Figure 9.15C) compared to the simulations without feed spacer (Figure 9.15F) with the same accumulated biomass amount (Figure 9.11B). Comparison of the average velocity field of the model and determined with MRI showed the same distortion of flow field and channeling (Figure 9.6). Also the spatial distributions of pressure and pressure drop are different with or without spacer. The pressure drop over the model domain with spacer declines stepwise over the spacer strands, even when biomass is not present (Figure 9.16A). The differences in pressure between the spacer diamonds are more accentuated by the biofilm formation (Figure 9.16BC). Within the spacer diamonds the pressure does not change significantly, showing that the main pressure drop is not due to the resistance to flow induced by the membrane walls but due to the spacer. In contrast, without spacer the pressure drop decline is gradual over the domain length (Figure 9.16D). In time, biomass accumulation increased the pressure drop and affected the pressure drop distribution over the domain (Figure 9.16F).

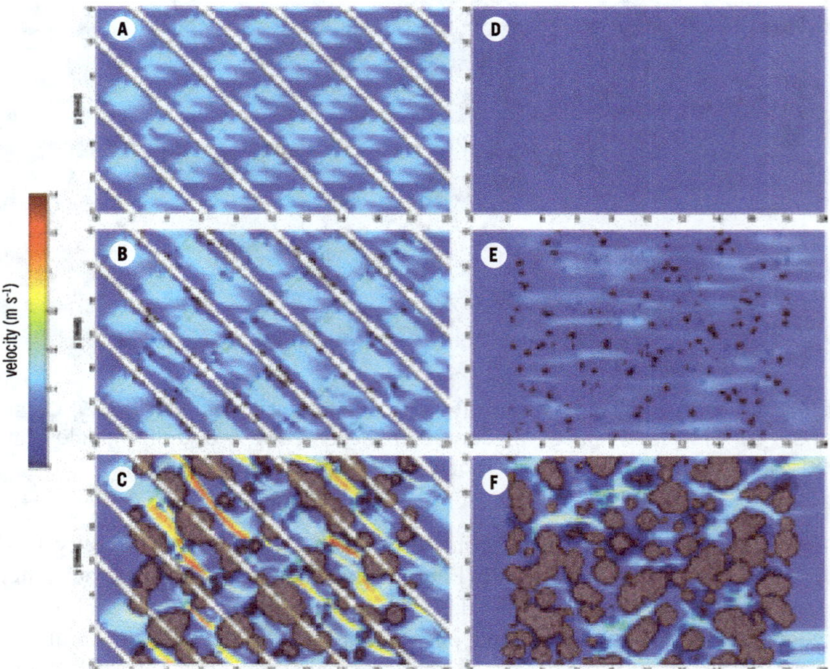

Figure 9.14 Calculated flow velocity field at different times with (**A, B, C**) and without (**D, E, F**) feed spacer, presented in sections at z = 585μm (75% of the feed channel height). The times are: (A,D) 6, (B,E) 42 and (C,F) 80 hours

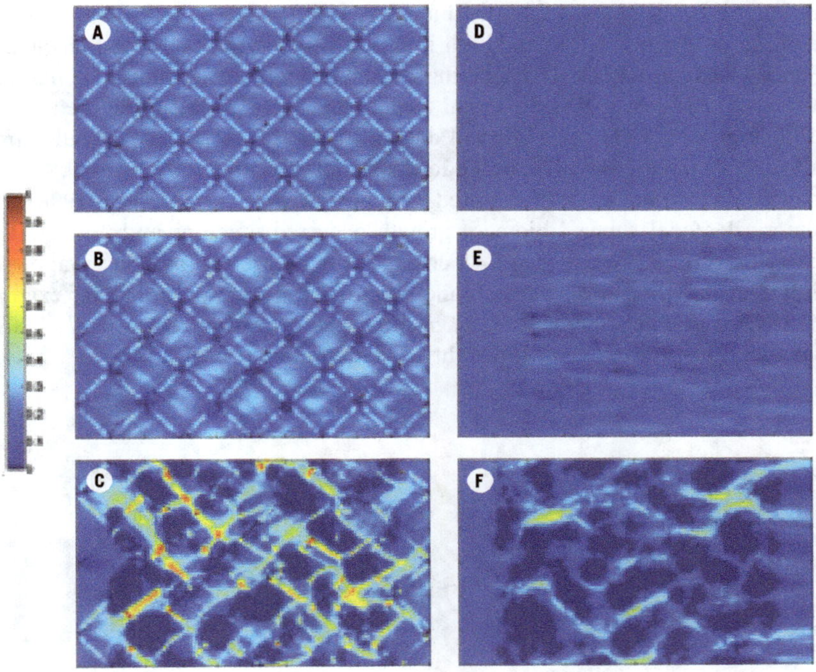

Figure 9.15 Calculated flow velocity field averaged over the channel height, at different times, with feed spacer (**A, B, C**) and without feed spacer (**D, E, F**). The times are: (A,D) 6, (B,E) 42 and (C,F) 80 hours

In conclusion, experiments showed that (i) the FCP was strongly affected by the feed spacer (Figure 9.10), (ii) flow channel formation occurred in the presence of a feed spacer (Figure 9.6), (iii) biomass accumulated predominantly on the spacer (Figure 9.9), and (iv) the absolute FCP was clearly higher with feed spacer compared to the monitor without spacer (Figure 9.10), all consistent with results of the 3-d model studies. 3-d modeling and experimental studies with and without feed spacer unambiguously illustrate that feed spacers play a crucial role in biofouling.

DISCUSSION

Comparison model with practice

Compared to experimental studies, the 3-d mathematical model developed by Picioreanu *et al.* (2009) showed the same development for (i) feed channel pressure drop (Figure 9.5), (ii) biomass, (iii) velocity distribution profile, resulting in regions of low and high liquid flow velocity also named channeling (Figures 9.6D and 9.14C). Channeling caused by biomass accumulation has been measured in-situ with NMR in spiral wound membrane modules (Figure 9.7) and MFS (Figure 9.6GH) and observed during studies on membrane modules from full-scale installations with elevated feed channel pressure drop (Figure 9.8). Flow channel formation has been reported previously (Baker and Dudley, 1998). The numerical model showed biomass growth occurred predominantly on the feed spacer (Figures 9.9AB, 9.13, 9.14), consistent with direct in-situ observations on biofouling of spiral wound membrane modules and monitors using for example MRI (Graf von der Schulenburg *et al.*, 2008 and Vrouwenvelder *et al.*, 2009c) and observations using the MFS sight window (Figure 9.9CD).

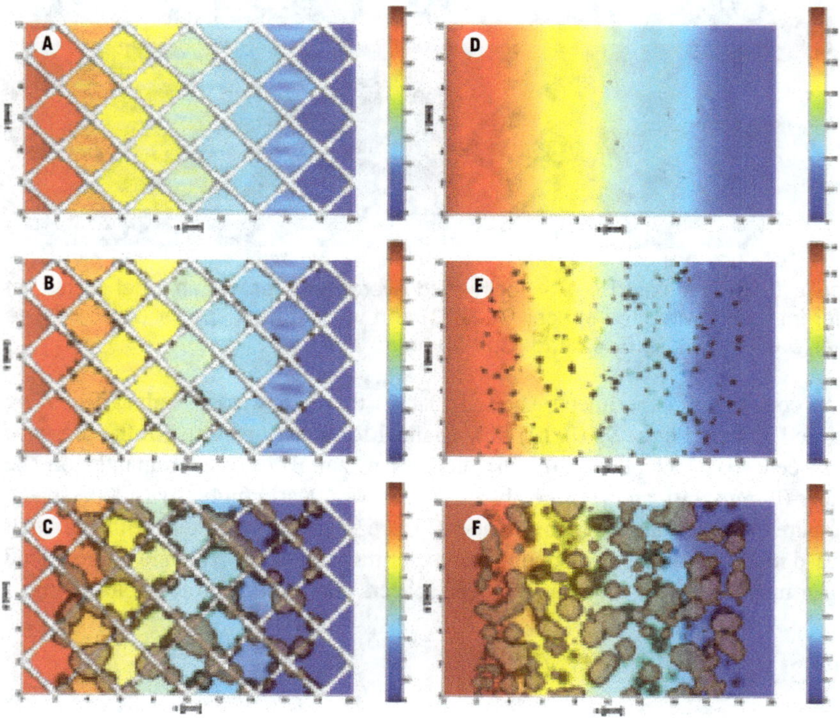

Figure 9.16 Calculated pressure distributions at different times with (**A, B, C**) and without (**D, E, F**) feed spacer, presented in sections at z = 390 μm (middle of the feed channel). The times are: (A,D) 6, (B,E) 42 and (C,F) 80 hours

The numerical model proves suitable to describe the overall biomass accumulation and pressure drop in time as observed during experiments. Moreover, the computed 3-d spatial distributions of biomass and flow velocity field agree qualitatively with the measurements. This 3-d model is based on generally accepted equations for the fluid flow and biofilm growth. The fact that some model results deviate from the experimental data is partially due to neglecting of potentially important processes, such as biomass detachment and substrate mass transport limitation. On the other hand, the quality of model predictions depends also on the choice of model parameters. No parameter optimization has been done in this study to fit the experimental data. All these facts underline that numerical tools have developed up to a stage where they can be used to get reliable information on such complex processes as the interaction between biofilm formation and hydraulics in RO and NF systems. This study and the study of Picioreanu *et al.* (2009) clearly show that in order to make reliable predictions the exact geometry of the spacer and flow channel and 3-d simulations need to be taken into account fully.

Spacer relevance

Spiral wound membrane modules need spacers to separate membrane envelopes and to promote the mass transfer along the membrane to minimize concentration polarization. The absolute FCP – sum of FCP and FCP increase – over the feed channel will determine the operational costs. Both our model and experimental studies with and without feed spacer showed that the absolute FCP is much higher in the presence of the feed spacer (Figure 9.10). Moreover, the increase in FCP due to biofouling is stronger and faster with spacer. Given this important increase in pressure drop, the feed spacer is therefore highly important for biofouling (Vrouwenvelder *et al.*, 2006b, 2008b, 2009c; Ho *et al.*, 2008 and Bartels *et al.*, 2008). The spacers not only influence the pressure drop but also the flow profile (Figure 9.15). This has a direct effect on the flow at the membrane surface and thereby will affect the extent of concentration polarization. Promoting mass transfer by using a spacer seems therefore to come at the expense of higher operational costs.

Most biofouling studies in laboratory systems used flow cells without spacers. Very often these studies come to the conclusion that biomass accumulated on the membrane, thus affecting the flux of permeate. The results we present here make it very likely that the findings of studies without spacers would have been different when feed spacers had been present.

In practice, the increased feed channel pressure drop over the membrane module is a dominant reason for operational problems caused by biofouling

(Vrouwenvelder *et al.*, 2008b, 2009c). An increased feed channel pressure drop or channeling over the membrane unit at constant inlet pressure will also lead to less permeate production. Studies without feed spacers are clearly not representative for biofouling studies in RO and NF systems. Not only is the flow profile (and thus mass transport to the membrane surface) different but also the amount of biomass growth on the membrane is strongly increased. In the absence of a spacer the effect of fluid flow channeling is much smaller. This again leads to an overemphasis of the role of concentration polarization and neglect of flow channeling.

Representative conditions for practice like feed spacers and hydrodynamics are considered essential for laboratory and modeling biofouling studies.

Future studies and practical implications

Recent biofouling papers (Herzberg and Elimelech, 2007 and Chong *et al.*, 2008) refer to a biofilm-enhanced osmotic pressure strongly contributing to a flux decline and salt passage increase. A flux decline was also found in a modeling study by Kim *et al.* (2006) on the physical presence of exopolymeric substances (EPS), which are produced and excreted by microorganisms. For a flux decline caused by biofouling, two mechanisms were identified: (i) an increase of the hydraulic resistance over the membrane and (ii) hindering the back diffusion of salts (Herzberg and Elimelech, 2007 and Chong *et al.*, 2008). Herzberg and Elimelech (2008) reported a higher distribution of active cells in a *Pseudomonas aeruginosa* biofilm close a RO membrane surface; likely due to higher substrate levels induced by concentration polarization effect caused by permeate production. In practice, increased feed channel pressure drop over the membrane module is a frequent dominant operational problem in RO and NF systems caused by biofouling (Vrouwenvelder *et al.*, 2008b, 2009b,c,d). Biofilm accumulation in NF and RO systems seem to effect both the biofilm-enhanced concentration polarization and the feed channel pressure drop, which may have synergistic negative effect on membrane performance. Evidently, there is need to unravel all the effects of (bio)fouling on performance of RO and NF systems (increase in feed spacer pressure drop, decrease in flux, and increase in salt passage) under representative conditions.

More insight in the biofouling process is needed to make progress in biofouling control. Direct in-situ spatially resolved information on biofilm accumulation, hydraulics and membrane performance is essential. The current 3-d model needs to be extended to contain all membrane performance indicators with feed channel pressure drop, permeate production (trans-membrane pressure drop) and salt passage. Concentration polarization has to be included as well since this may contribute to biofouling, salt passage and scaling. Additionally, biomass detachment has to be incorporated in the model. The influence of parameters like spacer geometries, hydraulics and feed water quality can be modeled to determine the effect on biomass accumulation, membrane performance indicators and cleanability. The nowadays commonly applied feed spacers restrict removal of biomass from the membrane module during cleaning. Direct in-situ observations with NMR and MFSs can be used to validate model results. The subsequent step may be pilot studies. Combination of 3-d mathematical modeling and direct in-situ observations using, for example, NMR and MFSs is the key to make progress in biofouling control. Moreover, the perspectives of combined modeling-measuring approach are integral fouling control and optimized membrane performance at minimal costs.

The development of membrane filtration processes less susceptible to (bio)fouling and optimized for performance may have consequences for module and total installation design. The viability of such (bio)fouling tolerant membrane system should be shown with pilot plant studies and cost based evaluations. Use of mathematical modeling techniques are expected to help with reducing the time and costs in developing new membrane and module configurations.

SUMMARY

A three-dimensional (3-d) computational model describing fluid dynamics and biofouling of feed channels of spiral wound reverse osmosis and nanofiltration membrane systems was developed based on results from practice and experimental studies. In the model simulations the same feed spacer geometry as applied in practice and the experimental studies was used. The 3-d mathematical model showed the same trends for (i) feed channel pressure drop, (ii) biomass accumulation, (iii) velocity distribution profile, resulting in regions of low and high liquid flow velocity also named channeling. The numerical model predicted a dominant biomass growth on the feed spacer, consistent with direct in-situ observations on biofouling of spiral wound membrane modules and monitors using Magnetic Resonance Imaging (MRI). The model confirms experimental results that feed spacer fouling is more important than membrane fouling. The paper shows that mathematical modeling techniques have evolved to a stage that they can be used hand-in-hand with experiments to understand the processes involved in membrane fouling.

Insights on Control Studies

10

Feed Water and Linear Flow Velocity: Significance of Substrate Load and Substrate Concentration

INTRODUCTION

The global drinking water demand is increasing and simultaneously drinking water regulations become stricter. This increases the demand for cost effective drinking water production technology. High pressure membrane filtration processes, reverse osmosis (RO) and nanofiltration (NF), produce high quality drinking water, virtually free of pathogenic microorganisms and inorganic and organic pollutants. Because of the decreasing costs of membrane applications, an increasingly important role is predicted for membrane filtration (Mallevialle *et al.*, 1996 and Shannon *et al.*, 2008).

A drawback of RO and NF applications is membrane fouling, resulting in a pressure drop increase over the membrane element and, therefore, higher plant operation costs. The major fouling mechanisms of RO and NF membranes are scaling, particulate and organic fouling and biofouling. Scaling by inorganic compounds is usually controlled using a scale inhibitor or an acid. Particulate fouling is controlled by extensive pretreatment, including ultrafiltration. Thus, all types of fouling except biofouling and organic fouling are controllable. Biofouling remains the major type of fouling in RO and NF membranes fed with extensively pretreated water, caused by biofilm formation in the membrane elements (Ridgway and Flemming, 1996; Patching and Fleming, 2003 and Shannon *et al.*, 2008).

Most biofouling studies are carried out on a trial and error approach in pilot and full-scale installations. Fouling is diagnosed on the basis of parameters such as pressure drop and from investigation of removed membranes (Tasaka *et al.*,

---

*This chapter is based, with permission from the copyright holder, on a paper previously published in Biofouling Vol. 25 No. 6 pp. 543–555 doi: 10.1080/08927010902972225

1994; Baker and Dudley, 1998; Vrouwenvelder *et al.*, 1998; Van Hoof *et al.*, 2002; Huisman and Feng Kong, 2004; Schneider *et al.*, 2005 and Karime *et al.*, 2008). Spiral wound RO and NF membrane elements have the same construction and size, complicating systematic studies on membrane fouling. Recently, developed tools such as the membrane fouling simulator (MFS) (Vrouwenvelder *et al.*, 2006b, 2007a) have facilitated methodological studies on biofouling. In a comparative study, the same pressure drop development and biofouling accumulation were observed in both spiral wound membrane elements and MFSs (Vrouwenvelder *et al.*, 2006b).

Insight into the factors influencing the development of biomass and pressure drop increase is needed to develop membrane systems, which are less susceptible to biofouling. Restricting the content of biodegradable compounds in the feed water by sand filtration pretreatment has been reported to lower biomass accumulation in a membrane test cell (Griebe and Flemming, 1998). Parallel test rig studies with spiral wound membrane elements operated with feed water with and without dosages of a biodegradable compound showed that substrate dosage stimulated the development of biofouling (Vrouwenvelder *et al.*, 2008b), supporting the findings of Griebe and Flemming (1998). Hydrodynamics and shear are important in biofilm development (Van Loosdrecht *et al.*, 1995; Kwok *et al.*, 1998 and Purevdorj-Gage and Stoodley, 2004). Biofilm formation is affected by substrate concentration, substrate load and hydrodynamic shear force. The thickness, structure, stability and density of the biofilm are influenced by substrate load and hydrodynamic shear stress/flow regime (Characklis and Marshall, 1990; Van Loosdrecht *et al.*, 1995; Peyton, 1996; Melo and Bott, 1997; Kwok *et al.*, 1998; Liu and Tay, 2002; Pereira *et al.*, 2002; Wäsche *et al.*, 2002 and Wijeyekoon *et al.*, 2004). Thicker biofilms were observed under high substrate load conditions (Peyton, 1996 and Wäsche *et al.*, 2002). Thinner biofilms were found during studies on *Pseudomonas fluorescens* biofilms grown under turbulent flow conditions compared with laminar flow conditions (Pereira *et al.*, 2002). Within membrane installations, the linear flow velocity over the pressure vessels reduces because of permeate production in the membrane elements. Between installations, the feed flow and the plant configuration may vary resulting in different flow profiles. Thus in practice, membranes are operated with different linear flow velocities resulting in differences in substrate load and hydrodynamic shear stress.

Increasing the linear flow velocity may reduce the accumulation of microorganisms on surfaces (Wills *et al.*, 2000). The effectiveness is attributed to the increased shear stress acting on the biofilm. Wills *et al.* (2000) reported that use of inserts reduced biofilm accumulation. It has been proposed that increasing the linear flow velocity may be a suitable method to reduce biofouling in tubular heat exchangers (Melo and Bott, 1997), as the shear stress is higher at elevated linear flow velocities. One consequence of increasing the linear flow velocity is an increase in the substrate loading rate.

The flux – water volume flowing through the membrane per unit area and time – as applied in practice in spiral wound NF and RO membranes has been shown to be unimportant for biofouling development after extensive pretreatment (Vrouwenvelder *et al.*, 2009b,c). In other words, the flux or permeate production was not contributing significantly to the amount of biofilm accumulation in the membrane elements. Furthermore, biofilm development did not influence the flux, as the biofilm predominantly occurred on the spacer structure. Irrespective of whether permeate was produced or not, the feed spacer channel pressure drop and biofilm concentration increased in RO and NF membranes in a monitor, test rigs, a pilot scale and a full-scale installation. Identical behavior with respect to biofouling and feed channel pressure drop development was observed in membrane elements in the same position in a NF installation operated with and without permeate production. Calculations to predict the influence of flux supported the experimental observations that the flux was playing an insignificant role in biofouling (Vrouwenvelder *et al.*, 2009b). Therefore, the experiments in this study were performed without flux. The feed-concentrate channel pressure drop increase problem is the consequence of biomass accumulation in the feed spacer channel.

Despite extensive biofilm literature, systematic studies on the influence of factors such as substrate load and linear flow velocity on the development of biofouling in spiral wound membrane systems are lacking. The main objective of this study was to determine the influence of substrate load, substrate concentration in the feed water and linear flow velocity on the development of pressure drop and biofilm in spiral wound NF and RO systems.

MATERIALS AND METHODS

Membrane fouling simulator

In all experiments, a MFS with external dimensions of $0.07\,m \times 0.30\,m \times 0.04\,m$ was used (Vrouwenvelder *et al.*, 2006b). Coupons of feed spacer, membrane and product spacer can be placed in the MFS resulting in the same spatial dimensions as in spiral wound membrane elements. The development of fouling was monitored by measuring the pressure drop increase over the feed spacer channel of the MFS, and visual observations using the sight glass and analysis of sheets of membrane and spacer taken from the monitor. During operation, the glass window was covered with a light tight lid to prevent growth of phototrophic organisms. Analyses performed on the sheets were total active biomass by adenosine-triphosphate (ATP) and dissolved organic carbon (DOC). Details of the methods have been described elsewhere (Vrouwenvelder *et al.*, 2006b, 2007a).

Membranes and spacer sheets used in the MFS research were taken from unused spiral wound membrane elements (Trisep TS 80, 8-inch diameter).

Experimental set-up

The MFS installation consisted of two pressure reducing valves, manometer, dosage point (for biodegradable compounds), MFS and flow controller (Vrouwenvelder *et al.*, 2007a). The MFS was fed with drinking water and operated at a pressure of $\sim 120\,\mathrm{kPa}$ to avoid degassing. Unless indicated otherwise, the feed water flow of the MFS was $16.0\,\mathrm{L\,h^{-1}}$ equal to a linear flow velocity of $0.163\,\mathrm{m\,s^{-1}}$. The MFS was operated cross-flow, without permeate production. Drinking water prepared from anaerobic groundwater (subsequently treated by aeration, rapid sand filtration, deacidification, softening and rapid sand filtration at treatment plant Spannenburg) was used as feed water of the MFS installation. The water had a DOC content of $8\,\mathrm{mg\,L^{-1}}$ and the temperature varied by a maximal 2°C during each experiment within a total range of 13 and 18°C. Acetate, a biodegradable compound, was added to the water. From a sterile vessel, containing a 5 L solution of concentrated sodium acetate, the substrate was dosed into the feed water prior to the MFS by a peristaltic pump (Masterflex) at a flow of $25\,\mathrm{mL\,h^{-1}}$. The dosage of substrate was checked periodically by measuring the weight of the dosing bottle. Pressure drop measurements were performed with a pressure difference transmitter (Endress and Hauser, type Deltabar S: PMD70-AAA7FKYAAA, Vrouwenvelder *et al.*, 2009d). The calibrated measuring range was 0–50 kPa. The differential pressure transmitter was connected onto the MFS with quick connectors.

Sampling and study of membranes

To quantify the accumulated fouling, sections of membrane, feed and product spacer were taken from the MFS. The sections ($4\,\mathrm{cm^2}$) were placed in a capped tube filled with 10 ml of autoclaved tap water. To determine the amount of biomass, the tubes with the membrane and spacer sections were placed in an ultrasonic cleaning bath (Bransonic). The low energy sonic treatment (2 min) followed by mixing on a Vortex (a few seconds) was repeated three times. Next, water volumes collected from the tubes were used to determine biomass parameters such as ATP. The choice for ATP was based on studies at full-scale installations to diagnose membrane biofouling (Vrouwenvelder *et al.*, 2008b). Additionally, total organic carbon (TOC) measurements were made.

Active biomass was estimated in duplicate by measuring the ATP concentration from $100\,\mu\mathrm{L}$ samples (Holm-Hansen and Booth, 1966). The luminometer (Celcis, model Advance) added $100\,\mu\mathrm{L}$ of LuminEX-B reagent (Celcis) to each sample to release ATP from the bacterial cells. Subsequently, $100\,\mu\mathrm{L}$ of LumATE PM (Celcis) was added for light production. The amount of light produced was measured with the luminometer (relative light units, RLU). The concentration of ATP was derived from the RLU values using the conversion factors of the linear relationship between RLU values and reference ATP concentrations. The ATP concentrations of autoclaved drinking water ($\leqslant 1\,\text{pg}\,\text{mL}^{-1}$) and two control solutions (2 and $100\,\text{pg}\,\text{mL}^{-1}$) were determined as quality control. TOC was determined as non-purgeable organic carbon by infra-red gas analysis. Calculation of the biomass load was derived from the biomass concentration and the surface area of the sections of membrane and feed spacer.

RESULTS

The development of feed spacer channel pressure drop and biofilm concentration in a MFS was investigated by (i) varying the substrate concentrations in the feed water at a constant linear flow velocity (Figures 10.1 and 10.2), (ii) varying the linear flow velocities at constant substrate concentration in the feed water (Figures 10.4 and 10.5) and (iii) varying the linear flow velocities and substrate concentrations at constant substrate load (Figures 10.6 and 10.7). The influence of flow direction change in a pressure vessel containing membrane elements on biofouling was simulated in the MFS by operating MFSs with varying linear flow velocities (Figures 10.8 and 10.9).

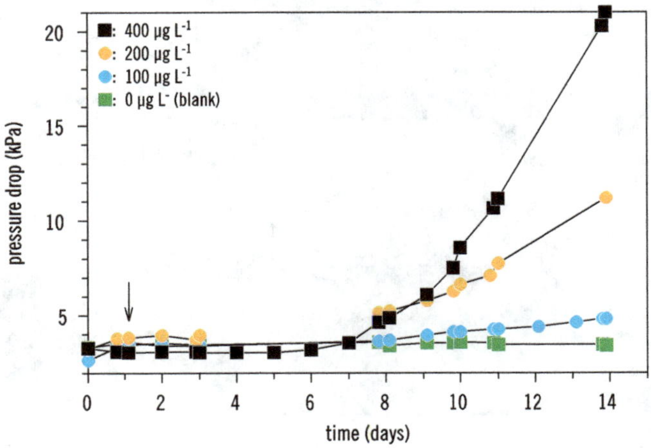

Figure 10.1 Pressure drop in time over the monitor with and without dosage of different concentrations of a biodegradable compound (acetate) to the feed water of the monitor. The arrow indicates the starting time of the dosage

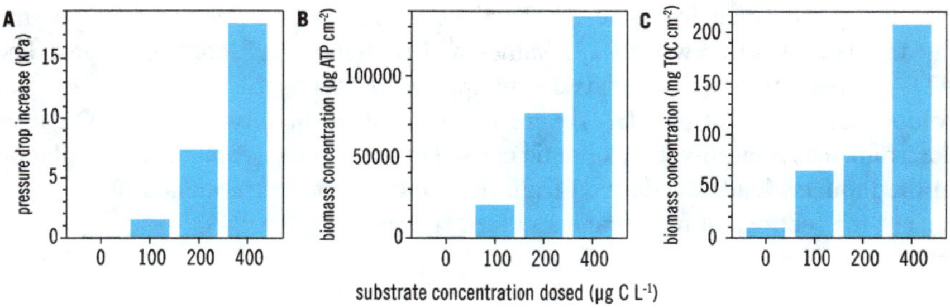

Figure 10.2 Pressure drop increase (**A**) and amount of accumulated biomass (**B** and **C**) on the membrane and spacer in the monitor after 14 days of operation with and without dosage of different acetate concentrations

Figure 10.3 Visual observations of the feed spacer and membrane in the monitor supplied 0, 100, 200 and 400 $\mu g\,C\,L^{-1}$ as acetate carbon (**A**, **B**, **C** and **D**) after 13 days of operation. The sight glass of the monitor was used for visual observations during operation of the monitor

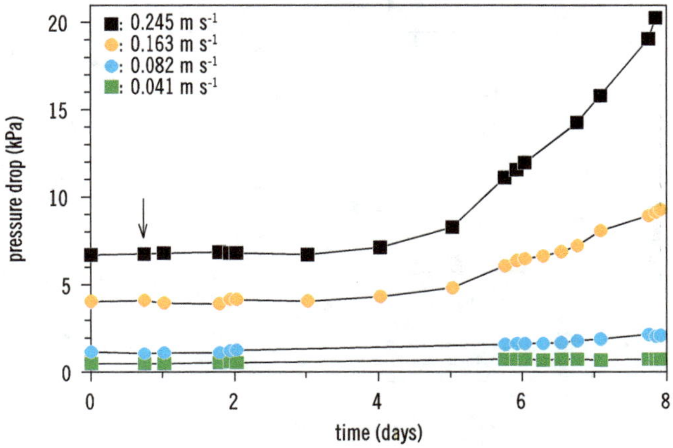

Figure 10.4 Pressure drop in time over the monitor operated with different linear flow velocities and the same concentration of a biodegradable compound ($200\,\mu g\,C\,L^{-1}$) in the feed water. The arrow indicates the starting time of the dosage

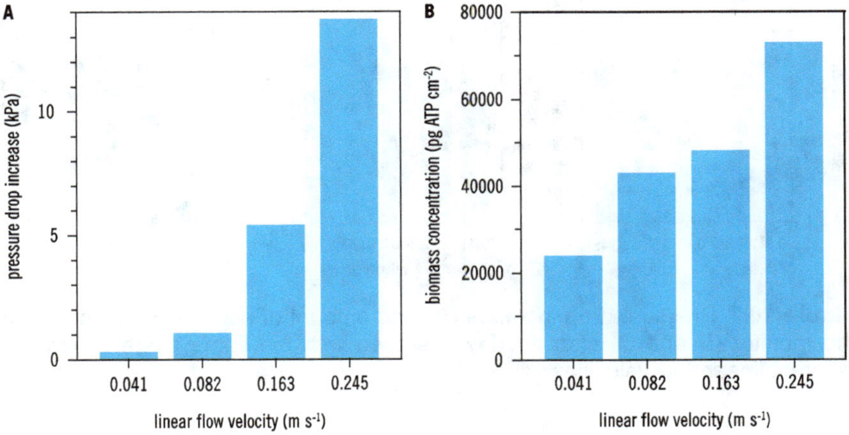

Figure 10.5 Pressure drop increase (**A**) and accumulated biomass in the monitor (**B**) after 8 day operation with different linear flow velocities and with the same concentration of a biodegradable compound ($200\,\mu g\,C\,L^{-1}$) in the feed water

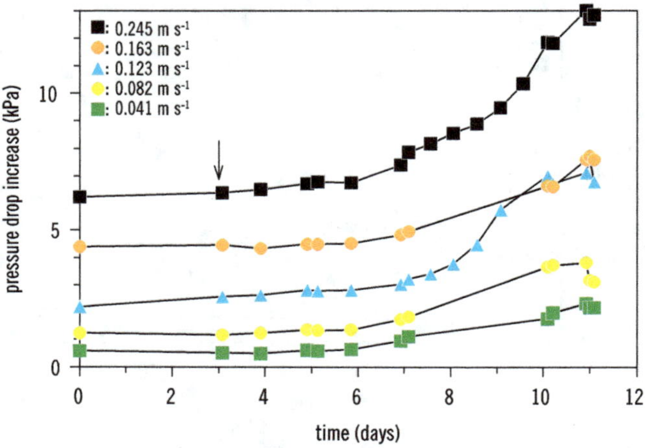

Figure 10.6 Pressure drop in time over the monitor with the same acetate load and different linear flow velocities. The arrow indicates the dosage start time

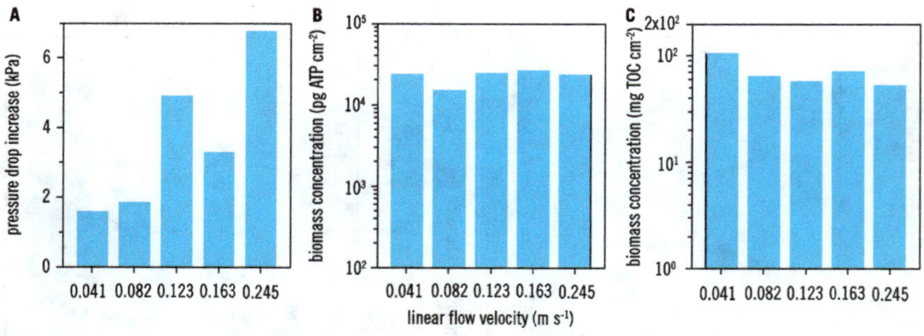

Figure 10.7 Pressure drop increase (**A**) and amount of accumulated biomass in the monitor (**B** and **C**) after 11 day operation with the same acetate load and different linear flow velocities

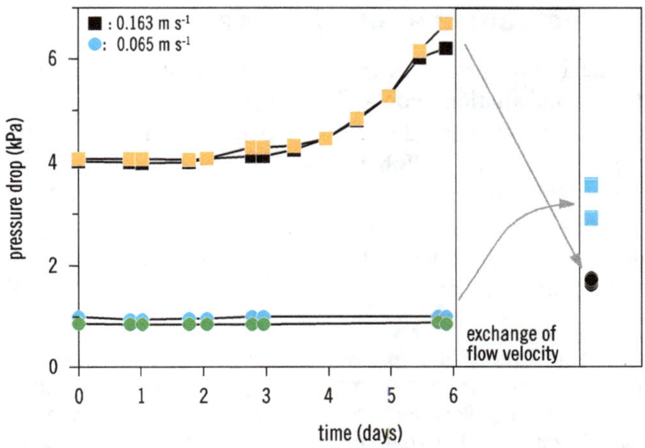

Figure 10.8 Pressure drop in time over the monitor at a high and low linear flow velocity (0.163 and $0.065\,\mathrm{m\,s^{-1}}$) before and after changing of the linear flow velocity on day 6. The arrows indicate the effect of flow velocity change on the pressure drop

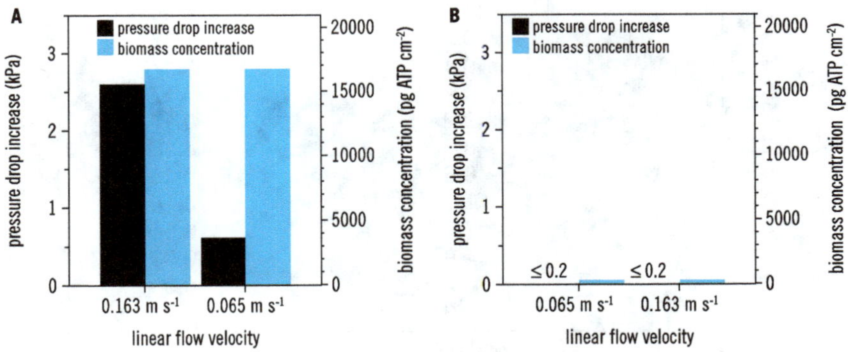

Figure 10.9 Pressure drop increase and accumulated biomass in the monitor after 6 day operation before and after reduction of the linear flow velocity (**A**) and before and after increasing the linear flow velocity (**B**). The pressure drop increase after adjustment of the linear flow velocity was calculated in comparison with pressure drop data determined at the same linear flow velocity prior to the experimental start

Linear flow velocities applied in practice

To quantify linear flow velocities in lead elements, an inventory was made at several full-scale installations containing 8-inch diameter spiral wound NF or RO membrane elements. The linear flow velocities in the lead elements ranged between 0.07 and 0.20 $m\,s^{-1}$ (Table 10.1). The same range of velocities was found for installations containing 4-inch diameter lead elements. Biofouling was predominantly observed in lead elements with a high linear velocity. Therefore, a relatively high linear flow velocity (0.163 $m\,s^{-1}$) was selected for monitor studies described in the following sections.

Table 10.1 Inventory of feed flows used in practice per lead membrane element (8 inch diameter) at the membrane installation feed side and linear flow velocity at inlet side of the lead membrane element. 13 different full-scale installations treating different water types were inventoried

NF/RO installation	Feed flow per lead element ($m^3\,h^{-1}$)	Linear flow velocity inlet side lead element ($m\,s^{-1}$)
1	11.2	0.20
2	11.1	0.20
3	10.0	0.18
4	9.4	0.17
5	9.4	0.17
6	9.2	0.17
7	8.8	0.16
8	8.6	0.16
9	7.9	0.14
10	7.7	0.14
11*	4.9	0.09
12*	4.2	0.08
13*	3.6	0.07

**Full-scale installations using the hydraulic optimized pressure vessel.*

Effect of substrate concentration at constant linear velocity

Four monitors were operated in a parallel study at constant linear water flow velocity (0.163 $m\,s^{-1}$). Water with different concentrations of a biodegradable compound (acetate carbon: 100, 200 and 400 $\mu g\,C\,L^{-1}$) or a blank solution without acetate were fed into the individual monitors. The pressure drop increased in time over the monitors fed with water containing acetate, whereas the pressure drop over the monitor fed with a blank solution remained constant (Figure 10.1). An increase in acetate concentration in the feed water resulted in a faster and greater pressure drop increase over the monitor (Figures 10.1 and 10.2A). At the end of the research period, the monitors were opened for analysis of the accumulated material. On the membrane and feed spacer taken from the monitor, the biomass load (ATP and TOC) increased with the acetate concentration in the feed water of the MFS (Figure 10.2BC). The TOC and ATP data showed the same trend. With increasing substrate concentration both the pressure drop and biomass load increased (Figure 10.2). Visual observations indicated that more material was present in the monitors supplied with higher substrate concentrations (Figure 10.3), supporting the pressure drop results and active biomass measurements.

A higher substrate concentration resulted in a faster and greater pressure drop increase and a higher amount of accumulated amount of biomass.

Effect of linear flow velocity at constant substrate concentration

Four monitors were operated with different linear flow velocities (0.041, 0.082, 0.163 and 0.245 $m\,s^{-1}$) at the same acetate concentration (acetate carbon: 200 $\mu g\,C\,L^{-1}$) in the feed water.

At the start of the experiment, a higher pressure drop was observed at higher linear flow velocities (Figure 10.4). In time, the pressure drop over the monitors increased (Figure 10.4). The extent of the pressure drop increase was related to the linear flow velocity (Figures 10.4 and 10.5A). The accumulated biomass (ATP) in the monitors (determined at the end of the research period) related with the pressure drop increase, although the relationship was not linear (Figure 10.5).

A higher linear flow velocity resulted in an initial higher pressure drop and a more rapid and heavier development of pressure drop increase and biomass accumulation.

Effect of linear velocity and substrate concentration at constant substrate load

The studies on the effect of substrate concentration (Figure 10.2) and linear flow velocity (Figure 10.5) were performed with different substrate loads ($\mu g\,C\,m^{-1}\,s^{-1}$). Differences in loading rate of biodegradable compounds may affect biofouling. To evaluate the influence of substrate load on biofouling development, comparative studies with monitors were performed with the same substrate load and different linear flow velocities and substrate concentrations (Table 10.2).

Table 10.2 Experimental conditions for the monitors with the same substrate load, as acetate carbon, and different linear flow velocities and substrate concentrations

Monitor no.	Linear velocity ($m\,s^{-1}$)	Feed ($L\,h^{-1}$)	Acetate concentration ($\mu g\,C\,L^{-1}$)	Acetate load ($\mu g\,C\,m^{-2}s^{-1}$)
1	0.041	4	400	56
2	0.082	8	200	56
3	0.123	12	133	56
4	0.163	16	100	56
5	0.245	24	66	56

The initial pressure drop was related to the linear velocity (Figure 10.6). The pressure drop increased during the run of 11 days and was lowest for the lowest linear velocity and highest for the highest linear flow velocity (Figures 10.6 and 10.7A). The accumulated amount of biomass determined in the monitors at the end of the research period was not significantly different (Figure 10.7BC). The same amounts of accumulated biomass in the monitors operated with different linear flow velocities and acetate concentrations but the same acetate load indicate that biomass accumulation was substrate load-related. Replicate experiments showed similar results (data not shown). Linear flow velocity had a greater impact on pressure drop increase than substrate concentration. In the monitors supplied with the same substrate load, the amount of accumulated biomass was constant but the pressure drop increase was a function of the linear flow velocity (Figure 10.7).

The biomass accumulation related to the substrate load. The effect of accumulated biomass on the pressure drop increase was affected by the linear flow velocity.

Effect of flow velocity

In the monitors supplied with the same substrate load, the amount of accumulated biomass was constant but the pressure drop increase was a function of the linear flow velocity (Figure 10.7). Therefore, lowering the linear flow velocity may be an approach to reduce the pressure drop increase.

In practice, the linear flow velocity of the water decreased during passage of the membrane elements in a pressure vessel. The reduction of linear flow velocity may be 64% over a pressure vessel containing eight membrane elements, each membrane element operating at 12% recovery. Usually, most biofouling is observed in the first membrane element at the feed side of the pressure vessels (Carnahan *et al.*, 1995 and Vrouwenvelder *et al.*, 1998, 2008b, 2009d). Therefore, changing the flow direction in a pressure vessel may be another option to reduce the pressure drop increase.

To investigate the influence of a flow direction change in a pressure vessel on biofouling, a simulation study was performed with four monitors. Two monitors were operated at a high feed flow velocity ($0.163\,m\,s^{-1}$) supplied with water containing acetate ($100\,\mu g\,C\,L^{-1}$) and two monitors were operated at a lower feed flow velocity ($0.065\,m\,s^{-1}$) supplied with water without acetate dosage. After a pressure drop increase in the monitors operated at high velocity ($0.163\,m\,s^{-1}$) and high substrate ($100\,\mu g\,C\,L^{-1}$) concentration (after 6 days), the flow rates in the monitors were exchanged to simulate a flow direction change in a pressure vessel.

The monitors with a high linear flow velocity ($0.163\,m\,s^{-1}$) showed a large pressure drop increase after 6 days (Figure 10.8). Reducing the linear flow velocity from 0.163 to $0.065\,m\,s^{-1}$ after 6 days resulted in a pressure drop only slightly higher as the original pressure drop for the monitors at a flow velocity of $0.065\,m\,s^{-1}$ directly prior to the experiment (Figure 10.9A). The decrease of flow velocity resulted in a 73% reduction of the pressure drop increase compared with the pressure drop at the same flow velocity prior to the start of the experiment.

The monitors operated with a low linear flow velocity ($0.065\,m\,s^{-1}$) for 6 days showed no detectable increase of pressure drop in time (Figure 10.8). Increasing the linear flow velocity to $0.163\,m\,s^{-1}$ resulted in the same pressure drop as determined at $0.163\,m\,s^{-1}$ directly prior to the experiment, indicating that the pressure drop of the monitors at a low linear flow velocity was not changed during operation (Figure 10.9B).

Reduction of the linear flow velocity resulted in a clear and instant reduction of pressure drop increase caused by biofilm accumulation, without affecting the amount of accumulated biomass.

Effect of substrate load reduction

The pressure drop increase and biomass concentration may decline in time after lowering the substrate load. To evaluate the influence of substrate load reduction on the development of pressure drop and biomass concentration, a simulation study was performed with four monitors. All monitors were operated at the same linear velocity ($0.163\,m\,s^{-1}$) supplied with water containing acetate ($100\,\mu g\,C\,L^{-1}$), simulating the lead element. At elevated pressure drop, the linear flow velocity was reduced to ($0.065\,m\,s^{-1}$) and the acetate dosage was stopped, simulating conversion of the lead element to the last element.

In time, the pressure drop over the monitors declined (Figure 10.10). The accumulated amounts of biomass determined after 0, 5, 10 and 12 days operation at $0.063\,m\,s^{-1}$ related with the pressure drop decline (Figures 10.10 and 10.11).

A reduction of substrate load resulted in a gradual decline of pressure drop increase and biomass amount.

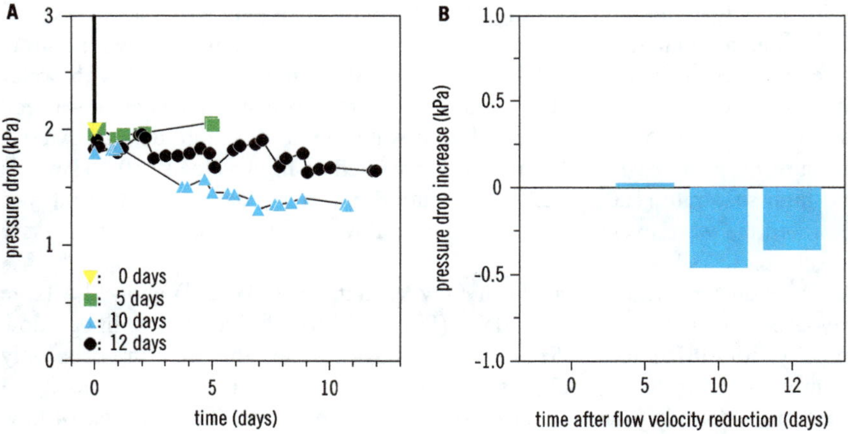

Figure 10.10 Pressure drop after reduction of the substrate load and linear velocity from 0.163 to 0.065 m s^{-1} **(A)** and pressure drop decline after 0, 5, 10 and 12 days of operation at 0.065 m s^{-1} **(B)**

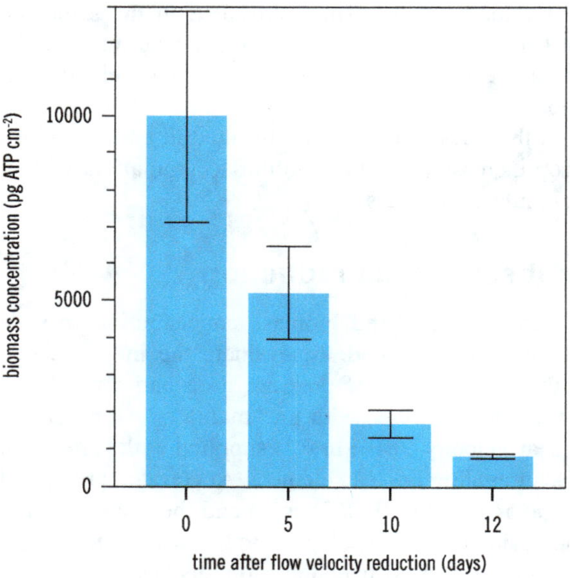

Figure 10.11 Biomass load in the monitors after 0, 5, 10 and 12 days operation at a reduced substrate load and a linear velocity of 0.065 m s^{-1}. The error bars indicate the standard deviation of the biomass load measured over the length of the MFS

DISCUSSION

Plant performance

Biofilm accumulation in NF and RO systems can reduce plant performance by increasing the feed channel pressure drop, reducing the flux and/or increasing salt passage. Previous biofouling studies over a wide range of scales (from full-scale installations to small flow cells) have shown that in spiral wound NF and RO systems with extensive pretreatment, like ultrafiltration, the feed channel pressure drop increase is more important than biofouling of the membrane (Vrouwenvelder *et al.*, 2009c). Although NF and RO studies on the influence of trans membrane flux under operating conditions showed that the flux did not influence biofouling (Vrouwenvelder *et al.*, 2009b,c), there may be cases where this flux does influence biofouling. To unravel the causes of reduced membrane performance, integral research is needed into the effect of both spacers and membranes under operating conditions.

Biomass parameters

Determining the influence of biomass accumulation in membrane elements on performance requires *in situ* spatially resolved information on biomass thickness (volume), structure (morphology, porosity and roughness), density, hydrodynamics and performance data. For biomass parameters, this information is difficult to acquire and interpret. The next best thing is to use biomass parameters that relate to operational problems. A broad range of biomass parameters and analytical tools are available (Ridgway and Flemming, 1996 and Speth *et al.*, 2000). To quantify biofouling, the biomass parameters ATP, TOC, total cell counts, colony counts, proteins, carbohydrates and dry weight were investigated in several membrane studies and compared with feed channel pressure drop increase in test rigs and full-scale investigations with different types of feed water (Vrouwenvelder *et al.*, 2008b). ATP was the biomass parameter that correlated best with the feed channel pressure drop increase. Other biofilm parameters such as TOC and total direct cell counts showed the same trend as ATP but were less sensitive and accurate. Therefore, in this present study biomass parameter measurements were restricted to ATP and TOC.

In situ observations on spiral wound membrane elements and (S-)MFSs during studies with magnetic resonance imaging (MRI) and the MFS sight window showed that most biomass was present on the feed spacer (Figure 10.12, Graf von der Schulenburg *et al.*, 2008 and Vrouwenvelder *et al.*, 2009c, 2010b). Discriminating between membrane and spacer fouling by analysis of separated spacer and membrane sheets is a challenge. Separating the stacked spacer and membrane during sampling disturbs the spatial biomass distribution. Therefore, separate biomass analysis of spacer and membrane showed a broad range of distributions over the two materials (Vrouwenvelder *et al.*, 2008b). *In situ* biomass measurements are strongly preferred over analysis of separated spacers and membranes.

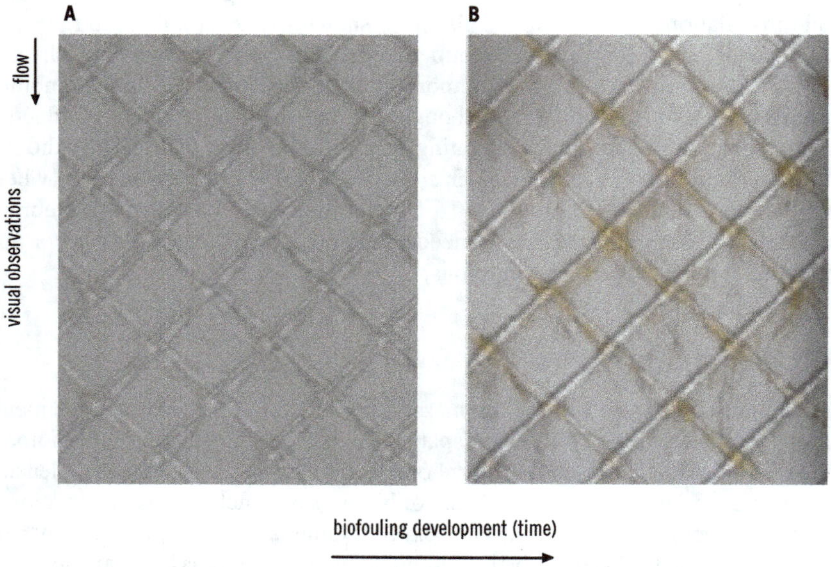

Figure 10.12 *In-situ* visual observations of the feed spacer and membrane in the MFS – using the sight window during operation – fed with water without (**A**) and with chemical containing substrate (**B**). Biofilm streamers attached to the feed spacer were observed

Recent studies indicate that spatially resolved data on biomass volume and structure can be measured and quantified using *in situ* MRI and MFS sight window imaging. It is expected that combining results from *in situ* MRI and MFS measurements and 3-dimensional modeling of membrane biofouling will enable progress to be made in determining the exact location of biofouling.

Linear flow velocities applied in practice

In full-scale NF/RO installations, the linear flow velocity varies greatly over membrane elements. In pressure vessels, membrane elements are placed in series of up to eight elements. A strong decline in flow velocity occurs over the membrane elements in pressure vessels with increasing distance from the feed side as a consequence of permeate production, reducing the water flow in the membrane element feed spacer channel. The lead element in the pressure vessel has the highest flow rate. In practice, in most cases biofouling was observed at the feed side of the installation, especially in the lead element (Carnahan *et al.*, 1995 and Vrouwenvelder *et al.*, 1998, 2006b, 2008b, 2009a).

An inventory made at several full-scale installations containing spiral wound NF or RO membrane elements showed that linear flow velocities in lead elements ranged between 0.14 and 0.20 $m\,s^{-1}$ (Table 10.1: installations 1–10).

Membrane element manufacturer specifications for a maximum feed flow are ~16 to 20 $m^3\,h^{-1}$ for 8-inch elements, a factor 2 higher than commonly applied in practice: ~8 to 11 $m^3\,h^{-1}$ (see Table 10.1).

Studies to determine a hydraulic optimized design of staging and the number of membrane elements per pressure vessel resulted in the development of a hydraulic optimized pressure vessel (Optiflux), which reduces pressure loss and increases membrane productivity (Van der Meer, 2003a,b). For the hydraulic optimized concept, the calculated concentration polarization was higher than prescribed by membrane manufacturers when using membrane suppliers software (Van der Meer, 2003a). Several full-scale installations using this optimized concept have been in operation since 1999 to 2005 without problems (Van Paassen *et al.*, 2005). Using the optimized pressure vessel, the number of lead elements is doubled and consequently the linear flow velocity is reduced (Van der Meer *et al.*, 2003a). Low linear flow velocities (0.07–0.09 $m\,s^{-1}$, Table 10.1: installations 11–13) are applied in lead elements of full-scale installations using the optimized concept.

Over pressure vessels, the linear flow velocity decreased with increasing distance from the feed side as a consequence of permeate production. Figure 10.13 illustrates the velocity profile in installations, without taking osmotic pressure and pressure drop into account. For example, in a pressure vessel with eight staged elements each performing with 12% recovery, the outlet side of membrane element at the longest distance from the feed side (position 8) has a linear fluid flow velocity that equals 36% of the lead element feed side velocity (position 1).

Biofilm accumulation

Biofilm accumulation is determined by the substrate loading rate. It is a matter of time before biofilms accumulate on surfaces under non-sterile conditions. Biofilm accumulation can be retarded by reducing the concentration of easily

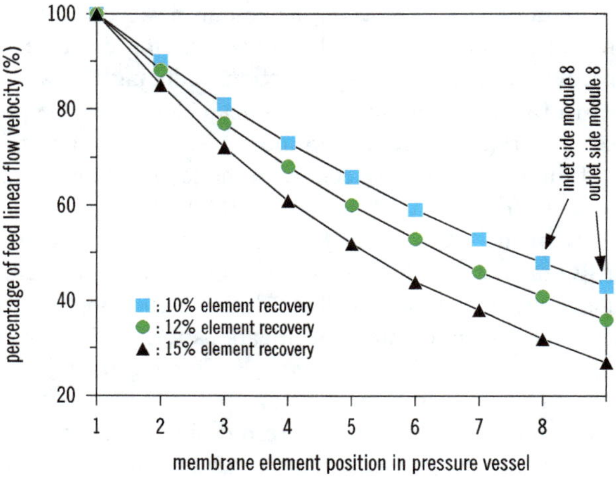

Figure 10.13 Linear flow velocity in membrane elements in pressure vessel expressed as percentage of the lead element feed side linear flow velocity at different element recoveries (10, 12 and 15%). The element position number increases with increasing distance to the feed side

biodegradable compounds in the feed water (Figure 10.2B), by lowering the linear flow velocity (Figure 10.5B) and combinations thereof. Lowering the velocity may be a way to restrict biofilm accumulation. Under low velocity conditions and low substrate conditions, the maintenance energy of the accumulated biomass may be sufficient to obtain a steady state situation in which the biomass amount is both low and stable in time with low impact on the pressure drop increase. Long term studies are needed to evaluate such an approach to 'live with' biofouling. Spatially resolved information on biofilm volume and structure may provide more insight into the biofouling process.

Membrane filtration installations with extended and robust pretreatment involving biological activity experienced relatively low pressure drop values for running times up to 1,000 days. Biologically active pretreatment steps are able to remove biodegradable compounds to a large extent. Griebe and Flemming (1998) reported that the use of biofiltration as a pretreatment resulted in a reduction in the biomass concentration in membrane elements, and this is supported by the findings of this study. However, biological pretreatment before membrane processes is no guarantee for preventing biofouling. Even when biological pretreatment is used, chemicals dosed (such as scale inhibitors) and seasonal variations in the water quality ask for attention (Carnahan *et al.*,

1995; Vrouwenvelder *et al.*, 1998, 2000 and Van der Hoek *et al.*, 2000). Testing of 14 commercially available polymer based scale inhibitors and dose–effect studies with test rigs demonstrated that chemicals used to prevent scaling differ greatly in their ability to promote the growth of microorganisms and confirmed experiences in practice (Vrouwenvelder *et al.*, 2000).

The linear flow velocity influences biofouling of membranes (Figure 10.5). In the literature, it has been proposed that increasing the linear flow velocity may be a suitable method for reducing biofouling in tubular heat exchangers (Melo and Bott, 1997), as the shear force is higher at an elevated linear flow velocity. The present results suggest that increasing the linear flow velocity results in a higher pressure drop increase (Figure 10.7), illustrating that the consequences of biofouling may be worse, even at a higher hydrodynamic shear force. The pressure drop is proportional to the square of the linear flow velocity:

$$\Delta p = \lambda \cdot \frac{\rho \cdot v^2}{2} \cdot \frac{L}{d_h} \qquad (10.1)$$

where λ is the friction coefficient, ρ the specific liquid density, v the linear velocity, L the length of the membrane or MFS and d_h the hydraulic diameter.

It has been observed in several biofilm studies that at high hydrodynamic shear stresses the biofilm becomes denser and stronger (Van Loosdrecht *et al.*, 1995; Kwok *et al.*, 1998; Pereira *et al.*, 2002 and Wäsche *et al.*, 2002). The influence of hydrodynamics on biofilm density and morphology is currently under investigation.

Pressure drop increase monitoring

A higher linear flow velocity enables early detection of a pressure drop increase (Figure 10.7) and stimulates the biomass accumulation rate (Figure 10.5B). Membrane fouling can be assessed more accurately and more early when the feed flow velocity is temporarily increased during sensitive pressure drop measurements. The effect of accumulated material on the pressure drop is greater at high feed flow velocities (Figure 10.7 and Equation 10.1). Also, studies can be shortened when membrane elements in parallel pilot installations or MFSs are continuously operated at higher linear flow velocities (Figure 10.4).

Biofouling analysis

The feed channel pressure drop increase caused by biofouling is affected by the accumulated biomass and the linear flow velocity (Figure 10.7). The lead element in the pressure vessel contributed most to the total pressure drop over the membrane installation because it had the highest linear flow velocity (Figure 10.13). In practice, in most cases biofouling is observed in the lead element. The accumulated biomass volume increases the effective water flow velocity, resulting in a more than proportional pressure drop increase (Figure 10.7 and Equation 10.1). At locations where biofilm accumulation in membrane systems occurs, the influence on the pressure drop is highest.

Comparison of biofouling data in different membrane installations is a challenge. Data on local accumulated biomass, linear flow velocities and pressure drop development are required. These data were not found in literature. Comparison of biofouling in membrane installations is hampered because of differences in linear flow velocities, feed spacer channel characteristics and pressure vessel loading program (number of elements in series) and plant staging and total design.

The absolute feed channel pressure drop (increase) caused by biofouling is a practical problem as this affects water production costs. The membrane manufacturers recommend corrective actions and restrict guarantees when the total pressure drop increase is $\geqslant$15% of the start-up value. This percentage pressure drop increase criterion, used in practice, is not a well defined guideline to take corrective action. The pressure drop increase over a lead membrane element can account for the total pressure drop increase over the pressure vessel. In other words, the pressure drop increase over a lead element can be around 100% whereas the pressure drop increase over the other elements in the pressure vessel can be much lower than 15% or even close to zero.

There is a need for membrane systems operating at a low pressure drop, tolerating biomass accumulation.

Biofouling control

In the late 1990s, two principal strategies were proposed to prevent and control membrane biofouling. These strategies were (i) physical removal of bacteria from the feed water of membrane systems (for example using microfiltration or ultrafiltration as a pretreatment) and (ii) metabolic inactivation of the bacteria by application of biocide dosage or UV irradiation (Ridgway, 1997). At present, the main research focus is substrate removal by biological pretreatment (e.g. sand filtration), adaptation of membranes (disinfectant resistant and low fouling) and, to a lesser extent, membrane cleaning. The present study shows that insight into the biofouling process is essential for biofouling control. A recent study showed that biomass accumulation on the feed spacer affects the fluid flow distribution and the pressure drop increase (Vrouwenvelder *et al.*, 2009c). Adaptation of spacers and the hydrodynamics and/or pressure vessel configuration may be elegant methods to obtain robust low pressure drop membrane systems.

An additional approach to reduce the effect of biofouling can be replacing of lead (biofouled) elements by (clean) elements, i.e. from the last position in the last stage or by changing the flow direction in the stages. The lead membrane has a much higher linear flow velocity (and pressure drop) than the membrane element in the last position in the membrane installation. So, by changing the positions of the elements in the installation or changing the flow direction, the pressure drop over the total installation will be instantaneously reduced (Figures 10.8 and 10.9). In practice, the decline of biomass and pressure drop in time may be even greater (Figures 10.10 and 10.11) because of differences in the osmotic pressure of the water between the membrane elements at the feed and concentrate side of the pressure vessel, which was not included in the present study. The MFS equipped with sensitive pressure difference transmitters are suitable for studying the development of membrane systems less sensitive to membrane biofouling.

Feed flow reversal in stages can be efficient for simultaneous control of several fouling types, i.e. biofouling, particulate fouling and mineral scaling. The principles for control of the individual fouling types differ. For prevention of mineral scaling without the addition of chemicals or periods of system downtime, feed flow reversal can be applied resulting in redissolving of deposited scale into solution (Uchymiak *et al.*, 2008).

Linear flow velocity adaptation: possible consequences

The linear flow velocity effects (i) plant performance in systems without fouling, (ii) the biofouling accumulation rate and (iii) the impact of accumulated biomass on plant performance. Plant performance is determined by feed channel pressure drop, flux and salt passage. In systems without fouling, a low cross flow velocity reduces the pressure drop. A low cross flow velocity reduces both the biofouling accumulation rate (Figure 10.5) and the impact of accumulated biomass on feed channel pressure drop increase (Figure 10.7).

In membrane installations, cross flow velocities in spiral wound membrane elements vary greatly (from ~0.07 to 0.20 $m\,s^{-1}$), depending on element position in the pressure vessel (Figure 10.13). In lead membrane elements in most current installations, the cross flow velocity ranges between 0.14 and 0.20 $m\,s^{-1}$ (Table 10.1). However, several full-scale installations are operated with lower linear flow velocities in the lead elements (0.07–0.09 $m\,s^{-1}$) without fouling problems (Van Paassen *et al.*, 2005; Table 10.1). The low cross flow velocity in the lead elements in these installations is the consequence of hydraulic optimized pressure vessel use (Van der Meer, 2003a). Originally, this adapted pressure design was developed to reduce pressure loss and improve membrane productivity. Low linear flow velocities in the lead elements obtained with these hydraulic optimized vessels will be suitable for reducing the pressure drop increase caused by biomass accumulation.

SUMMARY

In an earlier study, it was shown that biofouling predominantly is a feed spacer channel problem. In this chapter, pressure drop development and biofilm accumulation in membrane fouling simulators have been studied without permeate production as a function of the process parameters substrate concentration, linear flow velocity, substrate load and flow direction. At the applied substrate concentration range, 100–400 μg C L^{-1} as acetate carbon, a higher concentration caused a faster and greater pressure drop increase and a greater accumulation of biomass. Within the range of linear flow velocities as applied in practice, a higher linear flow velocity resulted in a higher initial pressure drop in addition to a more rapid and greater pressure drop increase and biomass accumulation. Reduction of the linear flow velocity resulted in an instantaneous reduction of the pressure drop caused by the accumulated biomass, without changing the biofilm concentration. A higher substrate load (product of substrate concentration and flow velocity) was related to biomass accumulation. The effect of the same amount of accumulated biomass on the pressure drop increase was related to the linear flow velocity. A decrease of substrate load caused a gradual decline in time of both biomass concentration and pressure drop increase.

It was concluded that the pressure drop increase over spiral wound reverse osmosis and nanofiltration membrane systems can be reduced by lowering both substrate load and linear flow velocity. There is a need for reverse osmosis and nanofiltration systems with a low pressure drop increase irrespective of the biomass formation. Current efforts to control biofouling of spiral wound membranes focus in addition to pretreatment on membrane improvement. According to these authors, adaptation of the hydrodynamics, spacers and pressure vessel configuration offer promising alternatives. Additional approaches may be replacing heavily biofouled elements and flow direction reversal.

11

Biofilm Accumulation and Morphology: Role of Flow Regime

INTRODUCTION

Biofouling – excessive growth of biomass, i.e. biofilms – is the major fouling type in spiral-wound nanofiltration (NF) and reverse osmosis (RO) systems, resulting in a pressure drop increase (Characklis and Marshal, 1990; Ridgway and Flemming, 1996 and Patching and Fleming, 2003).

In spiral wound membrane elements, two types of pressure drops can be discriminated: the trans-membrane pressure drop and the feed spacer channel pressure drop. The trans-membrane pressure drop is the differential pressure between feed and permeate lines, describing the resistance over the membrane. The trans-membrane pressure drop is related to the membrane flux (permeation rate). When the trans-membrane pressure is increased, the membrane flux is declined. The feed spacer channel pressure drop is the pressure drop between the feed and concentrate lines (Flemming *et al.*, 1994 and Patching and Fleming, 2003). In practice, the feed channel pressure drop is the most serious problem related to biofouling (Baker and Dudley, 1998 and Vrouwenvelder *et al.*, 2008b, 2009c).

The flux – the water volume flowing through the membrane per unit area and time – as applied in practice in spiral wound NF and RO membranes has been shown irrelevant for biofouling development in extensively pretreated water (Vrouwenvelder *et al.*, 2009b). In other words, presence of a flux (permeate production) was not contributing significantly to the biofilm accumulation in membrane elements. Furthermore, biofilm development was not influencing the

*This chapter is based, with permission from the copyright holder, on a paper previously published in Water Research Vol. 44 No. 3 pp. 689–702 doi: 10.1016/j.watres.2009.09.054

flux since the biofilm was predominantly formed onto the feed spacer structure. Irrespective whether a flux was applied or not, the feed spacer channel pressure drop and biofilm concentration increased in the same order of magnitude in reverse osmosis and nanofiltration membranes in a monitor, test rigs, a pilot scale and a full-scale installation. Identical behavior with respect to biofouling and feed channel pressure drop development was observed in membrane elements in the same position in a nanofiltration installation operated with and without flux. Calculations to predict the influence of flux confirmed the experimental observations that the flux can only be playing an insignificant role in biofouling (Vrouwenvelder *et al.*, 2009b). Therefore, the experiments in this study were performed without flux.

Under practical conditions in NF and RO systems biomass accumulates in the membrane elements causing a feed channel pressure drop increase, which is a problem for process operation. RO plants treating water with a high biofouling potential are applying successfully (intermittent) dosing of biocides e.g. monochloramine or DBNPA. A commonly applied approach to control biofouling is curative chemical cleaning. Currently applied chemical cleanings are (i) not effective to remove (dead) biomass from the elements (Whittaker *et al.*, 1984; Baker and Dudley, 1998; Vrouwenvelder and Van der Kooij, 2001; Al-Amoudi and Lovitt, 2007 and Zondervan and Roffel, 2007), (ii) increasing the costs of membrane operation and (iii) environmentally unfriendly (discharge regulations). Therefore, curative cleanings should be avoided as much as possible. An alternative approach to control biofouling is extensive pretreatment of the feed water, reducing the substrate concentration of the feed water (Griebe and Flemming, 1998; Kamp *et al.*, 2000; Van der Hoek *et al.*, 2000 and Speth *et al.*, 2000). In addition to pretreatment, variation of NF/RO installation operational parameters such as flow regime or linear flow velocity may result in biofouling control.

Increasing the linear flow velocity may reduce the accumulation of microorganisms on surfaces, attributed to the increased shear on the biofilm (Wills *et al.*, 2000). They reported that inserts reduced biofilm accumulation. It has been proposed that increasing the linear flow velocity may be a suitable method to reduce biofouling in tubular heat exchangers (Melo and Bott, 1997), since the shear force is increased by elevation of the linear flow velocity. In addition to the biomass accumulation, the flow regime influences the biofilm morphology: A compact smooth biofilm is formed at high shear force, while a thick, fluffy biofilm is produced at low shear force (Van Loosdrecht *et al.*, 1995; Tijhuis *et al.*, 1996 and Beun *et al.*, 2002).

The objective of this study was to establish the influence of flow regime on biofilm accumulation and morphology in spiral wound nanofiltration and reverse osmosis membranes. Monitor studies were performed with single phase flow (water) and two phase flow (water with air sparging: bubble flow).

MATERIALS AND METHODS

Experimental set-up

An overview of the experimental conditions is shown in Table 11.1. The experimental conditions for the studies on effect of varying linear flow velocities at constant substrate load and biofilm cohesion strength are shown in Table 11.2. The experimental conditions for the two studies on the effect of bubble flow at constant substrate load and linear flow velocity are shown in Table 11.3.

Table 11.1 Scheme experimental conditions of the biofouling studies

Study	Linear flow velocity $m\ s^{-1}$	Substrate concentration in feed water $mg\ C\ L^{-1}$	Substrate load $mg\ C\ m^{-2}\ s^{-1}$	Bubble flow
Effect substrate concentration at constant linear flow velocity	*constant*	varying	varying	no
Effect linear flow velocity at constant substrate concentration	varying	*constant*	varying	no
Effect linear flow velocity at constant substrate load	varying	varying	*constant*	no
Effect bubble flow at constant substrate load and linear flow velocity	*constant*	*constant*	*constant*	yes
Effect flow regime on biofilm cohesion strength	varying	varying	*constant*	no

Membrane fouling simulator (MFS)

In all experiments, a Membrane Fouling Simulator (MFS) with external dimensions of $0.07\,m \times 0.30\,m \times 0.04\,m$ was used (Vrouwenvelder *et al.*, 2006b). Coupons of feed spacer, membrane and product spacer can be placed in the MFS resulting in the same spatial dimensions as in spiral wound membrane elements. The development of fouling was monitored by measuring the pressure drop increase over the feed spacer channel of the MFS, photographic observations using the sight glass and analysis of sheets of membrane and spacer taken from the monitor. During operation, the glass window was covered with a light tight lid to prevent growth of phototrophic organisms. Analyses performed on the sheets were total active biomass (ATP) and dissolved organic carbon (DOC). Details of the methods have been described elsewhere (Vrouwenvelder *et al.*, 2006b, 2007a).

Table 11.2 Experimental conditions for the monitors during study on effect of linear flow velocity at constant substrate load and effect of flow regime on biofilm cohesion strength

Study	Monitor no	Linear velocity $m\ s^{-1}$	Feed $L\ h^{-1}$	Acetate concentration $mg\ C\ L^{-1}$	Acetate load $mg\ C\ m^{-2}\ s^{-1}$
Effect linear flow velocity at constant substrate load	1	0.041	4	0.40	0.056
	2	0.082	8	0.20	0.056
	3	0.123	12	0.13	0.056
	4	0.163	16	0.10	0.056
	5	0.245	24	0.066	0.056
Effect flow regime on biofilm cohesion strength	1	0.06	6	1.00	0.21
	2	0.06*	6	1.00	0.21
	3	0.31	30	0.20	0.21
	4	0.31*	30	0.20	0.21

**After a pressure drop increase caused by biomass accumulation was observed the linear flow velocity was increased until 0.42 $m s^{-1}$ to determine the biofilm cohesion strength.*

Table 11.3 Experimental set-up for monitors operated in parallel during studies on effect of bubble flow at constant substrate load and linear flow velocity

Exp.	Code	Linear flow velocity $m s^{-1}$	Water feed flow $L h^{-1}$	Air feed flow $L h^{-1}$	Acetate concentration $mg C L^{-1}$	Acetate load $mg C m^{-2} s^{-1}$
1	-S	0.16	16	0	0	0
	+S	0.16	16	0	0.20	0.10
	bubble flow+S	0.16	16	~100	0.20	0.10
	bubble flow-S	0.16	16	~100	0	0
2	-S	0.20	20	0	0	0
	+S	0.20	20	0	0.30	0.20
	bubble flow+S	0.20	20	~36	0.30	0.20

Membranes and spacer sheets used in the MFS research were taken from new and unused spiral wound membrane elements (Trisep TS 80, 8 inch diameter). The feed spacer taken from the element was a 31 mil thick diamond-shaped polypropylene spacer with porosity ~0.85. The feed spacers in the MFS had the same spatial orientation as in spiral-wound membrane elements (45° rotation).

The installation consisted of two pressure reducing valves, manometer, dosage point (for biodegradable compounds), MFS and flow controller (Vrouwenvelder *et al.*, 2007a). The MFS was fed with drinking water and operated at a pressure of ~120 kPa to avoid degassing. Unless mentioned otherwise, the feed water flow was $16\,L\,h^{-1}$ equal to a linear flow velocity of $0.16\,m\,s^{-1}$, which is a representative linear flow velocity for practice (Vrouwenvelder *et al.*, 2009a). The MFS was operated cross-flow, without permeate production.

Pressure drop

Pressure drop measurements were performed with a pressure difference transmitter (Endress & Hauser, type Deltabar S: PMD70-AAA7FKYAAA). The calibrated measuring range was 0 to 50 kPa (Vrouwenvelder *et al.*, 2009d).

Bubble flow studies

In the experiments, the MFSs were installed in a vertical position and operated in parallel (Figure 11.1). Compressed air at a pressure of 150 kPa was fed into the feed water of the MFS between the substrate dosing point and the MFS. The air supply tube contained a diaphragm pressure relief valve (KNF flodos diaphragm valve, Type: FDV 30 KPZ) to prevent back flow of water into the air system. A mass flow controller (5850TR, Brooks Instruments, The Netherlands) was used to regulate air supply. The linear velocity was calculated for the water flow.

Biofouling development was observed with a CCD camera (Pixelfly VGA, PCO AG, Kelheim, Germany) connected to a personal computer equipped with camera software (Camware, PCO, Germany). The camera was fitted with a fixed focal length lens (Edmund Optics, York, UK) with a minimum working distance of 30 cm. This system allowed a maximum frame rate of 50 Hz at a resolution of 640 × 480 pixels, which resulted in 3–4 pixels per millimeter of the flow cell at the applied working distance. In order to obtain constant light intensity in the MFS, it was illuminated from above with a cold light source (Euromex microscopen BV, Arnhem, The Netherlands). More details on the camera system are described elsewhere (Willems *et al.*, 2009).

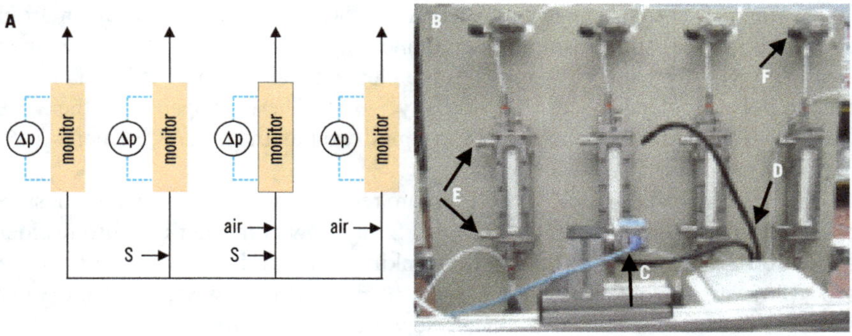

Figure 11.1 Set-up scheme (**A**) and photograph (**B**). A rail was used to position the camera (**C**) and lights (**D**). The pressure drop was measured over the monitor spacer channel (**E**). The flow rate was kept constant using a flow controller (**F**). Between imaging the monitor sight window was covered to prevent growth of phototrophic organisms. Air = water with air sparging: bubble flow; S = substrate dosage; Δp = differential pressure transmitter

Pressure drop measurements were performed in the monitors supplied with air during the experiment under two conditions: (i) continued air supply and (ii) shortly discontinued air supply. The pressure drop measurement during bubble flow resulted in a higher pressure drop (1.5 kPa) compared to the pressure drop measured during temporarily interruption of air supply (Figure 11.2). The same pressure drop increase by air supply was observed in the monitors supplied with and without substrate. Accurate monitoring of the pressure drop can be realized during bubble flow.

Feed water and substrate dosage

Drinking water prepared from anaerobic groundwater (subsequently treated by aeration, rapid sand filtration, deacidification, softening and rapid sand filtration at treatment plant Spannenburg in the Netherlands) was used as feed water of the MFS installation. The water had a dissolved organic carbon (DOC) content of 8 $mg\,L^{-1}$ and temperature varied maximal 2°C during each experiment within a total range of 13°C and 18°C.

Substrate was dosed to the feed water of MFSs to stimulate biomass growth. From a sterile vessel, containing a 5 L solution of concentrated substrate, the substrate was dosed into the feed water prior to the MFS by a peristaltic pump (Masterflex) at a flow of 0.03 $L\,h^{-1}$. The dosage of substrate was checked periodically by measuring the weight of the dosing bottle. The chemicals $NaCH_3COO$, $NaNO_3$ and NaH_2PO_4 were used with a ratio C:N:P of 100:20:10 for the dosage solution. C was the growth limiting compound. N and P were dosed to eliminate growth limiting conditions caused by potentially low N and P concentrations in the feed water. The substrates were dissolved in ultrapure water. To restrict bacterial growth in the substrate dosage bottle, the solution pH was set to 10.5 using NaOH. Dosage bottles were replaced every 5 days. The substrate dosage ($0.03\,L\,h^{-1}$) was low compared to the feed water flow rate ($16\,L\,h^{-1}$). Thus, the pH of the feed water was not measurably influenced by substrate dosage.

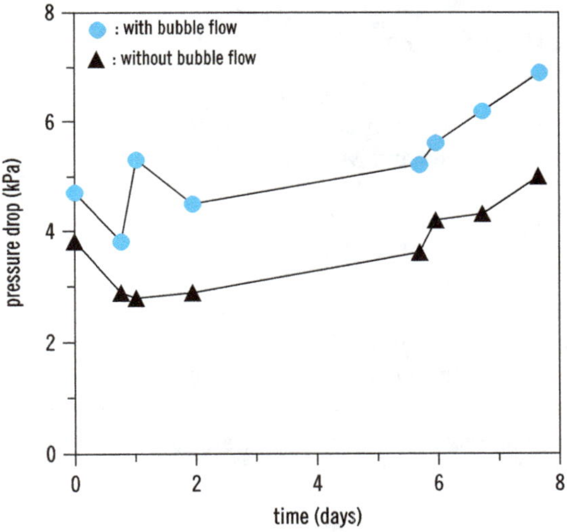

Figure 11.2 Pressure drop in time over the monitor with and without air supply to feed water containing substrate. Without air supply means that air supply was shortly interrupted for pressure drop measurements only

The effect of substrate concentration at constant linear flow velocity on biofouling (Table 11.1) was investigated with four monitors, which were operated in a parallel study at constant linear water flow velocity ($0.16\,m\,s^{-1}$). A blank solution without acetate and water with three acetate concentrations (0.10, 0.20 and $0.40\,mg\,C\,L^{-1}$) were fed into the individual monitors.

The effect of linear flow velocity at constant substrate concentration on biofouling (Table 11.1) was studied with four monitors operated with varying linear flow velocities (0.041, 0.082, 0.163, and 0.245 $m\,s^{-1}$) at constant acetate concentration (0.20 $mg\,C\,L^{-1}$) in the feed water.

The effect of linear flow velocity at constant substrate load on biofouling was performed with five monitors (Table 11.2).

The effect of bubble flow on biofouling (Table 11.3) was determined with two studies using monitors at constant substrate load and linear flow velocity.

The effect of flow regime on biofilm cohesion strength (Table 11.2) was performed with five monitors. Four monitors were operated at a low (0.06 $m\,s^{-1}$) and high (0.31 $m\,s^{-1}$) linear flow velocity supplied with substrate. A blank monitor operated at 0.31 $m\,s^{-1}$ without substrate dosage showed no fouling accumulation, indicating that the water was not contributing significantly to the fouling.

Relative friction factor

The friction coefficient λ is a function of the pressure drop (Δp), specific liquid density (ρ), linear velocity (v), membrane length (L) and the hydraulic diameter (d_h), see equation 1 (Schock and Miquel, 1987).

$$\lambda = 2\frac{\Delta p}{\rho v^2} \cdot \frac{d_h}{L} \tag{11.1}$$

The relative friction factors (λ/λ_o) of MFSs operated in parallel were determined. λ_o is the friction coefficient of the blank, so the blank measurement had a relative friction factor of 1. The purpose of using relative friction factors was to determine changes in biofilm volume and/or roughness.

RESULTS

The development of feed channel pressure drop and biofilm concentration and morphology in membrane fouling simulators operated in parallel were investigated with (i) varying substrate concentrations in the feed water at constant linear flow velocity (Figure 11.3ABC), (ii) varying linear flow velocities at constant substrate concentrations in the feed water (Figure 11.3DEF), (iii) varying linear flow velocities at constant substrate load (Figure 11.3GHK) and (iv) different flow regimes with and without bubble flow at constant substrate load and linear water flow velocity (Figures 11.4 and 11.5). The development of biofilm morphology was observed in the different experiments (Figures 11.6 and 11.7) and the effect of flow regime on biofilm cohesion strength was investigated (Figures 11.8 and 11.9).

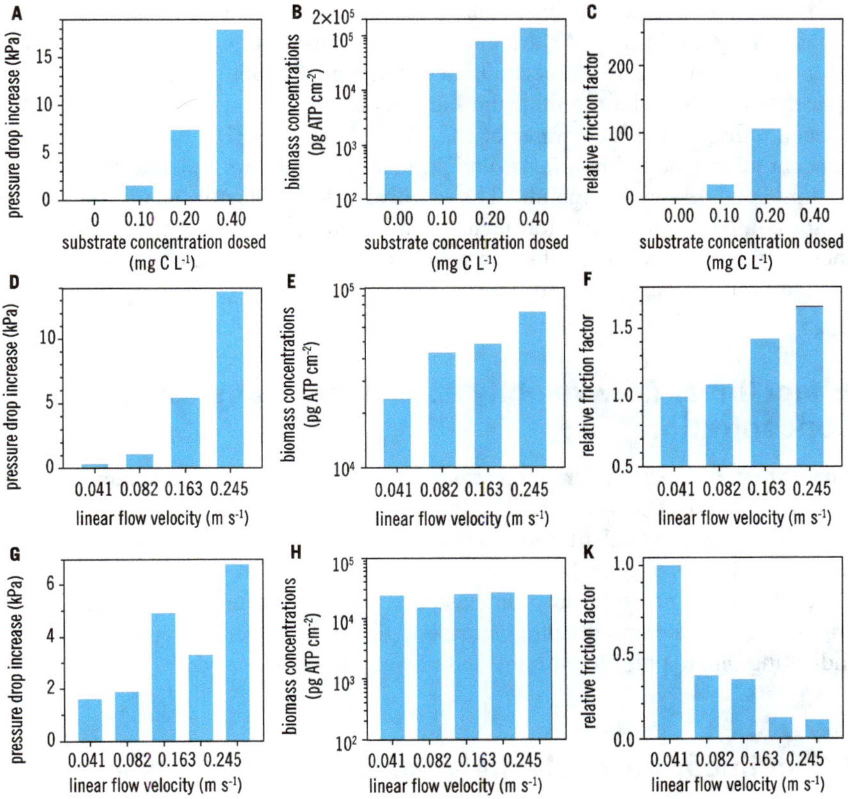

Figure 11.3 Observations on biofouling accumulation in monitors under varying conditions. Monitors were operated in parallel. **(A,B,C)** With varying substrate concentrations (0.00, 0.10, 0.20 and 0.40 mg acetate-C L^{-1}) and constant linear flow velocity. **(D,E,F)** With varying linear flow velocities and constant substrate concentration (0.20 mg acetate-C L^{-1}) in the feed water. **(G,H,K)** With varying linear flow velocities and the same substrate load (0.056 mg acetate-C m^{-2} s^{-1}). Adapted from Vrouwenvelder *et al.*, 2009a

Effect substrate concentration at constant linear flow velocity

The pressure drop increased in time over the monitors fed with water containing acetate, while the pressure drop over the monitor fed with a blank solution remained constant (Figure 11.3A). An increase of acetate concentration in the feed water resulted in a faster and stronger pressure drop increase over the monitor (Figure 11.3A). At the end of the research period, the monitors were opened for analysis of the accumulated material. On both membrane and feed spacer taken from the monitor, the biomass amount increased with the acetate concentration in the feed water of the MFS (Figure 11.3B). A higher substrate concentration resulted in a faster and stronger pressure drop increase and higher accumulated biomass amount. The relative friction factor (ratio of friction factor compared to the friction factor of the blank measurement, equation 11.1) increased with increasing substrate concentration (Figure 11.3C), indicating that the biofilm volume and/or roughness increased.

Effect linear flow velocity at constant substrate concentration

The extent of pressure drop increase was related to the linear flow velocity (Figure 11.3D). The accumulated biomass in the monitors (Figure 11.3E) – determined at the end of the research period – was related to the pressure drop increase. A higher linear flow velocity resulted in a faster and stronger development of pressure drop increase and biomass accumulation. The relative friction factor increased with increasing linear flow velocity (Figure 11.3F), indicating that the biofilm volume and/or roughness increased.

Effect linear flow velocity at constant substrate load

The load of biodegradable compounds may affect biofouling. To evaluate the influence of substrate load on biofouling development, comparative studies with monitors were performed with varying linear flow velocities at constant substrate load (Table 11.2).

The pressure drop increase increased with linear flow velocity (Figure 11.3G). The accumulated amount of biomass in the monitors at the end of the research period was not significantly different (Figure 11.3H). The same amounts of accumulated biomass in the monitors operated with different linear flow velocities and acetate concentrations but the same acetate load indicate that biomass accumulation was substrate load related under the applied conditions. Replicated experiments showed similar results (data not shown). In other words, in the monitors supplied with the same substrate load, the amount of accumulated biomass was constant but the pressure drop increase was a function of the linear flow velocity (Figure 11.3G).

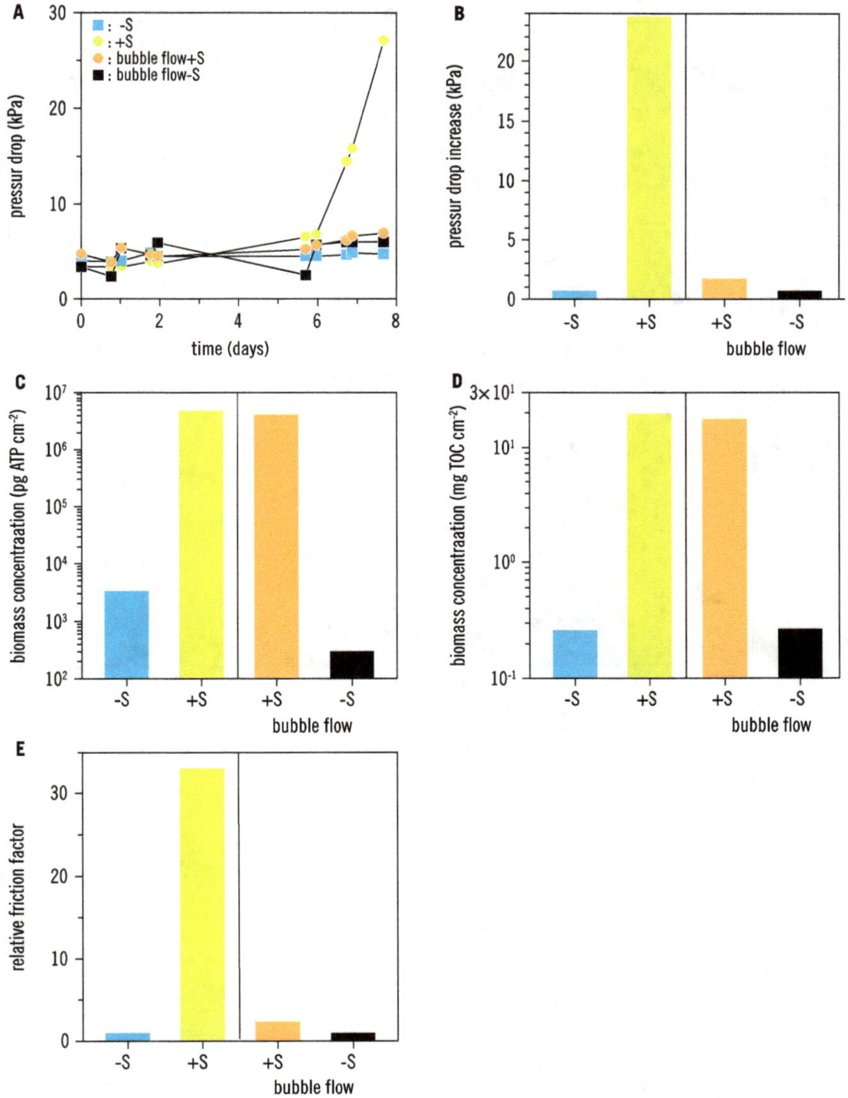

Figure 11.4 Pressure drop in time (**A**), pressure drop increase (**B**), biomass concentration ATP (**C**) and TOC (**D**) after 7.8 day MFS operation and the relative friction factor (E). The monitors were operated with and without dosage of a biodegradable compound (0.20 mg acetate-C L^{-1}) to the feed water of the monitor without and with bubble flow. -S = without bubble flow without substrate dosage; +S = without bubble flow with substrate; bubble flow+S = bubble flow with substrate; bubble flow-S = bubble flow without substrate

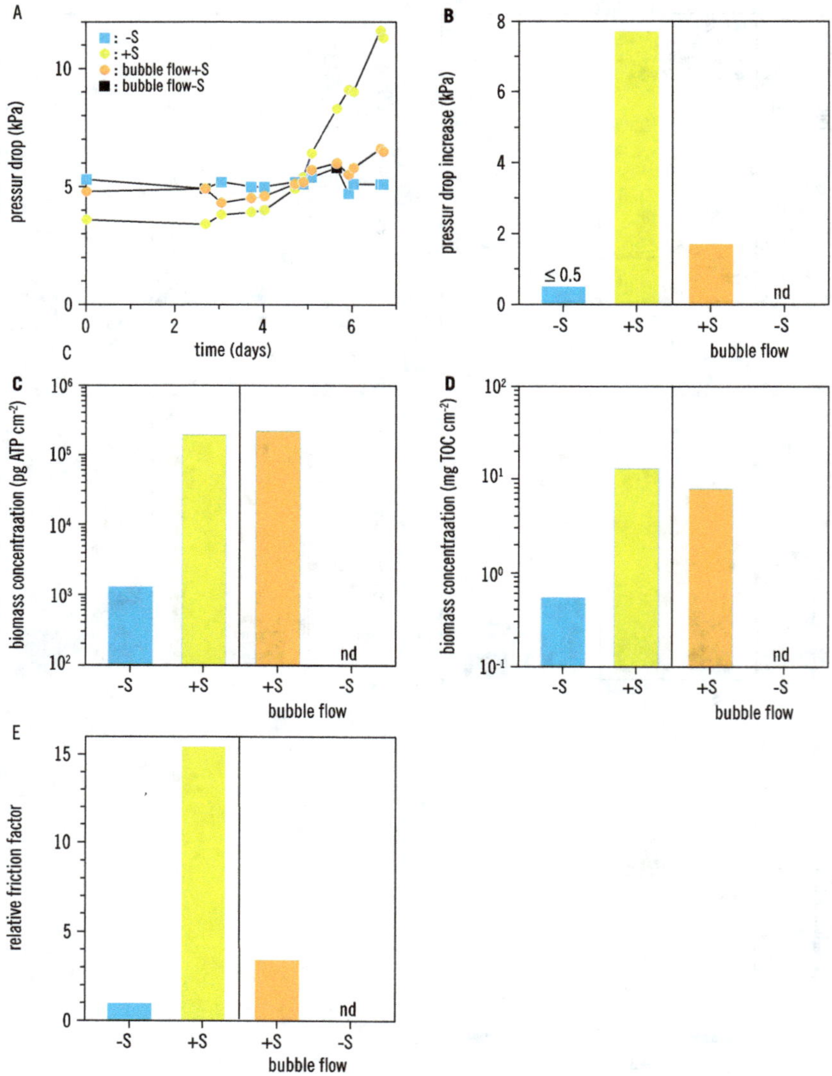

Figure 11.5 Pressure drop in time (**A**), pressure drop increase (**B**), biomass concentration ATP (**C**) and TOC (**D**) after 6.7 day MFS operation and the relative friction factor (**E**). The monitors were operated with and without dosage of a biodegradable compound (0.30 mg acetate-C L^{-1}) to the feed water of the monitor without and with bubble flow. -S = without bubble flow without substrate dosage; +S = without bubble flow with substrate; bubble flow+S = bubble flow with substrate; bubble flow-S = bubble flow without substrate was not determined (nd)

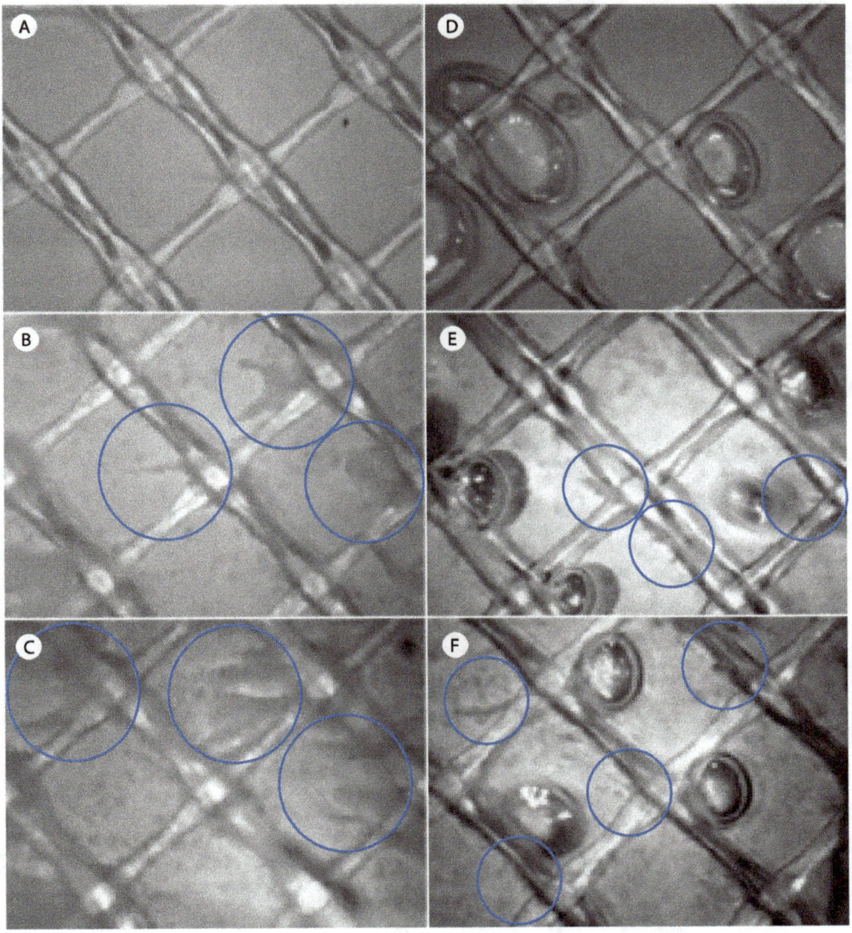

Figure 11.6 Biofilm characterization in time with substrate dosage without (**A,B,C**) and with bubble flow (**E,F,G**). A and D = day 0; B and E = day 5; C and F = day 7. The flow direction is from right to left. The circles indicate biofilm streamer locations

The relative friction factor decreased with increasing linear flow velocity (Figure 11.3K), indicating that the biofilm volume and/or roughness decreased. The opaque nature of the biofilm in the MFSs hampered direct observations of biofilm volume and roughness.

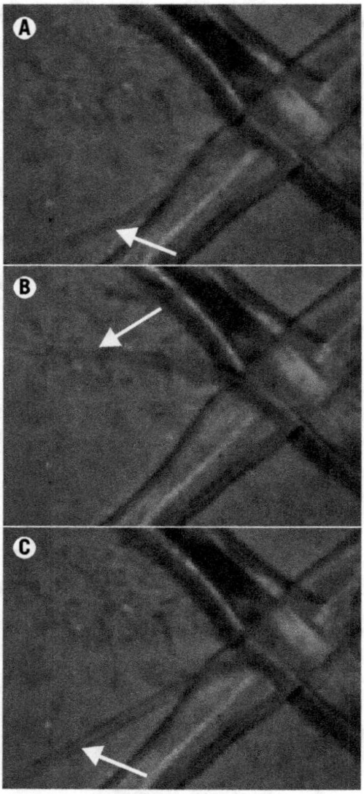

Figure 11.7 Filamentous biofilm structure oscillating in time in the water stream. The time between the images is 0.2 s. The flow direction is from right to left. The arrow indicates the biofilm streamer location

Effect bubble flow at constant substrate load and linear flow velocity

Pressure drop and biomass amount

The substrate concentration, linear flow velocity and substrate load affect the pressure drop (increase), biofilm accumulation and relative friction factor (Figure 11.3). To evaluate the influence of flow regime on biofouling development and morphology, comparative studies with monitors were performed with varying flow regimes (single and two phase flow) with constant substrate load and linear flow velocity (Figure 11.1 and Table 11.3: exp.1). As two phase flow bubble flow, water sparged with air, was applied while for single phase flow water only was used. Monitors without substrate dosage (-S and bubble flow -S) were operated under the same conditions as the monitors with substrate feed (+S and bubble flow+S).

Figure 11.8 Effect of flushing (at 0.42 $m s^{-1}$) on biofouling developed at low and high linear flow velocity (0.06 and 0.31 $m s^{-1}$). Biofilm developed at low linear flow velocity (0.06 $m s^{-1}$) before (**A**) and after flush (**B**), and developed at high linear flow velocity (0.31 $m s^{-1}$) before (**C**) and after flush (**D**)

The pressure drop increased strongly in the MFS supplied with substrate (+S) with water flow only, while the pressure drop remained low in the MFS supplied with substrate with bubble flow (bubble flow+S). As expected, the blank MFSs without substrate without and with bubble flow (-S and bubble flow-S) remained low as well, see Figure 11.4AB. At the end of the research period, the monitors were opened for analysis of the accumulated material. On the feed spacer and membrane sheets taken from the monitors with substrate addition without (+S) and with bubble flow (bubble flow+S), the same high concentrations of biomass (ATP, TOC) were observed, while the monitors without substrate dosage without (-S) and with bubble flow (bubble flow-S) had low biomass concentrations (Figure 11.4CD).

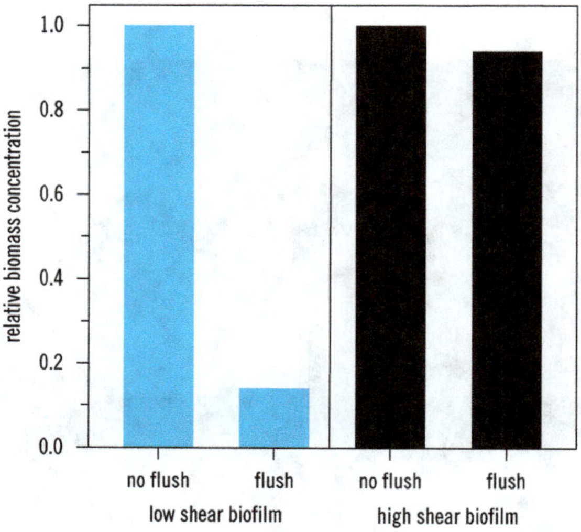

Figure 11.9 Effect of flushing (at 0.42 m s^{-1}) on biofouling developed at low and high linear flow velocity (0.06 and 0.31 m s^{-1})

Apparently, contrary to the pressure drop increase biomass growth was not influenced by the flow regime. In other words: with and without bubble flow the same biomass concentrations were found while the pressure drop increase differed strongly (see Figure 11.4, +S and bubble flow+S). The relative friction factor was high for the monitor supplied with substrate (+S) and single phase flow while the monitor with substrate and bubble flow (bubble flow+S) had a low relative friction factor (Figure 11.4E), indicating a higher biofilm volume and/or roughness in the monitor with substrate and single phase flow (+S) compared to the monitor with substrate and bubble flow (bubble flow+S).

A second study (Figure 11.5) performed with a similar set-up (Table 11.3: exp. 2) confirmed these results (Figure 11.4). A high pressure drop increase and high relative friction factor were observed for the monitor with substrate dosage without bubble flow (+S) while both values were low for the monitor with substrate and bubble flow (bubble flow+S, Figure 11.5ABE). The biomass concentrations had the same order of magnitude (Figure 11.5CD).

Biofilm compactness and morphology

In the blank monitors without substrate dosage (-S and bubble flow-S) no biofilm accumulation in time was observed. Accumulation of material was only observed in the monitors supplied with substrate.

Without bubble flow, the development of filamentous biofilm structures and voluminous biofilms was observed (Figure 11.6ABC). The filamentous biofilm structures were mainly attached to the feed spacer (Figure 11.6BC), and oscillated in the water stream, sweeping the open area between the spacer strands (Figure 11.7). Biofilm filaments lengths was up to ~6 mm, exceeding the distance between adjoining feed spacer intersections by a factor two (Figures 11.6 and 11.7). Biomass accumulation seemed to start predominantly on the feed spacer strand downstream side (Figure 11.6BE), in a later phase followed by a more generally spread biomass accumulation on all materials (spacer and membrane).

With bubble flow, development of a thin biofilm with incidental biofilm streamers of short length (~0.5 mm) and short diameter was observed (Figure 11.6EF). Biofilm streamers (indicated with circles) and the presence of the air bubbles increasing the shear force in the monitor are shown in Figure 11.6DEF. Biofilm compactness and morphology differed strongly between monitors operated with the different two flow regimes.

Effect flow regime on biofilm cohesion strength

To evaluate the effect of flow regime on biofilm cohesion strength, studies were performed with biofilms developed in monitors with low and high linear flow velocities at constant substrate load without and with a flush (Table 11.2).

Under low shear conditions much more biomass volume accumulated as compared to high shear conditions (Figure 11.8AC). The biofilm grown under low shear conditions was much more easy to remove during water flushing as more than 85% of the biomass detached compared to virtually no detachment from the biofilm grown under high shear conditions (Figures 11.8BD and 11.9). In the system grown under high shear even more biomass remained after the flush as compared to the system grown under low shear after the water flush (Figure 11.8). Clearly, the biofilm cohesion strength was influenced by the applied shear rate during biofilm growth.

DISCUSSION

Analogy biofilm formation in RO/NF and other systems

The same biomass amount (Figures 11.4CD and 11.5CD) was observed in membrane fouling simulators supplied with substrate operated with and without bubble flow. However, a heavy pressure drop increase was observed under 'standard' shear force conditions only while a low pressure drop increase was observed under 'enhanced' shear force conditions caused by bubble flow (Figures 11.4B and 11.5B). In other words, the same biomass concentration caused a significantly different pressure drop increase. The differences in effect on pressure drop are related to the structure of the biofilm: a thin compact biofilm developed under enhanced shear conditions and a thick fluffy biofilm developed under standard shear conditions (Figure 11.6).

The influence of hydrodynamic conditions on biofilm structure and compactness is referenced broadly. In several biofilm studies it has been observed that at high hydrodynamic shear force the biofilm becomes more compact, stable and dense (Van Loosdrecht *et al.*, 1995; Kwok *et al.*, 1998; Pereira *et al.*, 2002 and Wäsche *et al.*, 2002). The thickness, structure, stability, adhesive strength and density of the biofilm were influenced by the substrate load and hydrodynamic shear stress/flow regime (Characklis and Marshall, 1990; Van Loosdrecht *et al.*, 1995; Peyton, 1996; Kwok *et al.*, 1998; Melo and Bott, 1997; Pereira *et al.*, 2002; Beun *et al.*, 2002; Wäsche *et al.*, 2002; Wijeyekoon *et al.*, 2004; Purevdorj-Gage and Stoodley, 2004 and Chen *et al.*, 2005). Thinner biofilms were found during growth of *Pseudomonas fluorescens* biofilms under turbulent flow conditions compared to laminar flow conditions (Pereira *et al.*, 2002). Thicker biofilms were observed at increasing substrate load (Peyton, 1996 and Wäsche *et al.*, 2002). Filamentous biofilm streamers and hydrodynamics were described in a review of Purevdorj-Gage and Stoodley (2004) and other studies (Stoodley *et al.*, 1999a,b, 2002, 2005; Hille *et al.*, 2009 and Garny *et al.*, 2009). Garny *et al.* (2009) reported that filamentous bacterial growth was due to the combination of limited substrate availability and high flow rates. Evidently, biofilm formation in spiral wound NF and RO systems is analogous to biofilm formation in other systems. A high shear force results in a thin compact biofilm while a low shear force results in a thick fluffy biofilm.

Manipulation of biofilm morphology

Biofilm compactness and morphology can be manipulated ('biofilm morphology engineering') by the flow regime (Figure 11.6). In reactor design this has regularly been used to obtain a desired biofilm morphology e.g. in biofilm airlift reactors (Tijhuis *et al.*, 1996) and aerobic granular sludge (Beun *et al.*, 2002). Manipulation of the biofilm morphology influencing membrane performance can be a suitable approach to control NF/RO biofouling as well.

Quantification of biofouling effect

Characklis and Marshall (1990) defined biofouling as 'biofilm formation causing unacceptable operational problems'. In this context 'unacceptable' means that operational guidelines are exceeded for e.g. pressure drop increase. A dominant operational problem in practice is the feed spacer pressure drop increase caused by biomass accumulation in the feed spacer channel (Baker and Dudley, 1998 and Vrouwenvelder *et al.*, 2008b, 2009c,d). A study on quantifying biofouling using membrane elements from practice showed that ATP is a suitable parameter to measure the amount of biomass in order to diagnose biofouling (Vrouwenvelder *et al.*, 2008b). However, this study shows that analysis of biomass parameters as ATP and TOC only in membrane elements operated at a high shear force is not sufficient to predict the pressure drop increase. In addition to the biomass amount, the biofilm compactness and morphology contributes strongly to the pressure drop increase.

Since the linear flow velocity in the membrane element influences the extent of pressure drop increase (Vrouwenvelder *et al.*, 2009a,d), the pressure drop has to be measured under standard conditions (fixed linear flow velocity). Measuring biomass concentrations in membrane elements only is not sufficient to quantify and diagnose biofouling. Quantifying and diagnosis of biofouling requires determining the (i) pressure drop increase under a standard linear flow velocity, (ii) biomass amount, and (iii) biofilm morphology and thickness.

Future studies and practical implications

Studies over a long period of time are needed to determine how the biofilm grown under high substrate conditions is comparable to a biofilm grown under low substrate conditions, including varying hydrodynamic conditions and biomass removal potential from membrane systems. Comparison with practice is considered a key aspect for combined studies on biofilms, hydrodynamics and concentration polarization, e.g. at low shear operation. Evidently, there is need to integrate fouling studies to unravel all the effects of (bio)fouling on performance of RO and NF systems (increase in feed spacer pressure drop, decrease in flux, and increase in salt passage) under representative conditions. Flow regimes manipulate biofilm morphology affecting membrane performance. This opens new ways to control biofouling. The question is how to control biofilm morphology.

Growth of thin compact biofilms may be preferable from a pressure drop increase point of view. An elevated shear force resulted in a low pressure drop increase (Figures 11.4B and 11.5B) because of the compact biofilm (Figure 11.6).

Therefore, high shear forces seem to be suitable to realize a stable pressure drop. However, the amount of biomass growing at elevated shear force was the same as under standard shear force (Figures 11.4CD and 11.5CD), suggesting that even at elevated shear force biomass growth will progress in time resulting in a thick compact biofilm that will increase the pressure drop as well (Table 11.4). A biofilm developed under high shear is compact, strongly attached (Van Loosdrecht *et al.*, 1995; Tijhuis *et al.*, 1996; Beun *et al.*, 2002 and Chen *et al.*, 2005) and probably more resistant to cleanings (Table 11.4). In other words, the short term effect may be beneficial, but the long term effect of compact biofilm formation may be resilience to biofouling control actions.

Table 11.4 Relative biofilm characteristics developed under high and low shear and considerations for biofouling control

	High shear	**Low shear**
biofilm structure	compact and dense	fluffy and open
biofilm cohesion strength	high	low
impact biomass accumulation on pressure drop increase		
short term	low	high
long term	high?[a]	high
risk of biofouling on long term	high	low?[b]
risk on concentration polarization (scaling)	low	high?
energy requirements for shear	high[c]	low

[a] *May be more difficult to control on long term than a biofilm developed under low shear conditions because of the strong attachment and compact nature of biofilms developed under high shear.*
[b] *Biofilm may be removed during short reverse flush with high shear (Cornelissen* et al., *2007). Biofilms at low linear flow velocity have low impact on pressure drop increase (Vrouwenvelder* et al., *2009a,d).*
[c] *A high shear caused by a high linear flow velocity requires in a higher energy input, increasing the operational cost.*

Thick fluffy biofilms with low cohesion strength produced at low velocity/shear force may be easier to remove from the attached materials (Figures 11.8 and 11.9). However, already during operation sloughing of streamers may occur and because of the streamer size and spacer construction there is a high risk of entrapment (blocking). Also, removal of biomass from the membrane element during cleaning may be a problem. Generally, biofouling is observed in lead membrane elements where the linear flow velocity and the shear force are highest (Carnahan *et al.*, 1995 and Vrouwenvelder *et al.*, 2008b, 2009a). Reducing the shear force in lead membrane elements may lead to a fluffy biofilm. Methods to reduce the shear force can be (i) spacer geometry adaptation and (ii) more elements in parallel, reducing the number of elements in a pressure vessel and reducing the linear flow velocity in lead elements (Van der Meer, 2003 and Van Paassen *et al.*, 2005). Reducing the linear flow velocity in lead elements will reduce the impact of biomass on the pressure drop increase (Vrouwenvelder *et al.*, 2009a,d). A potentially suitable method to control biofouling may be operation of membrane elements at standard or even lower shear force combined with a periodic reverse flush at elevated shear force. Adaptation of feed spacer geometry may facilitate removal of biomass from the element.

The viability to reduce biofouling in practice by manipulating biofilm morphology depends on the (i) risk of unmanageable biofouling, (ii) risk on other fouling types like scaling, and (iii) energy use (Table 11.4). These aspects reducing operational costs should be balanced with increased costs for more pressure vessels and membrane elements.

SUMMARY

Biomass accumulation and pressure drop development have been studied in membrane fouling simulators at different flow regimes. At linear flow velocities as applied in practice in spiral wound nanofiltration and reverse osmosis membranes, voluminous and filamentous biofilm structures developed in the feed spacer channel, causing a significant increase in feed channel pressure drop. Elevated shear by both single phase flow (water) and two phase flow (water with air sparging: bubble flow) caused biofilm filaments and a pressure drop increase. The amount of accumulated biomass was independent of the applied shear, depending on the substrate loading rate (product of substrate concentration and linear flow velocity) only. The biofilm streamers oscillated in the passing water. Bubble flow resulted in a more compact and less filamentous biofilm structure than single phase flow, causing a much lower pressure drop increase. The biofilm grown under low shear conditions was more easy to remove during water flushing compared to a biofilm grown under high shear. To control biofouling, biofilm structure may be adjusted using biofilm morphology engineering combined with biomass removal from membrane elements by periodic reverse flushing using modified feed spacers. Potential long and short term consequences of flow regimes on biofilm development are discussed. Flow regimes manipulate biofilm morphology affecting membrane performance, enabling new approaches to control biofouling.

12

Reduction of Membrane Biofouling and Phosphate Limitation

INTRODUCTION

Microbial growth in drinking water distribution networks can be affected by various factors like disinfection, temperature and available nutrients such as organic carbon, nitrogen and phosphorous. Biofilm formation in the distribution network is hard to control by post chlorination because of the required residual concentration (LeChevalier *et al.*, 1987; Van der Wende *et al.*, 1989 and Block, 1992). In most cases, organic carbon is considered to be the limiting nutrient for microbial growth of drinking water biofilms (Van der Kooij *et al.*, 1982; LeChevalier, 1990; Joret *et al.*, 1991; Mathieu *et al.*, 1992; Servais *et al.*, 1995; Prévost *et al.*, 1998; Niquette *et al.*, 2001 and Hijnen *et al.*, 2009).

Phosphorous may be limiting microbial growth in aquatic systems. Phosphorous or phosphate limitation in relation with (micro)biological growth has been regularly reported within the field of waste water (Alphenaar *et al.*, 1993), rivers, surface and seawater (Paetsch *et al.*, 1980; Scavia and Laird, 1987; Harrison, 1990; Van Donk *et al.*, 1990; Toolan *et al.*, 1991; Petterson and Blomqvist, 1992; Mohamed *et al.*, 1998, 2003 and Zohary *et al.*, 2005) and drinking water (Miettinen, Vartiainen and Martikainen, 1997; Lehtola *et al.*, 1999, 2002a,b,c, 2003; Keinänen *et al.*, 2002; Torvinen *et al.*, 2007; Haas *et al.*, 1988; Sathasivan *et al.*, 1997, 1998, 1999; Nishijima *et al.*, 1997; Kors *et al.*, 1998; Van der Aa *et al.*, 2002 and Kasahara *et al.*, 2004).

In microbial biomass, the molar ratio for carbon (C), nitrogen (N) and phosphorous (P) is ~100:20:1.7 (Tchobanoglous *et al.*, 2003). This corresponds with a mass ratio of 100:23:4.3. Compared to carbon, lower phosphorous levels

*This chapter is based, with permission from the copyright holder, on a paper previously published in Water Research Vol. 44 No. 11 pp. 3454–3466 doi: 10.1016/j.watres.2010.03.026

are needed for microbial growth. In nature, phosphorous is present in a variety of chemical compounds (Holtan *et al.*, 1998). The soluble form of phosphorous normally consists of ortho-phosphate (H_3PO_4, $H_2PO_4^-$, HPO_4^{2-}), inorganic polyphosphates and dissolved organic phosphorous (Holtan *et al.*, 1998). Most of the total phosphorous in natural waters is not directly available for microbial growth. (Ortho)phosphate is the form in which phosphorous is most readily available for biological utilization (Maher and Woo, 1998). In this study, orthophosphate is referred to as phosphate.

Currently applied thin film composite NF and RO membranes are sensitive for free chlorine. Free chlorine damages the membrane structure causing decrease of membrane rejection. A limited number of plants apply successfully monochloramines in controlling biofouling. A much better membrane resistance to monochloramines compared to chlorine has been reported (DOW, 2009a): 300,000 ppm-hours for chloramine and up to ~1000 ppm-hours for free chlorine. Since monochloramine is formed by adding ammonia to chlorine, it is possible that free chlorine will be present (for e.g. the ammonia dosing is not correct or fails). Moreover, iron and manganese catalyzes membrane oxidation by monochloramines (Gabelich *et al.*, 2005 and Da Silva *et al.*, 2006). Another reason to avoid chloramination is the formation of N-nitrosodimethylamine (NDMA), a probable human carcinogen. The RO rejection capacities for NMDA are 10 to 50%. Recently, an alternative for chloramines, 2,2-Dibromo-3-Nitrilopropionamide (DBNPA. DOW, 2005, 2009b) is applied successfully in a limited number of plants.

At present, it is common practice for drinking water production to reduce the concentration of easily biodegradable organic carbon of water by pretreatment like biologically activated carbon filtration or (slow) sand filtration. However, membrane biofouling is still considered as a major problem. The combined search criteria 'biofouling' and '(phosphorous or phosphorus or phosphate) limitation' in article title, abstract and keywords in the Scopus database of July 2009 yielded no references. Recently, reports and conference papers suggesting that biofouling in a full-scale system might have been limited by phosphate have been published (Vrouwenvelder *et al.*, 2006a; Van der Kooij *et al.*, 2007 and Galjaard *et al.*, 2008). Jacobson *et al.* (2008, 2009) also suggest the potential of phosphate limitation as fouling control. Direct experimental evidence is lacking whether or not low phosphate concentrations in feed water may restrict biofouling.

The study of phosphate limitation was originated by observations made at two RO membrane installations with extensive pretreatment and a stable feed channel pressure drop, differing strongly in biomass amount compared to other installations. The biomass concentrations in the lead elements of the two RO installations were low compared to lead elements in other installations. It was hypothesized that these low biomass concentration were due to elimination of phosphate during pretreatment.

The goal of this study was to verify experimentally whether or not phosphate limitation may be a suitable approach to reduce membrane biofouling.

MATERIALS AND METHODS

Experimental set-up

Monitoring studies were performed with and without dosages (organic carbon and/or phosphate and/or antiscalants) at a full-scale RO installation with a very low phosphate and organic substrate concentration in the RO feed water.

An overview of the experimental conditions is shown in Table 12.1. The experimental conditions for the studies on 'proof of principle' and effect of antiscalant dosage are summarized in Table 12.2. The antiscalant coding and composition are presented in Table 12.3. The experimental conditions for the study at several stages of the membrane installation to determine the growth limiting conditions at these stages are shown in Table 12.4.

Plant description

PWN Water Supply Company North Holland has taken water treatment plant Jan Lagrand in Heemskerk (HMK) into production on 21 August 1999. HMK feeds pretreated surface water into ultrafiltration followed by RO. To produce drinking water the RO permeate of HMK is mixed with treated dune filtrate (Kamp, 1995, 2000). Figure 12.1 shows the treatment scheme and Figure 12.2A a picture of the RO installation. The objectives of the RO are desalination, softening, disinfection and organic contaminant control (e.g. pesticides etc.). The RO has a two-stage configuration with pressure vessels containing 7 elements each. The total amount of RO elements is 2016. The permeate production capacity is $1{,}800\,m^3\,h^{-1}$. Antiscalant dosage ($2.3\,mg\,L^{-1}$) to the RO feed water enables a RO permeate recovery of 0.82. Selected water quality parameters during several stages of water treatment are shown in Table 12.5. Monitoring studies described in this paper were performed at HMK at ambient water temperature (17–18°C).

The research facility of PWN in Andijk (ADK) has the same water treatment concept as HMK (Figure 12.1) with coagulation/sedimentation/filtration pretreatment, excluding biological activated carbon filtration.

Table 12.1 Experimental conditions for the studies on phosphate limitation to control biofouling and duration of study

Study	**Location**	**Monitor**	**Dosages#**			**Duration of study (days)**
	In installation*	**Study**	**Phosphate**	**Substrate**	**Antiscalant**	
RO installations: evaluation of practical data		no	no	no	no	
Proof of principle effect phosphate limitation	1	yes	yes/no	yes/no	no	7.7
Comparison of antiscalants	1	yes	yes/no	yes	yes/no	14.8
Growth limiting conditions in full-scale RO installation	1,2,3	yes	yes/no	yes/no	no	10.9

* = *coded locations are shown in Figure 12.1;* # = *dosages supplemented to the feed water of the monitor only.*

Table 12.2 Experimental conditions for study on the proof of principle and effect anti-scalant (AS) dosage. Feed water: ultrafiltrate before anti-scalant dosage (location 1 in Figure 12.1)

Study	MFS code*	Phosphate (μg P L^{-1})	Organic substrate (μg C L^{-1})	Nitrate (μg N L^{-1})	Anti-scalant (mg L^{-1})	Linear flow velocity (m s^{-1})
Proof of principle	blank	–	–	–	–	0.16
	P + S	20	200	40	–	0.16
	P	20	–	40	–	0.16
	S	–	200	40	–	0.16
Effect anti-scalant	S	–	200	40	–	0.16
	P + S	20	200	40	–	0.16
	AS1 + S#	–	200	40	2.5	0.16
	AS2 + S#	–	200	40	2.5	0.16
	AS3 + Sx	–	200	40	2.5	0.16
	AS4 + Sx	–	200	40	2.5	0.16

**Blank = no dosages; P + S = phosphate and substrate dosage; S = substrate dosage; P = phosphate dosage; AS = antiscalant dosage. # = AS1 and AS2 are phosphonate based anti-scalants. x = AS3 and AS4 are non-phosphonate based. – = no dosage.*

Table 12.3 Antiscalant coding and composition

Code	General composition	pH undiluted AS
AS1	phosphonate (sodium-l-hydroxyethylidene-l, l-diphosphonate – HEDP)	10.9
AS2	phosphonate (amino-tris-methylene-phosphonate) in sodiumhydroxide solution	11.0
AS3	carboxymethylinuline (derivate of Chicory root), no phosphate derivates	11.0
AS4	dendripolymer, no phosphates or phosphate derivates	4.4

Table 12.4 Experimental conditions for several stages of the membrane installation. Feed water: ultrafiltrate without and with antiscalant dosage and RO concentrate (locations 1, 2 and 3 in Figure 12.1) supplemented without or with chemicals as indicated in table

Water	MFS code	Phosphate (μg P L^{-1})	Organic substrate (μg C L^{-1})	Nitrate (μg N L^{-1})	Anti-scalant (mg L^{-1})	Linear flow velocity (m s^{-1})
RO feed water before AS dosage	B1	–	–	–	–	0.16
	S1	–	200	40	–	0.16
RO feed water with AS	AS2	–	–	–	2.3[a]	0.16
	AS + S2	–	200	40	2.3[a]	0.16
	AS + P + S2	20	200	40	2.3[a]	0.16
RO concentrate with AS	AS3	–	–	–	12.8[b]	0.16
	AS + S3	–	200	40	12.8[b]	0.16
	AS + P3	20	–	40	12.8[b]	0.16

[a] = *antiscalant dosage applied in the full-scale installation, antiscalant was not dosed to the MFS feed water in this study.* [b] = *antiscalant concentration calculated based on recovery RO installation 82%.*

Membrane fouling simulator

In all experiments, a Membrane Fouling Simulator (MFS) with external dimensions of 0.07 m × 0.30 m × 0.04 m was used (Vrouwenvelder *et al.*, 2006b). Coupons of feed spacer, membrane and product spacer can be placed in the MFS resulting in the same spatial dimensions as in spiral wound membrane elements. The development of fouling was monitored by measuring the pressure drop increase over the feed spacer channel of the MFS, photographic observations using the sight glass and analysis of sheets of membrane and spacer taken from the monitor. During operation, the glass window was covered with a light tight lid to prevent growth of phototrophic organisms.

In the MFS research, membranes and spacer sheets were taken from virgin spiral wound membrane elements. The feed spacer taken from the element was a 31 mil thick diamond-shaped polypropylene spacer with porosity ~0.85. The feed spacers in the MFS had the same spatial orientation as in spiral-wound membrane elements (45° rotation).

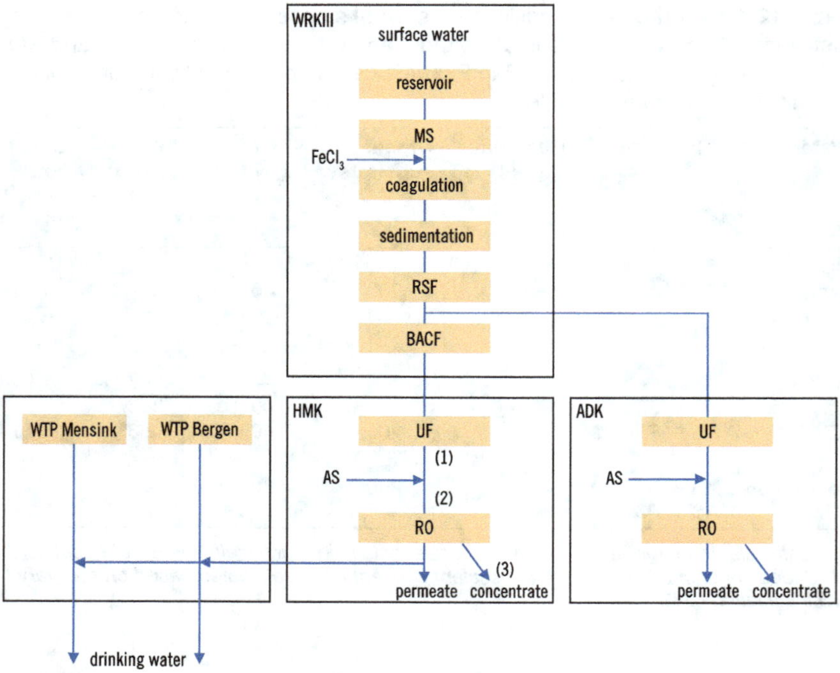

Figure 12.1 Plant treatment scheme with study locations (1), (2) and (3) at HMK. Preatreatment is performed at site WRKIII (lake IJssel) and UF/RO installations are in operation in Heemskerk (HMK) and Andijk (ADK). The subsequent treatments steps are reservoir storage, microstraining (MS), coagulation, sedimentation, rapid sand filtration (RSF), biological activated carbon filtration (BACF), ultrafiltration (UF) and reverse osmosis (RO). Iron (III) chloride ($FeCl_3$) is dosed as coagulant. Antiscalant (AS) is dosed to prevent scaling of RO. HMK permeate is mixed with dune filtrate from water treatment plants (WTP) Mensink and Bergen to produce drinking water

The installation consisted of two pressure reducing valves, manometer, dosage point (for biodegradable compounds), MFS and flow controller (Vrouwenvelder *et al.*, 2007a). The set-up with several MFSs during the studies at HMK is shown in Figure 12.2B and 12.2C. The MFSs were operated at a pressure of 120 kPa to avoid degassing. Unless mentioned otherwise, the feed water flow was $16\,L\,h^{-1}$ equal to a linear flow velocity of $0.16\,m\,s^{-1}$, representative for practice (Vrouwenvelder *et al.*, 2009a).

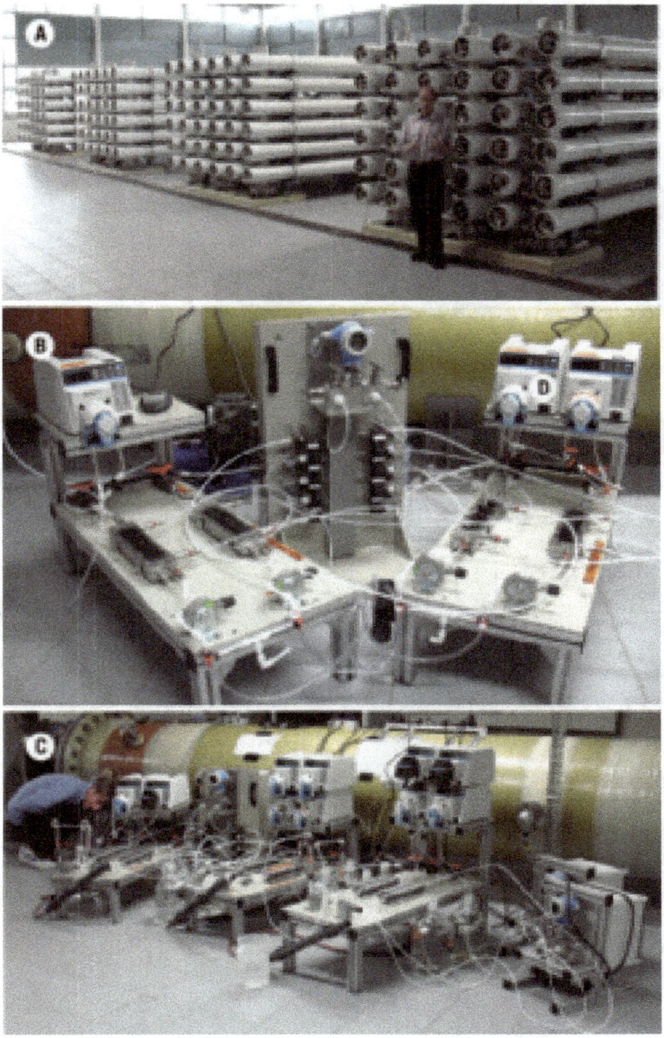

Figure 12.2 RO installation (**A**) and experimental set-up consisting of in parallel installed monitors and differential pressure transmitter during the proof of principle study (**B**) and antiscalant dosage (**C**). Pressure drop data were accessed using a modem. The horizontal tube on the background of (B) and (C) is the RO feed water pipe (Ø 1.0 m)

Table 12.5 Water quality parameters (average ± standard deviation) in the research period

Parameter	Unit	Raw water[a]	After RSF[b]	UF feed	RO feed	RO concentrate	RO permeate
conductivity	$mS\ m^{-1}$	60 ± 3	71 ± 1	nd	71 ± 3	336 ± 18	2.6 ± 0.2
suspended solids	$mg\ L^{-1}$	21 ± 20	<0.2	<0.2	nd	<0.2	nd
TOC	$mg\ C\ L^{-1}$	7.0 ± 1.5	4.1 ± 0.7	2.9 ± 0.3	2.9 ± 0.2	nd	<0.2
total phosphate	$mg\ P\ L^{-1}$	0.03 ± 0.01[c]	nd	nd	nd	nd	nd
ortho-phosphate	$mg\ P\ L^{-1}$	<0.02	nd	nd	nd	nd	nd

[a]*Raw water = intake point lake IJssel (before reservoir);* [b]*RSF = rapid sand filter;* [c] *= total phosphate values in the raw water ranged between 0.03 and 0.05 mg P L^{-1}. After reservoir storage the values ranged between 0.02 and 0.04 mg P L^{-1}; nd = not determined.*

The permeate flux – water volume passing the membrane per unit area and time – as applied in practice in spiral wound NF and RO membranes has been shown irrelevant for biofouling development in extensively pretreated water (Vrouwenvelder *et al.*, 2009b,c). Therefore, the MFS was operated cross-flow, without permeate production.

Chemicals were dosed to the MFS feed water to determine the effect on biomass growth. From a sterile vessel, containing a 5 L solution of concentrated substrate, substrate was dosed into the feed water prior to the MFS by a peristaltic pump (Masterflex) at a flow of $0.03\,L\,h^{-1}$. The dosage of substrate was checked periodically by measuring the weight of the dosing bottle. The chemicals $NaCH_3COO$, $NaNO_3$ and NaH_2PO_4 were dosed in a mass ratio C:N: P of 100:20:10 (Table 12.2). Dosed P was a factor 2.3 higher than the biomass mass ratio (~100:23:4.3, Tchobanoglous *et al.*, 2003) low enough to avoid growth by P-limitation, because P is used in other processes as well (Gagnon *et al.*, 2004 and Vrede *et al.*, 2002). The substrates were dissolved in ultrapure water. To restrict bacterial growth in the substrate dosage bottle, the pH was set at 10.5 by dosing NaOH. Stock solution bottles were replaced every 5 days.

Antiscalant was dosed from separate bottles (Table 12.2). The antiscalants were dissolved in ultrapure water. The pH of AS2 and AS3 was set at 11 by dosing NaOH. The chemical and antiscalant dosage flow rate (both $0.03\,L\,h^{-1}$) was low compared to the feed water flow rate ($16\,L\,h^{-1}$). Thus, the effect of the chemical dosage on pH of the feed water was insignificant.

Pressure drop

Pressure drop measurements were performed with a pressure difference transmitter (Endress & Hauser, type Deltabar S: PMD70-AAA7FKYAAA). The calibrated measuring range was 0 to 50 kPa (Vrouwenvelder *et al.*, 2009d). The pressure drop transmitter was combined with a valve system developed to enable measurement and logging the pressure drop over several monitors simultaneously (Figure 12.2B, centre). Pressure drop and water temperature data could be remotely accessed using Internet.

Membrane autopsy from elements and MFSs

Membrane elements from HMK and ADK were autopsied in 2001 when no feed channel pressure drop increase was observed. MFSs operated in parallel were autopsied at the same time. For both membrane elements and MFSs, the same sampling methods and analyses were performed. Analysis performed on stacked membrane and spacer sheets were total active biomass (ATP) and total organic carbon

(TOC). Details of the methods have been described elsewhere (Vrouwenvelder *et al.*, 2006b, 2008a). At the time of membrane autopsies, chemical dosage to the RO feed water consisted of pure hydrochloric acid and 0.9 $mg L^{-1}$ antiscalant (Galjaard *et al.*, 2008). The monitor studies and water quality parameters described in this paper were performed from early 2008 until July 2009 when only 2.3 $mg L^{-1}$ antiscalant was dosed into the feed water without a hydrochloric acid dosage.

RESULTS

Full-scale RO investigations

Usually, biofouling is observed at the feed side of the membrane installation (Carnahan *et al.*, 1995 and Vrouwenvelder *et al.*, 2008b, 2009c,d). Destructive studies on membrane elements taken from the feed side of the installation, have been performed to determine the cause(s) of the increased feed channel pressure drop at several full-scale installations. Investigations at most full-scale RO and NF installations suffering from an elevated feed channel pressure drop showed high biomass concentrations in the lead elements at the installation feed side (Table 12.6: A–F), supporting previous studies (Carnahan *et al.*, 1995 and Vrouwenvelder *et al.*, 2008b, 2009c,d).

Table 12.6 Biofilm concentrations (pg ATP cm^{-2}) in the lead element at the installation feed side and feed channel pressure drop increase for 8 installations

Membrane installation code	Biomass concentration (pg ATP cm^{-2}) lead element at installation feed side	Feed channel pressure drop increase (%)
A	46.000	>15
B	39.000	>15
C	31.000	>15
D	23.600	>15
E	15.600	>15
F	14.500	>15
ADK	27	<15
HMK	165	<15

Unusual observations were made at two RO membrane installations (a pilot and full-scale) with extensive pretreatment. Together with a constant feed channel pressure drop the biomass concentrations in the lead element were low compared to the lead elements from other installations (Table 12.6: *ADK, HMK*). The installations ADK and HMK applied extensive pretreatment removing biodegradable substrate to a great extent. When substrate would be the growth limiting factor than growth is expected especially in the lead elements. It was hypothesized that phosphate limitation may be the cause of the unique observation.

'Proof of principle' phosphate limitation

To determine whether low phosphate concentrations can be the reason for low biomass concentrations a study was performed at HMK with four monitors in parallel. The MFSs were operated under identical conditions except for a different supplementation to the feed water of phosphate, organic carbon and nitrate (Table 12.2). The feed water was ultrafiltrate (Figure 12.1 code 1).

Dosage of both phosphate and substrate (P + S; S = C + N) to the feed water caused a rapid increase of pressure drop over the monitor (Figure 12.3A and 12.3B). Within the research period, no pressure drop increase was observed in the monitors fed with water supplied with phosphate only (P), substrate only (S) or the blank without any dosage. After 7.7 days, all monitors were opened for visual observations and for measurements of the biomass parameters ATP and TOC. Biomass accumulation could only be seen in the monitor supplied with phosphate and substrate (P + S). This was confirmed by analysis of accumulated material in the monitors (Figure 12.3C and 12.3D). The biomass parameters ATP and TOC showed the same trend. The pressure drop development, visual observations and accumulated biomass analysis all unambiguously indicated that fouling accumulation occurred in the monitors supplied with both substrate and phosphate. The low phosphate concentration restricted substrate use for biomass growth (Figure 12.3, see S). Dosage of phosphate only caused no additional growth compared to substrate only (Figure 12.3), indicating that both substrate and phosphate concentrations were low in the ultrafiltrate of this installation.

This 'proof of principle' experiment clearly shows that phosphate limitation can restrict biomass growth. A high substrate concentration in the feed water was not causing biofouling in the monitor under phosphate limiting conditions, suggesting that substrate removal during pretreatment (current practice) to control biofouling can be avoided when the phosphate concentration is very low. In other words, when the phosphate concentration is very low or pretreatment removes phosphate to very low concentrations, biofouling can be restricted even in the presence of high substrate concentration.

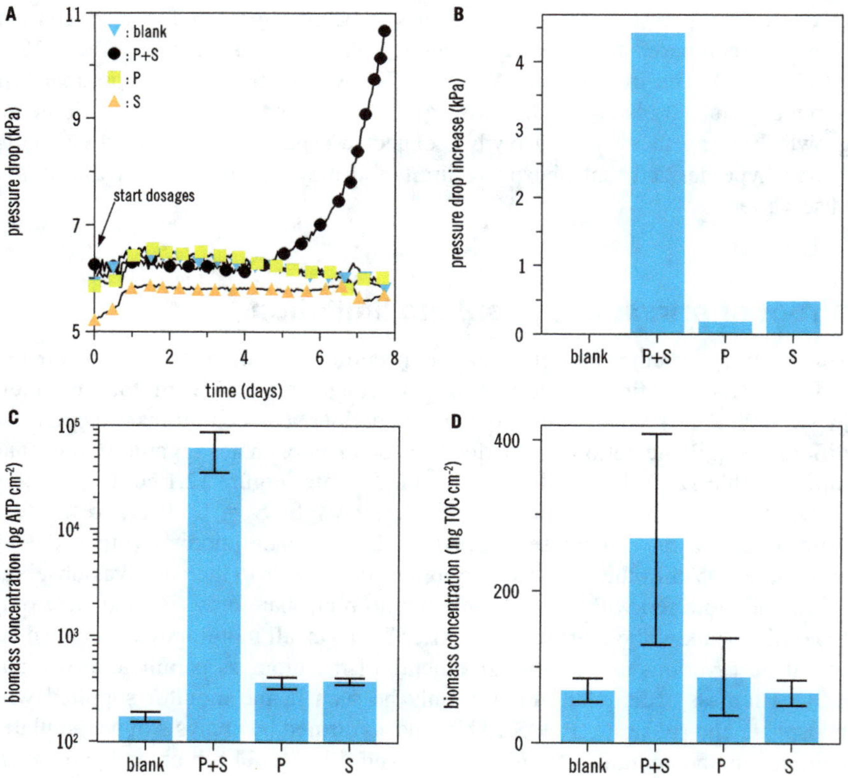

Figure 12.3 Pressure drop in time (**A**), pressure drop increase (**B**), biomass concentration ATP (**C**) and TOC (**D**) after 7.7 days MFS operation fed with water supplied with phosphate and substrate (P + S), phosphate (P), substrate (S) and without dosages (blank)

Comparison of antiscalants

The influence of four antiscalants on biofouling was studied with monitors fed with ultrafiltrate supplemented with substrate (S) and two blank measurements (Tables 12.3 and 12.4). A high pressure drop increase was observed with phosphonate based antiscalants (AS1 + S and AS2 + S) and with phosphate dosage (P + S) (Figure 12.4A and 12.4B). The phosphate-free antiscalants (AS3 + S and AS4 + S) and substrate only (S) showed a low pressure drop increase. Visual observations and biomass concentrations (Figure 12.4C and 12.4D) showed severe fouling of the monitors with phosphonate based antiscalants and phosphate, supporting the feed channel pressure drop increase data (Figure 12.4A and 12.4B). The phosphonate based antiscalants (AS1 + S and AS2 + S) showed the same or higher values for pressure drop increase and biomass concentrations compared to phosphate (P + S). The phosphonate based antiscalants increased besides phosphate also the substrate level but it is believed the main effect was caused by the phosphate in the antiscalant. Further studies were performed with a phosphonate based antiscalant.

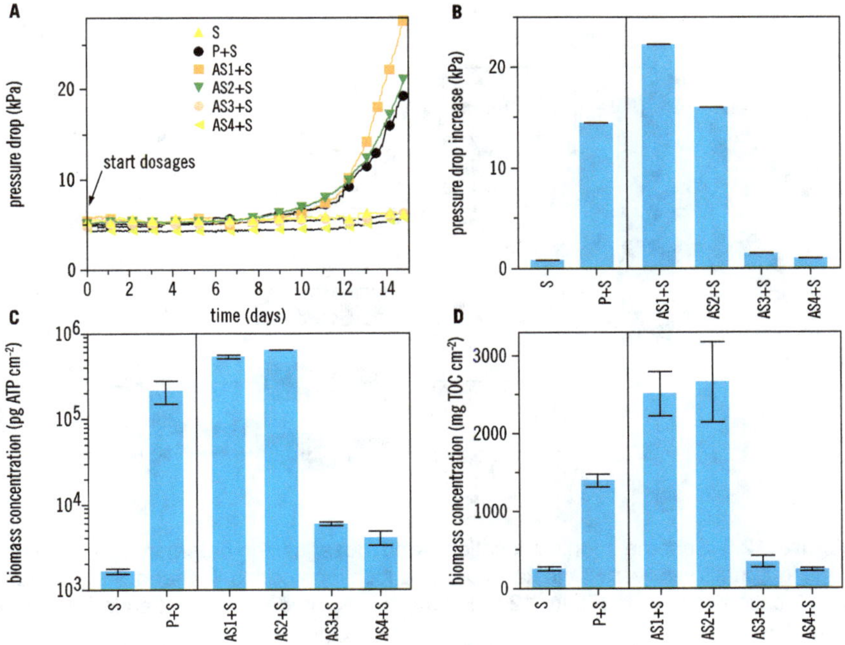

Figure 12.4 Pressure drop in time (A) and pressure drop increase (B) and biomass concentration ATP (C) and TOC (D) after 15 days MFS operation fed with water supplied with substrate (S), phosphate and substrate (P + S), and four antiscalants with substrate (AS1 + S, AS2 + S, AS3 + S, AS4 + S)

Growth limiting conditions in RO installation

To determine the growth limiting conditions in the RO installation (Table 12.6, HMK) studies were performed at several stages of the RO installation (Figures 12.5 and 12.1) using monitors with varying dosages (Table 12.4) and a phosphonate based antiscalant.

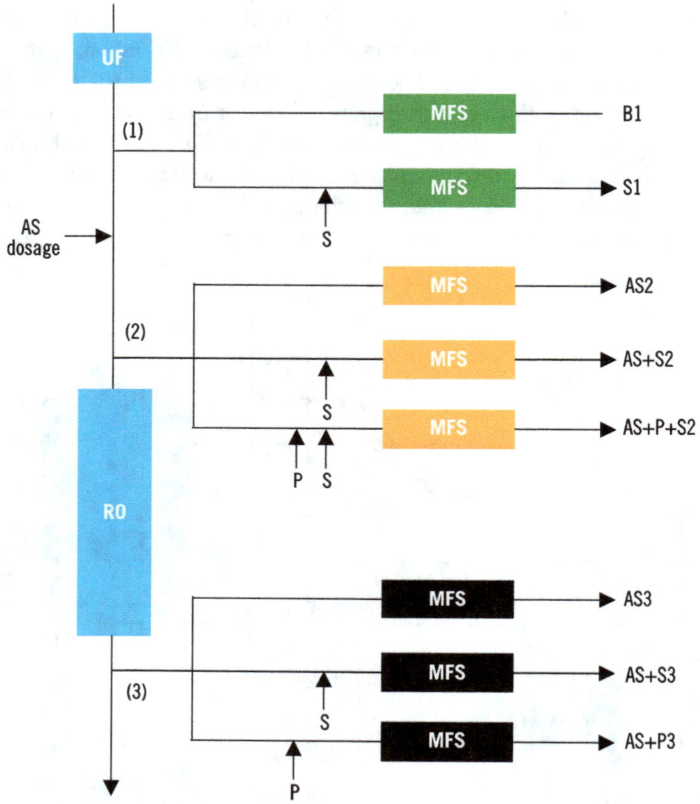

Figure 12.5 Scheme monitor locations and dosages for determining phosphate limiting conditions over RO installation. The locations codes 1, 2 and 3 correspond with locations shown in Figure 12.1 and the monitor codes are clarified in Table 12.4

A pressure drop increase was observed in the monitors fed with feed RO stage 1 and RO concentrate supplemented with substrate (AS + S2, AS + S3) and phosphate and substrate (AS + P + S2) (Figure 12.6A). No pressure drop increase was observed (i) without dosages to the ultrafiltrate (B1), feed RO stage 1 (AS2) and RO concentrate (AS3), (ii) with dosage of substrate only to the ultrafiltrate (S1), and (iii) with dosage of phosphate only to the RO concentrate (AS + P3). Visual observations (Figure 12.7) and analysis of the biomass parameters ATP and TOC (Figure 12.6C and 12.6D) supported the findings from the pressure drop measurements. In other words, a pressure drop increase, visual fouling and high biomass concentrations were found in monitors AS + S2, AS + S3 and AS + P + S2 only.

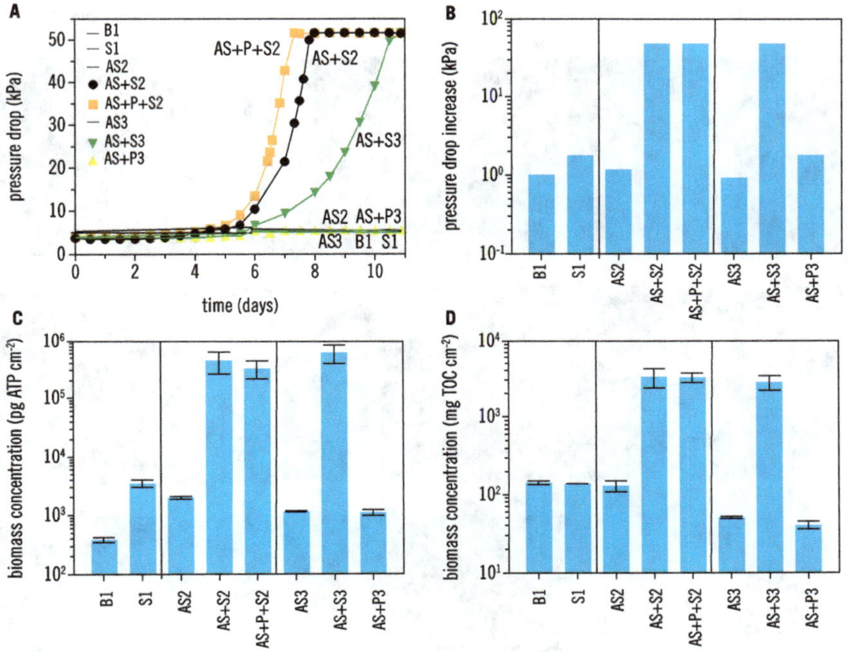

Figure 12.6 Pressure drop in time (**A**), pressure drop increase (**B**) and biomass concentration ATP (**C**) and TOC (**D**) after 11 days MFS operation fed with water from several stages of the RO installation supplied with substrate (S1, AS + S2, AS + S3), without substrate (B1, AS2, AS3) and with phosphate and substrate (AS + P + S2) and phosphate (AS + P3). Ultrafiltrate was the feed of B1 and S1. RO feed including AS was the feed of AS2, AS + S2, AS + P + S2. RO concentrate was the feed of AS3, AS + S3, AS + P3

Substrate dosage caused biomass accumulation in the RO feed water containing antiscalant (AS + S2) while no biomass accumulation was observed withouth antiscalant supplemention (S1), indicating that the antiscalant applied in practice contained microbially available phosphorous. Similar biomass concentrations were found in monitors AS + P + S2 and AS + S2 (Figure 12.6C and 12.6D), indicating that the antiscalant dosage ($2.3\,mg\,L^{-1}$) provided sufficient phosphate for extensive biofilm growth in the presence of substrate.

After antiscalant dosage to the RO installation feed water, phosphate was not longer the growth limiting compound. So, in this case the low substrate concentrations achieved by the extensive pretreatment restricted biomass accumulation in the RO elements.

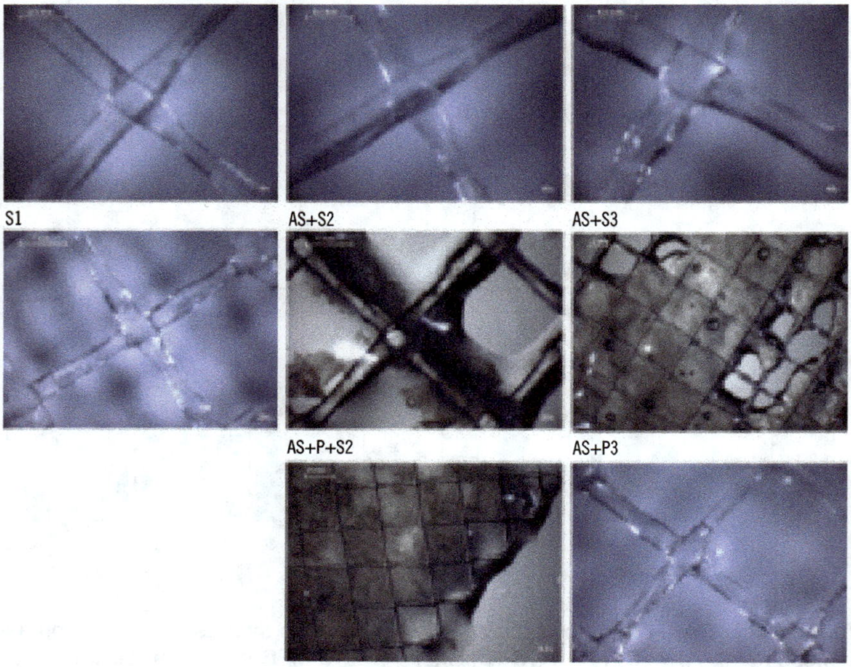

Figure 12.7 Visual observations of spacers from monitors used in this study at several locations in the RO installation (Figure 12.6). Biomass accumulation was observed in monitors coded AS + S2, AS + P + S2 and AS + S3 only. A larger magnification was used for AS + S3 and AS + P + S2 to show biofilm accumulation

Low phosphate concentrations during water treatment

The method for phosphate measurements applied during regular water quality monitoring program is not sensitive enough (detection limit of $0.02\,\mathrm{mg\,P\,L^{-1}}$) to provide detectable phosphate concentrations, since the phosphate concentrations in the raw water is much lower than $0.02\,\mathrm{mg\,P\,L^{-1}}$ (Table 12.5). Jacobson *et al.* (2008) reported for HMK distinguishable accurate quantitative phosphate concentrations in the range from <0.2 to $7.1\,\mu\mathrm{g\,P\,L^{-1}}$ (Figure 12.8). UF permeate, before antiscalant dosage, had a lower phosphate concentration than the RO feed including antiscalant dosage (Figure 12.8, Jacobson *et al.*, 2008). Apparently, the phosphonate based antiscalant increased the phosphate availability. The consequence of water permeation is that the phosphate concentration increases during passage of membrane modules in the pressure vessel. At a plant recovery of 80%, the phosphate concentration will be a factor five higher in the module at the installation concentrate side compared to the feed side.

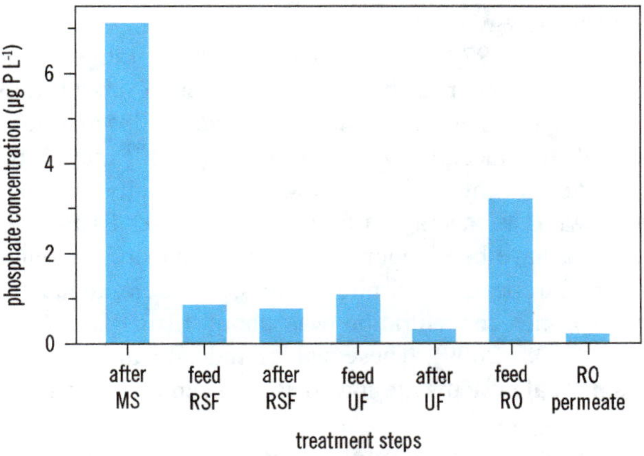

Figure 12.8 Phosphate concentration in water over several treatment steps of HMK (adapted from Jacobson *et al.*, 2008). For explanation of abbreviations see subscript Figure 12.1

DISCUSSION

Biofouling control

RO feed water with low phosphate concentrations showed a very restricted biomass accumulation in MFSs, even at high organic carbon substrate concentrations (Figure 12.3). Characklis and Marshall (1990) stated that 'nutrient limitation, aside from organic carbon, may limit microbial fouling accumulation. In such systems, chemical additives (e.g. corrosion and scale inhibitors) must be screened carefully lest they increase the biofouling'. Use of phosphonate based antiscalants increased the phosphate concentration causing biofouling in the presence of substrate (Figures 12.4 and 12.6). Antiscalant dosage may potentially increase the phosphate and substrate concentration (Figure 12.4). Therefore, antiscalant selection and a biofouling control strategy are linked actions. A previous study showed that antiscalants can contribute to biofouling (Vrouwenvelder *et al.*, 2000). However, the growth potential in that study was based on substrate limiting conditions only. For biofouling control under phosphate limiting conditions, an antiscalant with a high substrate growth potential but without any phosphate may be suitable for biofouling control while an antiscalant with a low substrate growth potential but containing some phosphate may be not. In other words, the requirements for antiscalant selection and dosage not contributing to biofouling are an important part of the overall biofouling control strategy.

Sathisvan *et al.* (1997, 1999) reported for drinking water distribution networks that phosphorous may become the microbial growth limiting nutrient at concentrations of 1 to 3 $\mu g\,P\,L^{-1}$. Jacobsen *et al.* (2008) mentioned that in RO feed water phosphate concentrations below 0.45 $\mu g\,P\,L^{-1}$ may lead to biofouling control. Phosphate concentrations in seawater are usually less than 33 $\mu g\,P\,L^{-1}$ (Garcia, 2006) with an average of 20 $\mu g\,P\,L^{-1}$ (OZ REEF, 2007). Similar values for seawater have been reported by the European Environmental Agency (EEA, 2007). In European rivers phosphate levels were about 45 $\mu g\,P\,L^{-1}$ and in European lakes the concentration was about 13 $\mu g\,P\,L^{-1}$ (EEA, 2005 and Jacobson *et al.*, 2008, 2009). These values indicate that although phosphate concentrations natural systems are low, these concentrations are not low enough to prevent microbial growth.

Phosphate concentrations are often already low in the feed of high pressure membrane systems by coagulation/sedimentation/filtration pretreatment (Figure 12.8, Jacobson *et al.*, 2008, 2009). This means that only a small additional phosphate removal is required for restricting biofilm growth in membrane elements. How low phosphate concentrations in the feed water should be to control biofouling is not clear yet. In membrane systems, the phosphate concentration increases as a consequence of (i) concentration polarization causing elevated phosphate concentrations near the membrane and (ii) the plant recovery increasing the phosphate concentration during passage of the installation (e.g. at 80% plant recovery, the phosphate concentration will be a factor five higher in the module at the installation concentrate side compared to the feed side).

Recently, for sustainable water reclamation and reuse desalination of waste water effluents has been introduced and has got more attention (Katz and Dosoretz, 2008). However, phosphate scaling was found to be the main obstacle for effluent desalination using RO (Katz and Dosoretz, 2008). Therefore, lowering phosphate concentrations in the feed water may therefore be an effective option to prevent biofouling and phosphate scaling facilitating higher water recoveries. An accurate and sensitive analytical method for phosphate is available to evaluate and monitor sufficiently low phosphate concentrations (Jacobson *et al.*, 2009).

Phosphate concentrations can already be reduced to low concentrations by conventional treatment processes (including coagulation). Often an overdosing of chemicals is needed due to the presence of humic substances. This leads to an increased salt load on the water. Thermostabile ferritin nanoparticles might present an alternative for conventional phosphate removal. Ferritin is an enzyme that can bind phosphate and iron. Advantage of ferritin is a high and selective phosphate removal rate even at very low concentrations. A feasibility study on ferritin use in practice for large scale water treatment showed that the costs are similar to the costs of currently applied chemical methods (Jacobs *et al.*, 2007).

Follow up

Phosphonate based antiscalants may increase phosphate levels (Figure 12.4). The origin of phosphate can be degrading of phosphonates, a by-product of the antiscalant production process or both. Studies are needed to elucidate the cause of phosphate availability for growth in phosphonate based antiscalants.

The influence of concentration polarization and plant recovery on phosphate limitation and plant performance needs study. Usually, biofouling is observed at the feed side of the membrane installation (Carnahan *et al.*, 1995 and Vrouwenvelder *et al.*, 2008b, 2009c,d). Because of the high linear flow velocity, biofouling of the lead module has a higher impact on plant performance than biofouling of modules on other locations in the installation (Vrouwenvelder *et al.*, 2009a, d). The phosphate concentration in the water increases during passage of the modules in the membrane filtration installation, which may lead to elevated biomass concentrations at the concentrate side, but with a low impact on the feed channel pressure drop, since the linear velocity is low (Vrouwenvelder *et al.*, 2009a).

Long term studies under phosphate limiting conditions are required. Phosphate limiting growth conditions to restrict biomass accumulation may have great implications for pretreatment and antiscalant selection. At present in practice, pretreatment aims at reducing substrate and most antiscalants are phosphonate based. The pretreatment for extensive substrate removal may not be needed when phosphate limiting conditions are achieved. Feasibility studies and cost balancing will elucidate the potential of phosphate limitation to control biofouling.

SUMMARY

Phosphate limitation as a method to control biofouling of spiral wound reverse osmosis (RO) membranes was studied at a full-scale installation fed with extensively pretreated water. The RO installation is characterized by (i) a low feed channel pressure drop increase and (ii) low biomass concentrations in membrane elements at the installation feed side. This installation contrasted sharply with installations fed with less extensively pretreated feed water (and therefore higher phosphate concentrations) experiencing a high pressure drop increase and high biomass concentrations in lead elements. Membrane fouling simulator (MFS) studies showed that low phosphate concentrations ($\sim 0.3\,\mu g\,P\,L^{-1}$) in the feed water restricted the pressure drop increase and biomass accumulation, even at high substrate (organic carbon) concentrations. In the MFS under ortho-phosphate limiting conditions, dosing phosphonate based antiscalants caused biofouling while no biofouling was observed when acids or phosphonate-free antiscalants were used. Antiscalant dosage could increase both phosphate and substrate concentrations of the water. Therefore, antiscalant selection may be critical for biofouling control. Since no biofouling was observed at low phosphate concentrations, restricting biomass growth by phosphate limitation may be a feasible approach to control biofouling, even in the presence of high organic carbon levels.

Future Perspectives

13

Biofouling Control: Potential Complimentary Perspectives

INTRODUCTION

High pressure membrane filtration processes such as nanofiltration (NF) and reverse osmosis (RO) produce water of high quality. A drawback of NF and RO applications is membrane fouling, resulting in a pressure drop increase over membrane modules in the membrane filtration installation causing technical problems and increasing operation cost. Biofouling (Figure 13.1) is the major type of fouling in NF and RO membranes fed with extensively pretreated water, caused by biofilm formation in membrane elements (Ridgway and Flemming, 1996; Patching and Fleming, 2003 and Shannon *et al.*, 2008). Numerous authors describe biofouling problems in membrane installations (Ridgway *et al.*, 1983, 1985; Flemming, 1993; Tasaka *et al.*, 1994; Ridgway and Flemming, 1996; Baker and Dudley, 1998; Van Hoof *et al.*, 2002; Schneider *et al.*, 2005 and Karime *et al.*, 2008). In the Middle East, about 70% of the seawater RO membrane installations suffer from biofouling problems (Gamal Khedr, 2000, 2002).

Over the years, strategies to control biofouling have not always been successful. In the late 1990s, two strategies were pursued to prevent and control membrane biofouling: (i) physical removal of bacteria from the feed water of membrane systems (for example by microfiltration or ultrafiltration pretreatment), and (ii) metabolic inactivation of bacteria by applying biocide dosage or UV irradiation (Ridgway, 1997). At present, the focus is on nutrient removal by biological pretreatment (e.g. sand filtration) and modification of membranes (disinfectant resistant and low fouling). In addition, membrane cleanings are

*This chapter is based, with permission from the copyright holder, on a paper previously published in Water Science & Technology Vol. 62 No. 11 pp. 2477–2490 doi: 10.2166/wst.2010.747

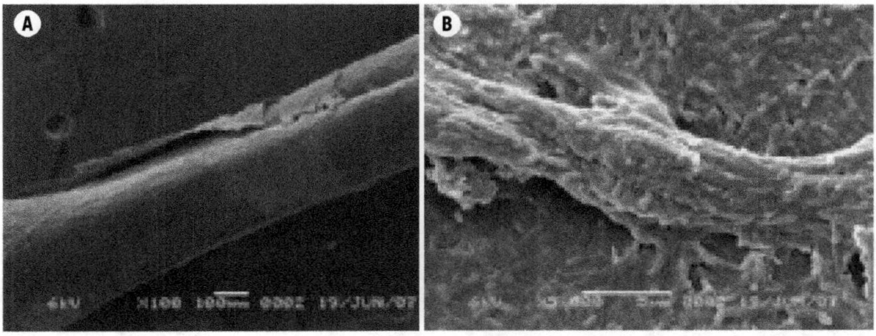

Figure 13.1 Scanning electron microscope images of feed spacer strand with biofilm **(A)** and biofilm on feed spacer **(B)** taken from a biofouled membrane module. B shows individual microorganisms embedded in – by microorganisms produced and excreted – extracellular polymer substances. The bars indicate the figurescale

commonly applied. Current cleaning strategies i.e. using cleaning chemicals are not very effective (Whittaker *et al.*, 1984; Baker and Dudley, 1998; Al-Amoudi and Lovitt, 2007; Zondervan and Roffel, 2007 and Creber *et al.*, 2010).

A rational approach is required to solve biofouling problems. Therefore, gaining insight in the biofouling process is a first priority.

PROBLEM ANALYSIS

Figure 13.2 shows the evolution of the annual number of publications satisfying the search criteria 'biofouling' and 'membrane' in the Scopus database, March 2010. Since the first paper was published in 1983 in total 2941 papers appeared until 2010, from which about 55% appeared within the last four years (2006–2009). The increasing number of papers illustrate that membrane biofouling is a major problem.

Ridgway (1998) summarized the effects of biofouling on RO membrane performance: (i) membrane flux decline, (ii) increase in trans membrane and differential pressure and (iii) permeability increase of dissolved material (decreased mineral rejection). Dissolved minerals may accumulate in the biofilm layer, increasing concentration polarization, thereby reducing the flux and rejection properties of the membrane. Characklis and Marshall (1990) (Figure 13.3) stated that biofouling causes energy losses due to increased fluid frictional resistance, indicating that in spiral-wound membrane systems biofouling may cause a pressure drop increase.

In spiral-wound membrane modules, two types of pressure drop can be distinguished: the trans-membrane pressure drop (TMP) and the feed channel pressure drop (FCP), the pressure drop between feed and concentrate lines

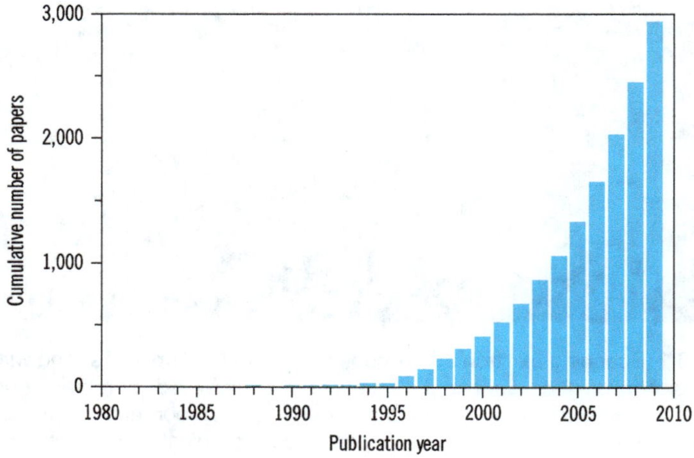

Figure 13.2 Cumulative number of publications on membrane biofouling in time

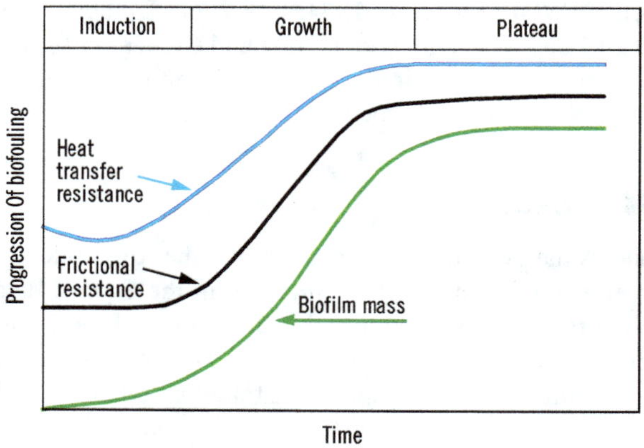

Figure 13.3 Progression of biofouling is generally described by a sigmoid function when expressed in terms of deposit accumulation, hydrodynamic frictional resistance, or heat resistance. Three phases can be arbitrarily defined: (1) the induction period, (2) the log growth rate period and (3) the plateau period. (adapted from Characklis and Marshall, 1990)

(Flemming *et al.*, 1994). The TMP is the differential pressure between feed and permeate lines, caused by the frictional resistance over the membrane. When the TMP is increased by biofouling, the membrane flux is declined.

Biofouling in spiral-wound nanofiltration and reverse osmosis membranes has been studied on monitor, test-rig, pilot and full-scale for extensively pretreated water (Vrouwenvelder *et al.*, 2009b,c,d). Identical behaviour with respect to biofouling and FCP development was observed in membrane elements in the same position in a NF installation operated with and without flux. Irrespective whether a flux was applied or not, the FCP and biofilm concentration increased. Calculations on mass transfer aspects supported the observations that the flux is not playing a significant role in substrate supply to the fouling layer. Also, test-rig and full-scale studies with different types of feed water showed that biofouling of membrane modules correlated very well with FCP-increase (Vrouwenvelder *et al.*, 2008b). Moreover, in systems suffering from biofouling cleaning cycles are governed by the pressure drop over the feed channel. Therefore, biofouling is considered as a FCP problem (Vrouwenvelder *et al.*, 2009c).

Biofouling is dominantly localized in the lead membrane module of the first stage in membrane filtration installations (Carnahan *et al.*, 1995 and Vrouwenvelder *et al.*, 1998, 2006b, 2007b, 2008b, 2009c,d). Monitoring the FCP over individual membrane modules and stages of a NF installation revealed that the FCP increased mainly over the lead element of the first stage (Vrouwenvelder *et al.*, 2009d). Destructive studies on full-scale membrane modules (autopsy) showed most biomass on the inlet side of lead membrane modules. Biofouling studies using a flat sheet monitor confirmed that biofouling was localised on the inlet side of lead membrane modules (Vrouwenvelder *et al.*, 2007a).

FCP development and biofilm accumulation have been studied in membrane fouling simulators, as a function of substrate concentration, linear flow velocity, substrate load and flow direction (Vrouwenvelder *et al.*, 2009a). At the low substrate concentrations applied, biofilm accumulation was related to substrate load (the product of substrate concentration and flow velocity). Reduction of the linear flow velocity resulted in an instantaneous reduction of FCP-increase caused by accumulated biomass, without affecting the biofilm amount present. The results indicate that the FCP-increase should be the most important parameter for process control.

Magnetic resonance imaging (MRI) studies showed that even restricted biofilm accumulation on the feed channel spacer influenced the velocity distribution profile strongly (Graf von der Schulenburg *et al.*, 2008 and Vrouwenvelder *et al.*, 2009c). The feed spacer presence strongly influenced the FCP-increase caused by biofilm accumulation: in both spiral-wound NF and RO systems biofouling is dominantly a feed spacer problem (Vrouwenvelder *et al.*, 2009c).

EARLY DETECTION

Early detection of biofouling enables timely actions to control biofouling (Flemming *et al.*, 1997b, 2003). Early detection of biofouling requires monitoring of the differential pressure with a sensitive differential pressure transmitter over a (i) lead membrane module of a first stage and/or a (ii) membrane fouling simulator supplied with feed water of the membrane filtration installation. Early detection of biofouling can be realized by operating the membrane fouling simulator under a higher linear flow compared to the lead membrane modules of the first stage (Vrouwenvelder *et al.*, 2010a, Figure 13.4). Monitoring the FCP over the total installation or complete stages is not suitable for early biofouling detection.

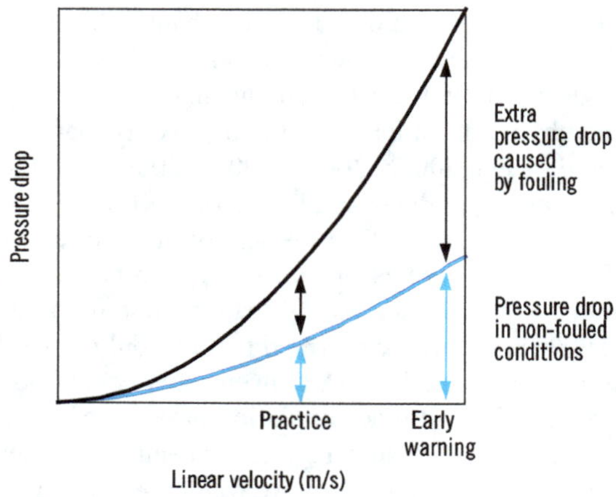

Figure 13.4 Pressure drop as function of the linear flow velocity, illustrating that the effect of biomass concentration on pressure drop increase is affected by linear flow velocity enabling early biofouling detection. The difference between the pressure drop in non-fouling conditions and pressure drop in fouling conditions is the pressure drop increase caused by accumulated fouling (adapted from Vrouwenvelder *et al.*, 2011)

BIOFOULING CONTROL

Strategy

In the past, biofouling control was pursued by microorganism removal, inactivation by pre-treatment and chemical dosages (Figure 13.5, (Ridgway, 1997)). In time, the approach shifted to substrate removal by pretreatment (Griebe and Flemming, 1998; Kruithof *et al.*, 1998; Kamp *et al.*, 2000; Ebrahim *et al.*, 2001; Visvanathan *et al.*, 2002; Wend *et al.*, 2003; Hu *et al.*, 2005 and Schneider *et al.*, 2005) and membrane modification (Wilf and Alt, 2000; Louie *et al.*, 2006 and Liu *et al.*, 2006). However, past and present strategies to control biofouling were not always successful to solve practical biofouling problems in spiral-wound NF and RO membranes (Bereschenko *et al.*, 2008; Chong *et al.*, 2008; Huertas *et al.*, 2008; Karime *et al.*, 2008; Shannon *et al.*, 2008 and Vrouwenvelder *et al.*, 2008b, etc).

Controlling biofouling may be achieved by chemical dosage to the feed water. Currently applied thin film composite NF and RO membranes are sensitive for free chlorine. Free chlorine damages the membrane structure causing decrease of membrane rejection. A limited number of plants apply successfully monochloramine in controlling biofouling. A much better membrane resistance to monochloramine compared to chlorine has been reported (DOW, 2009a): 300,000 ppm-hours for monochloramine compared to ~1,000 ppm-hours for free chlorine. Since monochloramine is formed by adding ammonia to chlorine, it is possible that free chlorine will be present (e.g. if the ammonia dosing is not correct or fails). Moreover, iron and manganese catalyzes membrane oxidation by monochloramine (Gabelich *et al.*, 2005 and Da Silva *et al.*, 2006). Another reason to avoid chloramination is the formation of N-nitrosodimethylamine (NDMA), a probable human carcinogen. The RO rejection capacities for NMDA are 10 to 50%. Recently, an alternative for monochloramine, 2,2-dibromo-3-nitrilopropionamide (DBNPA. DOW, 2005, 2009b and Bertheas *et al.*, 2009) is applied successfully in a limited number of plants.

A holistic Approach for Biofouling Control (ABC) is proposed to solve biofouling of high pressure membranes, symbolised by an umbrella shielding against biofouling (Figure 13.5). The ABC consists of three corner stones: (i) appropriate equipment design and operation, (ii) controlling biomass growth conditions, and (iii) application of cleaning agents. Combinations of approaches will in general be more efficient than a single approach. The proposed ABC methodology is developed to select the right methods for biofouling control. Although in this stage chemical cleaning and biofouling inhibitor dosages seem unavoidable, it is expected that in future – because of sustainability and costs reasons – adapted membrane systems will be operated without or with minimal chemical cleaning and dosages still avoiding biofouling (Figure 13.6). The ABC approach describes a trajectory to come to a comprehensive scenario for effective biofouling control.

Potential approaches

Proposed approaches to control biofouling based on our research efforts are nutrient removal by pre-treatment and module modification. Obviously, nutrient removal is essential to restrict and delay biofouling, but the effect of membrane module modification on biofouling is not clear yet. Extensive research efforts during the last decade into the modification of membrane properties have not resulted in solutions to control biofouling under all circumstances. Use of low fouling membranes did not cause an effective biofouling control (Figure 13.7). Our findings that biofouling in spiral-wound membrane modules is predominantly a feed spacer problem (Vrouwenvelder *et al.*, 2009c) may explain the ineffectiveness of membrane modification for biofouling control. Prevention of biofouling is clearly most successful by minimizing the availability of nutrients (similar to conditions that minimize regrowth of bacteria in drinking water lines without chlorination). Eventually, microbial growth will always happen, therefore it is recommended to design a system in such a way that biofilm formation doesn't lead to operational problems or that biomass can be easily removed. In the next sections of this paper several approaches to control biofouling are introduced.

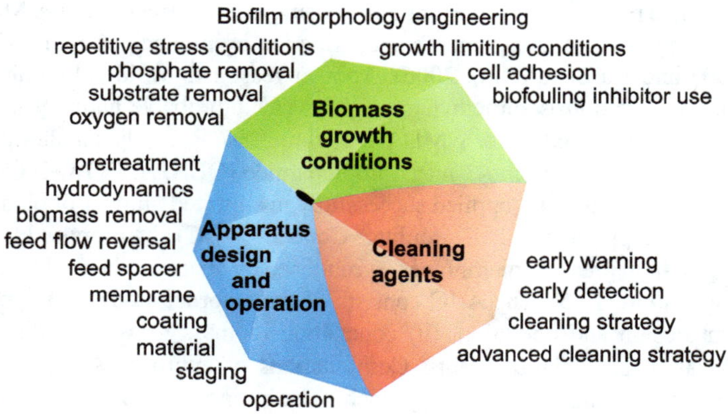

Figure 13.5 Approach for Biofouling Control (ABC) consisting of three corner stones: equipment design and operation, biomass growth conditions, and cleaning agents. Combination of actions to control biofouling is preferred over a single action. The ABC is symbolised by an umbrella shielding against biofouling

Cleaning strategies

The real problem related to biofouling is the FCP-increase caused by biofouling since this affects water production costs. Membrane manufacturers recommend corrective actions when the FCP-increase is 15% of the FCP start-up value determined under 'industrial' conditions and restrict guarantees when the pressure drop increase is higher. This percentage pressure drop increase criterion, used in practice, is not a well-defined guideline to take corrective actions. The

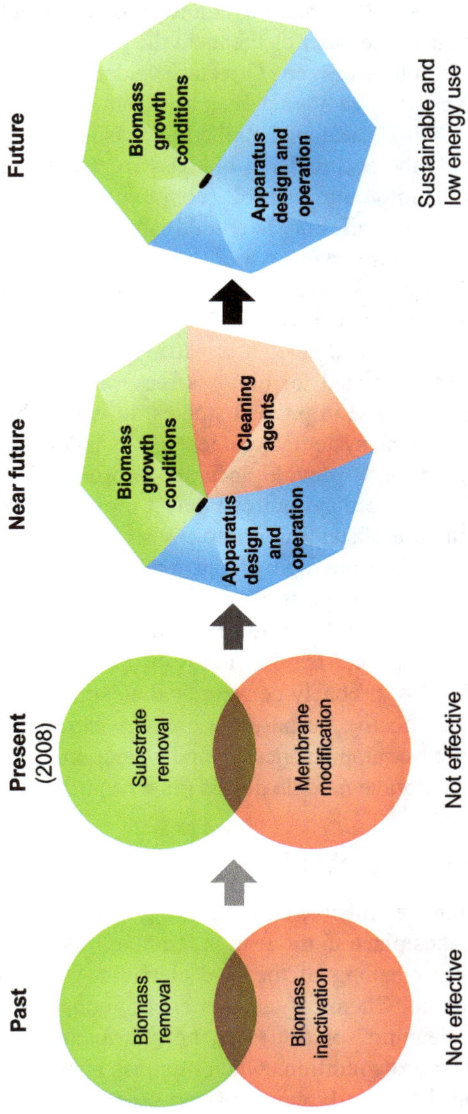

Figure 13.6 Past, present, near future and future approach for biofouling control of high pressure membranes

FCP-increase over a lead membrane module can account for the total FCP-increase over the pressure vessel. In other words, for a FCP-increase over the total pressure vessel of 15% the FCP-increase over a lead module can be around 100% while the FCP-increase over the other elements in the pressure vessel can be much lower than 15% or even (close to) zero (Vrouwenvelder *et al.*, 2009d). Cleaning strategies must be efficient in removing biomass from spiral-wound membrane modules (Flemming, 1997b). Cleanings applied in practice usually inactivate but do not remove (most of) the accumulated biomass from the membrane module, which is essential for biofouling control (Flemming, 1997b; Vrouwenvelder *et al.*, 1998; Zondervan and Roffel, 2007 and Al-Amoudi and Lovitt, 2007). The remaining inactivated biomass serves as food for surviving bacteria and bacteria in the feed water causing rapid regrowth in the cleaned membrane modules again (Vrouwenvelder and Van der Kooij, 2001; Vrouwenvelder *et al.*, 2008a; Kappelhof *et al.*, 2003 and Bereschenko *et al.*, 2011). The effectivity of cleanings should be evaluated and optimized for biomass removal. For this the membrane fouling simulator can be u sed in parallel studies, in which the FCP and direct observations using a camera system and the sight window of the membrane fouling simulator (Figure 13.8) provide accurate information on the efficiency of biomass removal. The camera system can detect biofilm thicknesses of a few micrometers, without the use of stains to improve visibility of the semi-transparent nature of the biofilm. Cleanings are more effective when the biofilm amount is still limited (Figure 13.8C) than when (parts of) the feed spacer channel are completely blocked by biomass (Figure 13.8D and 13.8E), limiting the transport of cleaning chemicals into the blocked spacer regions and restricts removal of biomass from the membrane module. Most probably conventional cleaning strategies as used in practice now can not be effective because the biomass can not be removed from the modules with current equipment design and operation. Therefore advanced cleaning strategies or membrane module design must be pursued.

Advanced cleaning strategies

Advanced cleaning strategies involve early cleaning of partially fouled modules. Biofouling primarily takes place in the lead membrane module of the first stage. Therefore with a timely cleaning of the lead element(s), cleaning of the total installation can be avoided. Chemical use and energy consumption (to heat and pump/recirculate the cleaning solution) can be restricted, lowering the costs. Additionally, under standard conditions when biomass is removed from the lead membrane module the detached biomass passes several less fouled membrane modules in the pressure vessel, which may negatively affects the cleaning performance. Isolating the lead membrane modules during cleaning from the rest of the membrane filtration installation prevent the possibility that detached biomass deposits in the rest of the installation. Early cleaning and isolating the lead membrane modules from the installation may have several advantages (Figure 13.9B): (i) reduction of cleaning chemical use and energy consumption and (ii) more accurate monitoring of biofouling. When isolated lead modules are installed vertically (Figure 13.9C and E), other types of cleanings such as enhanced reverse flushing with water containing air eventually in combination with a copper salt solution (Cornelissen *et al.*, 2007) can be applied as well. Application of cleaning by air sparging enhanced the flux in both ultrafiltration (Cabassud *et al.*, 1997) and flat sheet NF systems (Ducom *et al.*, 2002). Combining chemical cleaning with reverse flushing can be more effective than chemical cleaning only. Lead membrane modules can be adapted to improve (i) resistance to high shear forces caused by enhanced flushing and (ii) cleanability by the use of (thicker/coated/adapted geometry) feed spacers.

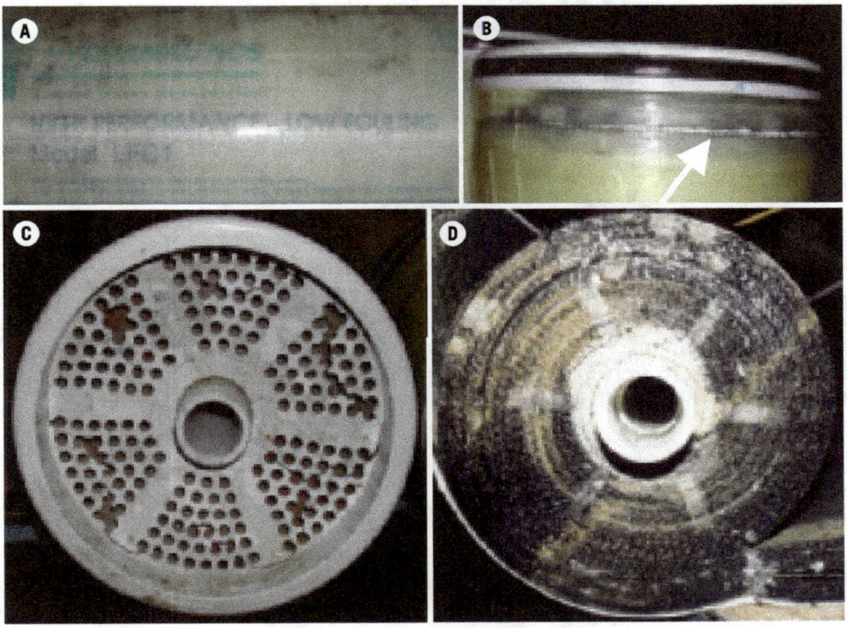

Figure 13.7 Low fouling membrane use **(A)** is no approach for biofouling prevention in most cases. The glass fibre casing (**B**, arrow) and the end cap **(C)** were damaged due to the high pressure drop over the feed spacer channel. Removal of the feed side end cap of the lead membrane showed biofouling of feed spacer channel **(D)**

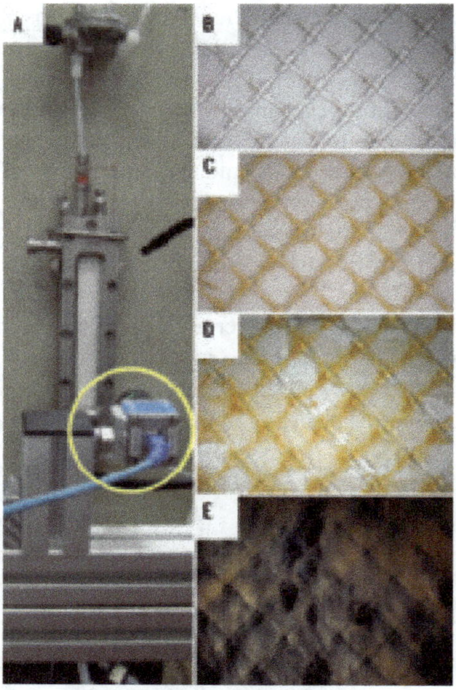

Figure 13.8 Direct in-situ observations of fouling amount in the feed spacer channel obtained with a camera system (yellow circle) and the sight window of the membrane fouling simulator **(A)**. The biofilm amount in the feed spacer channel can be monitored in time **(B, C** and **D)**. Shown for comparison is a fouled feed spacer taken from a membrane module from practice **(E)**, illustrating that the feed spacer channel is blocked. Analysis of (E) confirmed biofouling

Biofouling inhibitor dosage

Biofouling inhibitors are dosed to prevent biomass accumulation in spiral-wound membrane modules. The efficiency of dosages to inhibit biofouling can be evaluated using the membrane fouling simulator in parallel studies, in which the FCP (Figure 13.10) and direct observations using a camera system and the sight window of the membrane fouling simulator (Figure 13.8) provide accurate information about biomass accumulation. The membrane fouling simulator can also be used to optimize inhibitor concentration and dosing frequency to reduce costs and impact on the environment (discharge). From environmental and sustainability aspects in principle it is undesirable to use chemicals for biofouling control. Without chemical use to prevent fouling and to clean membranes, the use of high pressure membrane systems would be more sustainable and environmentally friendly. It is expected that chemical dosages will be restricted or even avoided in future because of environmental aspects and availability of adapted membrane systems.

Chemical selection and use

Wrong selection of chemicals like anti-scalants and biocides may result in a higher risk of biofouling since dosage of biodegradable chemicals will contribute to biofouling. Chemicals dosed to the feed water of membrane installations to prevent scaling (Van Paassen *et al.*, 1998; Van der Hoek *et al.*, 2000 and Vrouwenvelder *et al.*, 2000) or biofouling (Figure 13.10) may even cause biofouling. The effect of chemical dosage can be evaluated and optimized using the membrane fouling simulator (Vrouwenvelder *et al.*, 2008b). The efficacy of biocides can be also determined using microcalorimetric measurements of the microbial activity of biofilms (Von Rège and Sand, 1998) or oxygen imaging in biofilms (Kühl *et al.*, 2007). The chemicals used must be of a constant and high quality since already a temporarily low content of biodegradable impurities may cause rapid, severe biofouling (Vrouwenvelder *et al.*, 1998), illustrating the importance of chemical selection and quality control. To reduce the risk of fouling and meet probably promulgated stricter discharge regulations in the future, it is recommended to avoid chemical dosage as much as possible.

Low flow velocities

Operating membranes at low flow velocity reduces the effect of biomass accumulation on FCP-increase (Vrouwenvelder *et al.*, 2009a). Membrane systems could be operated at an optimal FCP, tolerating biomass accumulation. In other words: a high biomass concentration causing a low FCP-increase only is not a problem. Therefore, from a practical point of view the focus of research should be restricting the FCP and the FCP-increase rather than biomass accumulation (Vrouwenvelder *et al.*, 2009b). The Optiflux concept (Van der Meer, 2003a and Van Paassen *et al.*, 2005) reduces the operational costs of membrane filtration by applying less membrane modules in a pressure vessel than commonly applied (i.e. three modules instead of up to eight). The operational cost reduction is based on lower energy consumption. In addition, a lower flow velocity will impact the effect of the biomass amount on the FCP, so the Optiflux concept may reduce the FCP-increase caused by biofouling, reducing the membrane operational costs even further. Another approach to reduce the linear flow velocity is to increase the number of lead membrane modules (Figure 13.9D). Therefore, biomass accumulation is reduced on the location where impact on the FCP is highest (Vrouwenvelder *et al.*, 2009a).

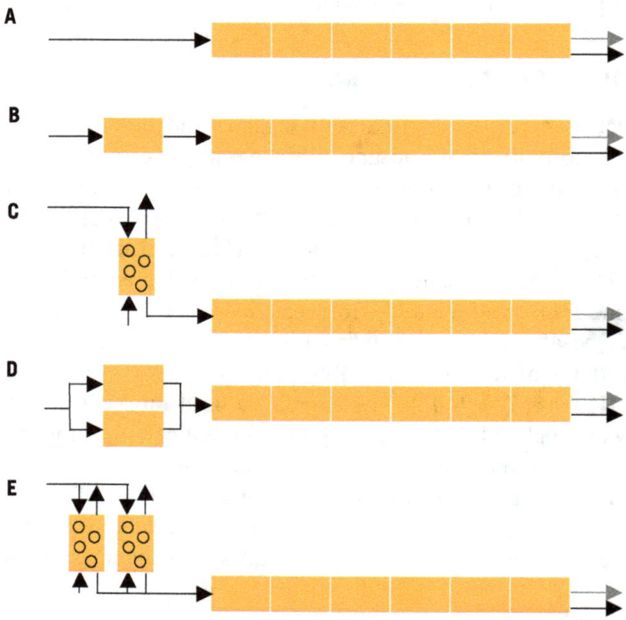

Figure 13.9 Scheme illustrating selected approaches for biofouling control. Pressure vessel configuration on feed side containing membrane modules as applied in practice **(A)**. A separate lead membrane module enables early biofouling detection and advanced cleaning strategies **(B)**, since only the fouled membrane is cleaned. Vertical positioning of the separate lead module enables alternative cleanings such as enhanced periodic back flushing with water containing gas **(C)**, increasing shear force. Increasing the number of lead modules by adding modules in parallel will lower the linear flow velocity, reducing the impact of accumulated biomass on the FCP **(D)**. **(E)** is a variation of (D) in a vertical position enabling alternative cleanings

Feed flow reversal

Feed flow reversal in pressure vessels with several modules can result in low FCP-levels. The effect of biofouling accumulation is strongly reduced by changing the flow direction in the pressure vessels. Before flow reversal, the lead membrane module has a much higher linear flow velocity (and pressure drop) than the membrane element in the last position of the pressure vessel. By changing the flow direction, the FCP-increase over the total installation will be instantly reduced. Simulation studies with the membrane fouling simulator support this theory (Vrouwenvelder *et al.*, 2009a). Also, a decline of biomass and FCP over a biofouled membrane fouling simulator in time was observed when the simulator was operated under the conditions of membrane modules at the last place in a pressure vessel (Vrouwenvelder *et al.*, 2009a). Feed flow reversal in stages may also be efficient for simultaneous control of other fouling types besides biofouling, i.e. particulate fouling and mineral scaling. The principles to control individual fouling types differ. For prevention of mineral scaling or periods of system downtime, feed flow reversal can be applied resulting in re-dissolving of deposited scale into solution avoiding chemical dosage (Uchymiak *et al.*, 2008). Practical consequences of feed flow reversal would be the use of adapted module sealing and some additional piping.

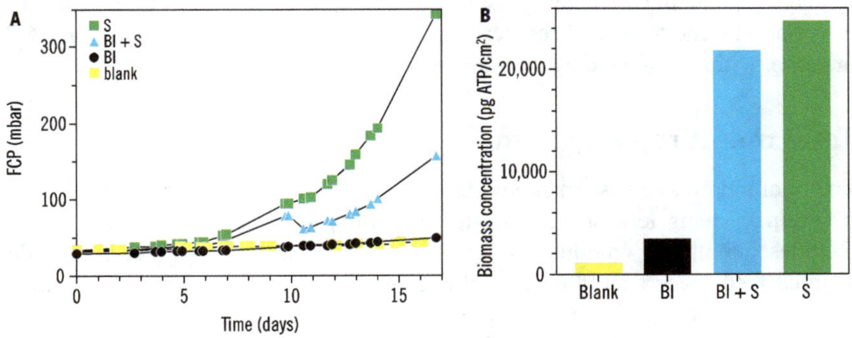

Figure 13.10 FCP development in time over membrane fouling simulators (**A**) and the biomass concentrations in membrane fouling simulators after 17 days operation (**B**). The membrane fouling simulators were fed with water with supplementing of substrate (S), biofouling inhibitor and substrate (BI + S), biofouling inhibitor (BI) and without dosage (blank). The monitors were operated in parallel. BI dosage resulted in a high FCP-increase and high biomass concentration compared to the blank: BI dosage can contribute to biofilm growth. The BI active compound is related to compounds used in toothpaste

Feed spacer modification

Modification of feed spacers may/could reduce the FCP, reduce the effect of biomass on the FCP-increase, and improve cleanability of the membrane module.

Ridgway (1997) already suggested adaptation of hydrodynamics of the feed spacer channel as an approach to control membrane biofouling. Nevertheless, still research to control biofouling is predominantly focused on development of low fouling membranes instead of feed channel and spacer modification. The number of publications in journals satisfying the search criteria 'biofouling' and 'modified and/or adapted membrane' in the article title, abstract and keywords in the scopus database of March 2008 amounted 59. The search criteria 'biofouling' and 'modified and/or adapted spacer' yielded no references at all. At the North American Membrane Society (NAMS) conference 2007, to control biofouling membrane modification was addressed in 6 presentations while feed spacer modification was addressed in 1 presentation only. Evidently, until now biofouling control was considered a membrane problem rather than a feed channel problem. One of our recent studies showed that biomass accumulation on the feed spacer affects both the fluid flow distribution and the FCP-increase (Vrouwenvelder *et al.*, 2009c). Biofouling may be affected by a number of spacer characteristics such as thickness, geometry, structure, material and coating. Thicker spacers with a unique geometry have been shown to have less pressure drop and reduce the frequency of chemical cleaning (Bartels *et al.,* 2008). The use of thicker spacers in membrane modules will restrict the effect of biomass accumulation on FCP-increase, without influencing membrane module construction.

Total membrane system

The total membrane system should be evaluated in order to develope membrane filtration systems less susceptible to biofouling. The total membrane system includes membrane module design, plant staging and operation. Besides adapted feed spacers and lower number of elements in pressure vessels enabling operation at lower flow velocities, additional adaptations may be considered to prevent biofouling problems. Capillary NF and RO membranes may be operated without elevated pressure drops caused by biomass accumulation. They might also be easier to clean. However, the use of capillary membranes may require a larger foot print. A large foot print should not be prohibitive when a stable and long term robust process is achieved. Balancing economics of investments versus operations to control biofouling is a logical step in developing membrane systems less susceptible to biofouling. It is anticipated that the costs of membrane process will be reduced significantly by rational biofouling control.

Growth limiting conditions

Biological pretreatment reduces the concentration of easily biodegradable compounds in the water fed into the membrane filtration installation, resulting in lower/delayed biomass accumulation in the membrane modules (Griebe and Flemming, 1998). However, biological pretreatment removing substrate is no guarantee for biofouling control in practice. Another approach to restrict microbiological growth conditions in membrane modules can be extensive removal of an essential biomass constituent from the water. For biomass synthesis besides carbon, nitrogen and phosphorous are essential. Very low phosphate levels will restrict biomass growth, even in the presence of high amounts of easily biodegradable compounds (Miettinen *et al.*, 1997; Mohamed *et al.*, 1998; Lehtola *et al.*, 1999, 2002a and Vrouwenvelder *et al.*, 2010c). The feed water of most membrane filtration installations already has low phosphate concentrations, which can be further reduced using for example thermostabile ferritin (Hasan *et al.*, 2007), (Figure 13.11), so that phosphate becomes the microbial growth limiting factor. Ferritin is an enzyme that can bind phosphate and iron. The advantage of ferritin use is the high phosphate removal rate even at very low concentrations. A feasibility study on ferritin use for large scale water treatment showed that the costs are similar to those of currently applied methods (Jacobs *et al.*, 2010). Phosphate limitation may be a suitable approach to control biofouling avoiding the need for pretreatment to achieve an extended removal of biodegradable compounds from the feed water. Obviously, with such an approach to control biofouling dosing of chemicals containing phosphate and/or materials releasing phosphate such as phosphonate based antiscalants should be avoided.

Repetitive stress conditions

Repetitive stress conditions may control the accumulation of biomass on surfaces. Instead of biocide dosing more environmentally friendly approaches may be available. The principle is to create conditions unfavourable for microorganisms to accumulate on a surface, so no biofouling will occur. The stress conditions may be created by alternating (i) high and low substrate concentrations, (ii) high and low salt concentrations, (iii) high and low flow velocities and (iv) feed flow reversal. Other stress conditions, such as flushing with permeate (osmotic shock) and/or with water containing copper (the copper solution is reused) can be applied as well.

Biofilm morphology engineering

Biofilm morphology engineering involves the manipulation of biofilms in such a way that system performance is preferentially improved or not negatively influenced. Biofilm morphology depends on the combined effect of growth rate of cells, substrate transport towards cells and detachment forces (Van Loosdrecht *et al.*, 1995; Kwok *et al.*, 1998 and Picioreanu *et al.*, 1998b, 2001). When there is a strong substrate limitation like often in NF/RO systems and a not too strong shear is applied to the biofilm an open, porous biofilm is formed (Figure 13.12C). Microbial growth will occur at the tips of the fingerlike structures or streamers. These tips regularly break off and leave the module while the open and porous structure is maintained (Figure 13.13). At the base of such an open and porous biofilm no substrate remains, i.e. preventing biofouling of the membrane itself (Figure 13.12C). The movement of the biofilm filaments (as e.g. observed by Stoodley *et al.*, 1998, 1999b; Purevdorj-Gage and Stoodley, 2004 and Vrouwenvelder *et al.*, 2009c) prevents strong concentration polarization due to the mixing effect a filamentous biofilm has on the mass transfer boundary layer. It is evident that such an approach would need a well balanced condition in the membrane module to really obtain the wanted biofilm type and avoid the formation of a dense biofilm layer covering the full membrane area (Figure 13.12B).

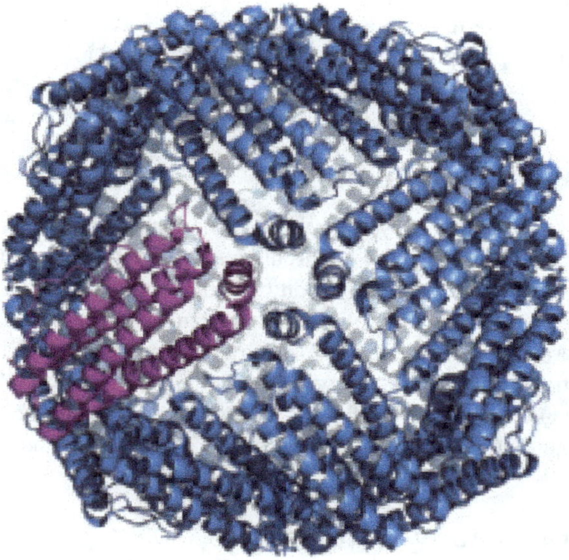

Figure 13.11 Ferritin structure, in which one of the 24 protein subunits is shown in purple. The nearly round-shaped protein has an outer diameter of ~12 nm. Adapted from Jacobs *et al.*, 2010

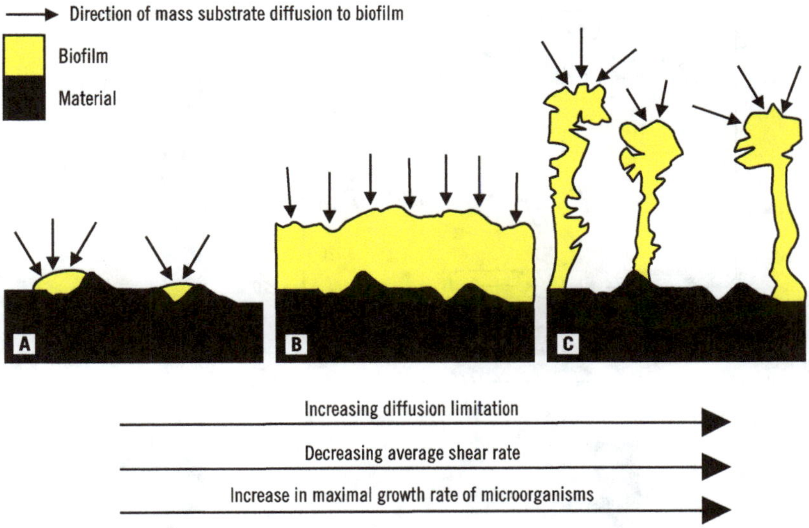

Figure 13.12 Schematic representation of the influence of substrate surface loading rate (increases from left to right) and detachment force on biofilm morphology (adapted from Van Loosdrecht *et al.*, 1995)

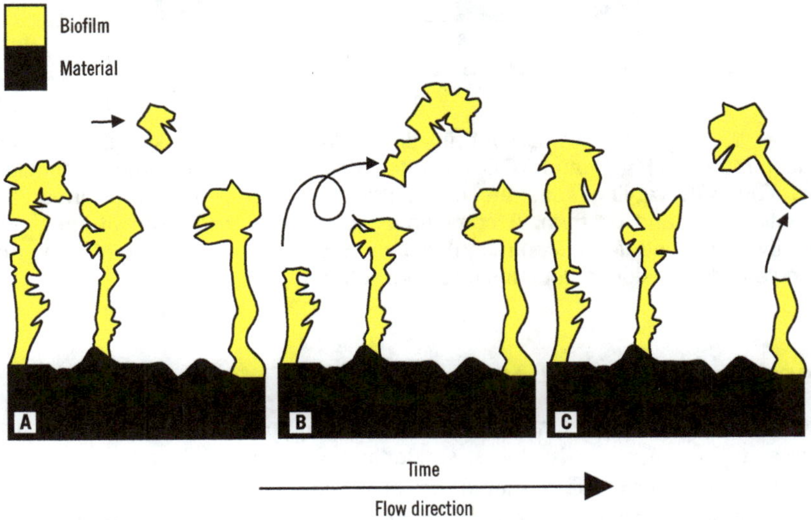

Figure 13.13 Schematic representation of sloughing and regrowth of fingerlike biofilm structures in time. The open and porous biofilm structure is maintained at the proper conditions

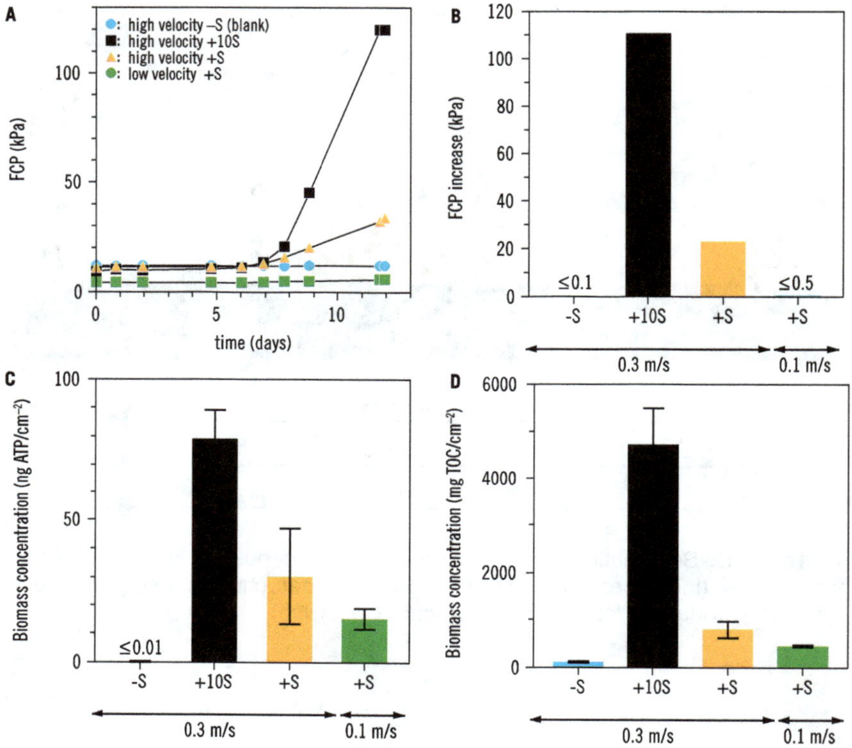

Figure 13.14 FCP development in time (**A**), FCP increase (**B**), biomass concentration ATP (**C**) and TOC (**D**) after 12 day membrane fouling simulator operation with varying substrate concentrations (10S and S) and linear flow velocities (0.3 and 0.1 m/s). A combination of actions (reduction of linear flow velocity and substrate concentration) was more efficient to control biofouling than a single action (reduction of substrate concentration)

Combined approaches

The combinations of several approaches may be more efficient than a single approach. In other words, by a single approach (i.e. substrate removal by pretreatment) 100% effect should be achieved, while by a combination of approaches (i.e. substrate removal, low flow velocities and adapted feed spacers, see Figure 13.14) each individual approach can be less effective but the combination is still efficient for biofouling control. A proposed combination is the use of adapted spacer design and lower linear flow velocities, addressed in the section promising scenarios for biofouling control.

MOST PROMISING SCENARIOS FOR BIOFOULING CONTROL

Biofouling tolerant conditions in spiral wound membrane systems are proposed as the highest potential scenario for biofouling control. Two additional alternative scenarios are proposed: (i) capillary membranes and/or (ii) phosphate limitation.

Biofouling tolerant conditions in spiral wound membrane systems

A lower linear flow velocity in spiral wound membrane modules will result in a (i) lower energy use, (ii) lower impact of biomass on the feed channel pressure drop, (iii) a more fluffy biofilm that may be easier to remove from the membrane module (Vrouwenvelder *et al.*, 2010b). Using biofilm morphology engineering may reduce the impact of accumulated biomass on plant performance and facilitate biomass removal from the modules using e.g. a reversed enhanced flow. Combining a lower linear flow velocity with adapted designs for feed spacer, membrane module and filtration plant will result in effective biofouling control at low energy requirements, minimal cost and minimal chemical use.

Capillary membranes

In the presence of a feed spacer the absolute feed channel pressure drop increase caused by biomass accumulation was much higher than when a feed spacer was absent (Vrouwenvelder *et al.*, 2009c). Already without biomass, the feed spacer presence caused a higher feed channel pressure drop compared to a flow channel without feed spacer (Vrouwenvelder *et al.*, 2009c). Spiral wound membrane modules without spacers are not feasible since spacers are needed for separating the membrane envelopes and creating turbulence. An approach to operate membranes without spacers is the use of capillary membranes. Laboratory studies using capillary membrane modules showed practical advantages for biofouling control compared to currently commercially available spiral wound membrane modules with feed spacers (unpublished data). Compared to spiral wound feed spacer membrane module systems, in capillary modules the feed channel pressure drop was lower, biomass accumulation had less impact on the feed channel pressure drop and the modules could be cleaned more efficiently.

Phosphate limitation

A low phosphate concentration (Miettinen *et al.*, 1997 and Lehtola *et al.*, 2002a) in the feed water of membrane filtration installations can restrict the pressure drop increase and biomass accumulation, even at high organic substrate concentrations. Phosphate limitation may therefore be suitable to reduce pre-treatment and chemical use, and can be combined with other complementary approaches like biofouling tolerant conditions to solve biofouling. In the membrane fouling simulator under phosphate limiting conditions, dosing phosphonate based antiscalants caused biofouling while no biofouling was observed when acids or phosphonate-free antiscalants were used. Antiscalant dosage may increase both phosphate and substrate concentrations of the water. Therefore, antiscalant selection can be critical for biofouling control (Vrouwenvelder *et al.*, 2010c).

SUMMARY

Despite extensive research efforts, past and present strategies to control biofouling problems in spiral-wound nanofiltration and reverse osmosis membranes have not been successful under all circumstances. Gaining insight in the biofouling process is a first necessity. Based on recent insights, an overview is given of 12 potential complementary approaches to solve biofouling. Combinations of approaches may be more efficient in biofouling control than a single approach. A single approach must be 100% effective, while in combination each individual approach can be partially effective while the combination is still efficient. An integrated Approach for Biofouling Control (ABC) is proposed, based on three corner stones: (i) equipment design and operation, (ii) biomass growth conditions, and (iii) cleaning agents as a framework to control biofouling. While past and present strategies addressed mainly membranes and microorganisms, i.e. removal or inactivation of biomass, this ABC-approach addresses the total membrane filtration system. It is anticipated that this integral approach will enable a more rational and effective control of biofouling. Although in this stage chemical cleaning and biofouling inhibitor dosage seem unavoidable to control biofouling, it is expected that in future – because of sustainability and costs reasons – membrane systems will be developed without or with minimal need for chemical cleaning and dosing. Three potential scenarios for biofouling control are proposed based on (i) biofouling tolerant spiral wound membrane systems, (ii) capillary membranes, and (iii) phosphate limitation.

References

Ahmad A. L. and Lau K. K. (2006). Impact of different spacer filaments geometries on 2D unsteady hydrodynamics and concentration polarization in spiral wound membrane channel. *J. Membr. Sci.*, **286**, 77–92.

Ahmad A. L., Lau K. K. and Abu Bakar M. Z. (2005). Impact of different spacer filament geometries on concentration polarization control in narrow membrane channel. *J. Membr. Sci.*, **262**, 138–152.

Al-Amoudi A. and Lovitt R. W. (2007). Fouling strategies and the cleaning system of NF membranes and factors affecting cleaning efficiency. *J. Membr. Sci.*, **303**, 4–28.

Allison D. G. and Sutherland I. W. (1984). A staining technique for attached bacteria and its correlation to extracellular carbohydrate production. *J. Microbiol. Methods*, **2**, 93–99.

Alphenaar P. A., Sleyster R., De Reuver P., Ligthart G.-J. and Lettinga G. (1993). Phosphorus requirement in high-rate anaerobic wastewater treatment. *Water Res.*, **27**, 749–756.

Amin M. H. G., Gibbs S. J., Chorley R. J., Richards K. S., Carpenter T. A. and Hall L. D. (1997). Study of flow and hydrodynamic dispersion in a porous medium using pulsed-field-gradient magnetic resonance. *Proceedings of The Royal Society of London, Series A: Mathematical and Physical Sciences*, **453**, 489–513.

Amjad Z. (1993). Reverse Osmosis: Membrane technology, water chemistry and industrial applications. Van Nostrand Reinhold, New York. ISBN 0-442-23964-5.

Argo D. G. and Ridgway H. F. (1982). Biological Fouling of Reverse Osmosis Membranes at water factory 21. Proceedings 10th Annual conference and trade fair of the water supply improvement association, vol. III, special interest presentations, July 1982, Honolulu, Hawaii.

Bacchin P., Aimar P. and Field R. W. (2006). Critical and sustainable fluxes: Theory, experiments and applications. *J. Membr. Sci.*, **281**, 42–69.

Bacchin P., Aimar P. and Sanchez V. (1995). Model for colloidal fouling of membranes. *AlChE J.*, **41**, 368–376.

Baker R. W. (2004). Membrane technology and applications. 2nd edition. John Wiley and Sons Ltd. ISBN: 978-0-470-85445-7.

Baker J. S. and Dudley L. Y. (1998). Biofouling in membrane systems–a review. *Desalination*, **118**, 81–90.

Bartels C., Hirose M. and Fujioka H. (2008) Performance advancement in the spiral wound RO/NF element design. *Desalination*, **221**(1–3), 207–214.

Belfer S., Gilron J., Daltrophe N. and Oren Y. (2005). Comparative study of biofouling of NF modified membrane at SHAFDAN. *Desalination*, **184**, 13–21.

Bereschenko L. A., Heilig G. H. J., Nederlof M. M., Van Loosdrecht M. C. M., Stams A. J. M. and Euverink G. J. W. (2008). Molecular characterization of the bacterial communities in the different compartments of a full-scale reverse-osmosis water purification plant. *Appl. Environ. Microbiol.*, **74**, 5297–5304.

Bereschenko L. A., Prummel H., Euverink G. J. W., Stams A. J. M. and van Loosdrecht M. C. M. (2011). Effect of conventional chemical treatment on the microbial population in a biofouling layer of reverse osmosis systems. *Water Res.*, **45**, 405–416.

Bertheas U., Majamaa K., Arzu A. and Pahnke R. (2009). Use of DBNPA to control biofouling in RO systems. *Desalination, Water Treat.*, **3**, 175–178.

Beun J. J., Van Loosdrecht M. C. M. and Heijnen J. J. (2002). Aerobic granulation in a sequencing batch airlift reactor. *Water Res.*, **36**, 702–712.

Block J. C. (1992). Biofilms in drinking water distribution systems. In: *Biofilms Science and Technology*, Melo L. F., Bott T. R., Fletcher M. and Capdeville B. (eds.). Kluwer Academic Publishers, The Netherlands.

Böddeker K. W. (1995). Commentary: Tracing membrane science. The early history of membrane science selected papers celebrating vol. 100. *J. Membr. Sci.*, **100**, 65–68.

Brandt D. C., Leitner G. F. and Leitner W. E. (1993). Reverse osmosis membranes state of the art. In: *Reverse Osmosis: Membrane technology, water chemistry and industrial applications*, Amjad Z. (ed.). Van Nostrand Reinhold, New York, ISBN 0-442-23964-5, 1–36.

Brouwer H., Meesters K. and Van Groenestijn J. (2006). Biofouling control in reverse osmosis membranes using rapid biofiltration technology. *Desalination*, **199**, 15–17.

Bryers J. D. (2000). Biofilms II: Process analysis and applications. John Wiley and Sons, New York. ISBN 0-471-29656-2.

Cabassud C., Laborie S. and Lainé J. M. (1997). How slug flow can improve ultrafiltration flux in organic hollow fibres. *J. Membr. Sci.*, **128**, 93–101.

Cao Z., Wiley D. E. and Fane A. G. (2001). CFD simulations of net-type turbulence promoters in a narrow channel. *J. Membr. Sci.*, **185**, 157–176.

Carnahan R. P., Bolin L. and Suratt W. (1995). Biofouling of PVD-1 reverse osmosis elements in the water treatment plant of the city of Dunedin, Florida. *Desalination*, **102**, 235–244.

Characklis W. G. (1973a). Attached microbial growths: I. Attachment and growth. *Water Res.*, **7**, 1113–1127.

Characklis W. G. (1973b). Attached microbial growths. II. Frictional resistance due to microbial slimes. *Water Res.*, **7**, 1249–1258.

Characklis W. G. and Marshall K. C. (1990). Biofilms. John Wiley & Sons, New York.

Chen C.-L., Liu W.-T., Chong M.-L., Wong M.-T., Ong S. L., Seah H. and Ng W. J. (2004a). Community structure of microbial biofilms associated with membrane-based water purification processes as revealed using a polyphasic approach. *Appl. Microbiol. Biotechnol.*, **63**, 466–473.

Chen K. L., Song L., Ong S. L. and Ng W. J. (2004b). The development of membrane fouling in full-scale RO processes. *J. Membr. Sci.*, **232**, 63–72.

Chen M. J., Zhang Z. and Bott T. R. (2005). Effects of operating conditions on the adhesive strength of Pseudomonas fluorescens biofilms in tubes. *Colloids Surf., B.*, **43**, 61–71.

Chiolle A., Gianotti G., Gramondo M. and Parrini G. (1978). Mathematical model of reverse osmosis in parallel-wall channels with turbulence promoting nets. *Desalination*, **26**, 3–16.

Chong T. H., Wong F. S. and Fane A. G. (2008). The effect of imposed flux on biofouling in reverse osmosis: Role of concentration polarisation and biofilm enhanced osmotic pressure phenomena. *J. Membr. Sci.*, **325**, 840–850.

Cornelissen E. R., Vrouwenvelder J. S., Heijman S. G. J., Viallefont X. D., Van der Kooij D. and Wessels L. P. (2007). Periodic air/water cleaning for control of biofouling in spiral wound membrane elements. *J. Membr. Sci.*, **287**, 94–101.

Cosgrove W. J. and Rijsberman F. R. (2000). World water vision: making water everybody's business. World Water Council. London, UK, Earthscan. http://watervision.cdinet.com/visioncontents.html.

Costerton J. W., Cheng K. J., Geesey G. G., Ladd T. I., Nickel J. C., Dasgupta M. and Marrie T. J. (1987). Bacterial biofilms in nature and disease. *Annu. Rev. Microbiol.*, **41**, 435–464.

Creber S. A., Vrouwenvelder J. S., Van Loosdrecht M. C. M. and Johns M. L. (2010). Chemical cleaning of biofouling in reverse osmosis membrances evaluating using Magn. Reson. Imaging. *J. Membr. Sci.*, **362**, 202–210.

Darcovich K., Dal-Cin M. M. and Gros B. (2009). Membrane mass transport modeling with the periodic boundary condition. *Comput. Chem. Eng.*, **33**, 213–224.

Da Silva M. K., Tessaro I. C. and Wada K. (2006). Investigation of oxidative degradation of polyamide reverse osmosis membranes by monochloramine solutions. *J. Membr. Sci.*, **282**, 375–382.

Dendukuri D., Karode S. K. and Kumar A. (2005). Flow visualization through spacer filled channels by computational fluid dynamics-II: Improved feed spacer designs. *J. Membr. Sci.*, **249**, 41–49.

Dirckx C. J., Clark S. A., Hall L. D., Antalek B., Tooma J., Hewitt J. M. and Kawaoka K. (2000). Magn. Reson. Imaging of the filtration process. *AlChE J.*, **46**, 6–14.

DiSalvo L. H. and Cobet A. B. (1974). Control of an estuarine microfouling sequence on optical surfaces using low intensity ultraviolet irradiation. *Appl. Microbiol.*, **27**, 172–178.

DOW. (2005). Micro-biocide prevents membrane biofouling. *Membr. Technol.*, **11**, 2.

DOW. (2009a). Form no. 609-02078-1204, 609-02034-1004 and 609-22010-604. 6 August.

DOW. (2009b). Form No. 609-02036-0808. 23 August.

Dubois M., Gilles K. A., Hamilton J. K., Rebers P. A. and Smith F. (1956). Colorimetric method for determination of sugars and related substances. *Anal. Chem.*, **28**, 350–356.

Ducom G., Matamoros H. and Cabassud C. (2002). Air sparging for flux enhancement in nanofiltration membranes: Application to O/W stabilised and non-stabilised emulsions. *J. Membr. Sci.*, **204**, 221–236.

Duranceau S. J. and Knippel G. G. (1996). Nanofiltration membrane biofouling control study for the city of Dunedin, Florida. *ACS Division of Environmental Chemistry, Preprints*, **36**, 101–104.

Eberl H. J., Picioreanu C., Heijnen J. J. and Van Loosdrecht M. C. M. (2000). Three-dimensional numerical study on the correlation of spatial structure, hydrodynamic conditions, and mass transfer and conversion in biofilms. *Chem. Eng. Sci.*, **55**, 6209–6222.

Ebrahim S. (1994). Cleaning and regeneration of membranes in desalination and wastewater applications: State-of-the-art. *Desalination*, **96**, 225–238.

Ebrahim S., Abdel-Jawad M., Bou-Hamad S. and Safar M. (2001). Fifteen years of R&D program in seawater desalination at KISR part I. Pretreatment technologies for RO systems. *Desalination*, **135**, 141–153.

EEA. (2005). Nutrients in Freshwater (CSI 020) – assessment published 30 November 2005. European Environment Agency.

EEA. (2007). Nutrients in Coastal Water. European Environment Agency.

El-Fadel M. and Alameddine I. (2005). Desalination in arid regions: Merits and concerns. *J. Wat. Sup.: Res. Technol. – AQUA*, **54**, 449–461.

ESCWA (Economic and Social Commission for Western Asia). (1997). Water desalination and wastewater reuse: Review of technology, economics and application in the ESCWA region. Expert Group Meeting on Development of Non-Conventional Water Resources and Appropriate Technologies for Groundwater Management in the ESCWA Member Countries, Manama, Bahrain, 27–30 October.

Field R. W., Wu D., Howell J. A. and Gupta B. B. (1995). Critical flux concept for microfiltration fouling. *J. Membr. Sci.*, **100**, 259–272.

Fimbres-Weihs G. A. and Wiley D. E. (2007). Numerical study of mass transfer in three-dimensional spacer-filled narrow channels with steady flow. *J. Membr. Sci.*, **306**, 228–243.

Fimbres-Weihs G. A., Wiley D. E. (2008). Numerical study of two-dimensional multi-layer spacer designs for minimum drag and maximum mass transfer. *J. Membr. Sci.*, **325**, 809–822.

Flemming H. C. (1993). Mechanistic aspects of reverse osmosis membrane biofouling and prevention. In: Amjad Z. (ed.), *Reverse Osmosis: Membrane Technology, Water Chemistry and Industrial Applications*. Van Nostrand Reinhold, New York. pp. 163–209.

Flemming H. C. (2003). Role and levels of real-time monitoring for successful anti-fouling strategies – An overview. *Wat. Sci. Technol.*, **47**, 1–8.

Flemming H. C., Schaule G., Griebe T., Schmitt J. and Tamachkiarowa A. (1997a). Biofouling–The Achilles heel of membrane processes. *Desalination*, **113**, 215–225.

Flemming H. C. (1997b). Reverse osmosis membrane biofouling. *Exp. Therm. Fluid Sci.*, **14**, 382–391.

Flemming H. C., Schaule G., McDonogh R. and Ridgway H. F. (1994). Effects and extent of biofilm accumulation in membrane systems. In: *Biofouling and biocorrosion in industrial water systems,* Geesey G. G., Lewandowski Z. and Flemming H. C. (eds.). CRC press, USA, 63–89.

Flemming H. C., Tamachkiarowa A., Klahre J. and Schmitt J. (1998). Monitoring of fouling and biofouling in technical systems. *Wat. Sci. Technol.*, **38**, 291–298.

Fowler J. D. and Robertson C. R. (1991). Hydraulic permeability of immobilized bacterial cell aggregates. *Appl. Environ. Microbiol.*, **57**, 102–113.

Frenkel V. (2004). Water desalination methods, technology, and economics. Kennedy/Jenks Consultants. Conference proceedings desalination conference Santa Barbara, California, April 16.

Gabelich C. J., Frankin J. C., Gerringer F. W., Ishida K. P. and Suffet I. H. (2005). Enhanced oxidation of polyamide membranes using monochloramine and ferrous iron. *J. Membr. Sci.*, **258**, 64–70.

Gagnon G. A., O'Leary K. C., Volk C. J., Chauret C., Stover L. and Andrews R. C. (2004). Comparative Analysis of Chlorine Dioxide, Free Chlorine and Chloramines on Bacterial Water Quality in Model Distribution Systems. *J. Envir. Engrg.*, **130**(11), 1269–1279.

Galjaard G., Kruithof J. C. and Kamp P. (2005). Recent and future advancements with UF membrane systems at NV PWN Water Supply Company North Holland. In: Proceedings 3rd IWA Leading Edge Conference on Water and Waste Water Treatment, Sapporo.

Galjaard G., Lampe M. and Kamp P. C. (2008). 8 years of RO-experience at WTP Heemskerk biofouling aspects. Proceedings American Membrane Technology Association (AMTA) 2008 conference, Naples, Florida.

Gamal Khedr M. (1998). A case study of RO plant failure due to membrane fouling, analysis and diagnosis. *Desalination*, **120**, 107–113.

Gamal Khedr M. (2000). Membrane fouling problems in reverse osmosis desalination applications. *Desalination and Water Reuse*, **10**, 8–17.

Gamal Khedr M. (2002). Personal communication.

Garcia H. E., Locarnini R. A., Boyer T. P. and Antonov J. I. (2006). World Ocean Atlas 2005, Volume 4: Nutrients (phosphate, nitrate, silicate). Levitus S. (ed.), NOAA Atlas NESDIS 64, U.S. Government Printing Office, Washington, D.C.

Garny K., Neu T. R. and Horn H. (2009). Sloughing and limited substrate conditions trigger filamentous growth in heterotrophic biofilms-Measurements in flow-through tube reactor. *Chem. Eng. Sci.*, **64**, 2723–2732.

Geesey G. G. (1982). Microbial exopolymers: Ecological and economic considerations. *ASM News*, **48**, 9–14.

Geesey G. G. and Bryers J. D. (2000). Biofouling of engineered materials and systems. In: *Biofilms II: Process analysis and applications*, Bryers J. D. (ed.). Wiley-Liss, New York, 242–245.

Geesey G. G., Lewandowski Z. and Flemming H.-C. (1993). Biofouling and biocorrosion in industrial water systems. CRC press, Boca Raton, Florida, ISBN 0-87371-928-X.

Geraldes V., Semião V. and De Pinho M. N. (2002a). Flow management in nanofiltration spiral wound modules with ladder-type spacers. *J. Membr. Sci.*, **203**, 87–102.

Geraldes V., Semião V. and De Pinho M. N. (2002b). The effect of the ladder-type spacers configuration in NF spiral-wound modules on the concentration boundary layers disruption. *Desalination*, **146**, 187–194.

Gimmelshtein M., Semiat R. (2005). Investigation of flow next to membrane walls. *J. Membr. Sci.*, **264**, 137–150.

Gjaltema A., Arts P. A. M., Van Loosdrecht M. C. M., Kuenen J. G. and Heijnen J. J. (1994). Heterogeneity of biofilms in rotating annular reactors: Occurrence, structure, and consequences. *Biotechnol. Bioeng.*, **44**, 194–204.

Gjaltema A., Van der Marel N., Van Loosdrecht M. C. M. and Heijnen J. J. (1997). Adhesion and biofilm development on suspended carriers in airlift reactors: Hydrodynamic conditions versus surface characteristics. *Biotechnol. Bioeng.*, **55**, 880–889.

Glater J. (1998). The early history of reverse osmosis membrane development. *Desalination*, **117**, 297–309.

Goosen M. F. A., Sablani S. S., Al-Hinai H., Al-Obeidani S., Al-Belushi R. and Jackson D. (2004). Fouling of reverse osmosis and ultrafiltration membranes: A critical review. *Sep. Sci. Technol.*, **39**, 2261–2297.

Graf von der Schulenburg D. A., Gladden L. F. and Johns M. L. (2007). Modelling biofilm-modified hydrodynamics in 3D. *Wat. Sci. Technol.*, **55**, 275–281.

Graf von der Schulenburg D. A., Pintelon T. R. R., Picioreanu C., Van Loosdrecht M. C. M. and Johns M. L. (2009). Three-dimensional simulations of biofilm growth in porous media. *AlChE J.*, **55**, 494–504.

Graf von der Schulenburg D. A., Vrouwenvelder J. S., Creber S. A., Van Loosdrecht M. C. M. and Johns M. L. (2008). Nuclear magnetic resonance microscopy studies of membrane biofouling. *J. Membr. Sci.*, **323**, 37–44.

Greenlee L. F., Lawler D. F., Freeman B. D., Marrot B. and Moulin P. (2009). Reverse osmosis desalination: Water sources, technology, and today's challenges. *Water Res.*, **43**, 2317–2348.

Griebe T. and Flemming H. C. (1998). Biocide-free antifouling strategy to protect RO membranes from biofouling. *Desalination*, **118**, 153–156.

Guillen G. and Hoek E. M. V. (2009). Modeling the impacts of feed spacer geometry on reverse osmosis and nanofiltration processes. *Chem. Eng. J.*, **149**, 221–231.

Haas C. N., Bitter P. and Scheff P. A. (1988). Preliminary determination of limiting nutrients for standard plate count organisms in Chicago intake water. *Wat. Air Soil Pollut.*, **37**, 65–72.

Hamida A. B. and Moch I. (1996). Controlling biological fouling in open sea intake RO plants without continuous chlorination. *Desalination and Water Reuse*, **6**, 40–45.

Hamilton W. A. (1995). Biofilms and microbially influenced corrosion. In: *Microbial biofilms*, Lappin-Scott H. M. and Costerton J. W. (eds.). Cambridge University Press. Cambridge, UK, 171–182.

Harrison P. J. (1990). Phosphate limitation in estuarine and coastal waters of China. *J. Exp. Mar. Biol. Ecol.*, **140**, 79–87.

Hasan M. N., Van Loosdrecht M. C. M. and Hagen W. R. (2007). New material comprising hyperthermophilic ferritin, useful for removing oxo-anions and metal cations from a liquid, preferably water. Patent Number(s): EP1764348-A1; WO2007031515-A1.

Hassler G. L. (1950). The use of molecular oil films in a pressure method for obtaining irrigation water from the sea, unpublished report. Gift to UCLA Engineering Library, July 1954.

Hassler G. L. (1964). Method and apparatus for separating solvents from solutions by distillation activated by pressure. patent US3129146.

Heath C. A., Belfort G., Hammer B. E., Mirer S. D. and Pimbley J. M. (1990). Magn. Reson. Imaging and modeling of flow in hollow–fiber bioreactors. *AlChE J.*, **36**, 547–558.

Hennig J., Nauerth A. and Friedburg H. (1986). RARE imaging: A fast imaging method for clinical MR. *Magn. Reson. Med.*, **3**, 823–833.

Henze M., Gujer W., Mino T. and Van Loosdrecht M. C. M. (2000). Activated sludge models ASM1, ASM2, ASM2d and ASM3 (IWA Task Group on Mathematical Modelling for Design and Operation of Biological Wastewater Treatment, Henze M., Gujer W., Mino T. and Van Loosdrecht M. C. M. (eds.). IWA Scientific & Technical Report, IWA Publishing, London, UK.

Herzberg M. and Elimelech M. (2007). Biofouling of reverse osmosis membranes: Role of biofilm-enhanced osmotic pressure. *J. Membr. Sci.*, **295**, 11–20.

Herzberg M. and Elimelech M. (2008). Physiology and genetic traits of reverse osmosis membrane biofilms: A case study with *Pseudomonas aeruginosa. ISME J.*, **2**, 180–194.

Heukelekian H. and Heller A. (1940). Relation between food concentration and surface for bacterial growth. *J. Bacteriol.*, **40**, 547–558.

Hille A., He M., Ochmann C., Neu T. R. and Horn H. (2009). Application of two component biodegradable carriers in a particle-fixed biofilm airlift suspension reactor: Development and structure of biofilms. *Bioprocess. Biosyst. Eng.*, **32**, 31–39.

Hijnen W. A. M., Biraud D., Cornelissen E. R. and Van der Kooij D. (2009). Threshold concentration of easily assimilable organic carbon in feedwater for biofouling of spiral-wound membranes. *Environ. Sci. Technol.*, **43**, 4890–4895.

Hinrichsen D., Robey B. and Upadhyay U. D. (1998). Solutions for a Water-Short World. Chapter 3: The coming era of water stress and scarcity. Population reports, series M, No. 14. Baltimore, Johns Hopkins School of Public Health. Population information program.

Ho C. K., Altman S. J., Jones H. D. T., Khalsa S. S., McGrath L. K. and Clem P. G. (2008). Analysis of micromixers to reduce biofouling on reverse osmosis membranes. *Environ. Prog.*, **27**, 195–203.

Hobbie J. E., Daley R. J. and Jasper S. (1977). Use of nuclepore filters for counting bacteria by fluorescence microscopy. *Appl. Environ. Microbiol.*, **33**, 1225–1228.

Holm-Hansen O. and Booth C. R. (1966). The measurement of adenosine triphosphate in the ocean and its ecological significance. *Limnol. Oceanogr.*, **11**, 510–519.

Holtan H., Kamp-Nielsen L. and Stuanes A. O. (1988). Phosphorus in soil, water and sediment: An overview. *Hydrobiologia*, **170**, 19–34.

Hoskins B. C., Fevang L., Majors P. D., Sharma M. M. and Georgiou G. (1999). Selective imaging of biofilms in porous media by NMR relaxation. *J. Magn. Reson.*, **139**, 67–73.

Howell J. A. (1995). Sub-critical flux operation of microfiltration. *J. Membr. Sci.*, **107**, 165–171.

Hu J. Y., Song L. F., Ong S. L., Phua E. T. and Ng W. J. (2005). Biofiltration pretreatment for reverse osmosis (RO) membrane in a water reclamation system. *Chemosphere*, **59**, 127–133.

Huertas E., Herzberg M., Oron G. and Elimelech M. (2008). Influence of biofouling on boron removal by nanofiltration and reverse osmosis membranes. *J. Membr. Sci.*, **318**, 264–270.

Huisman I. H. and Feng Kong J. (2004). Biological fouling of RO membranes: detection and control. *Asian water*, 16–18.

Huiting H., De Koning M. and Beerendonk E. F. (1999). Normalisatie van gegevens bij nanofiltratie en omgekeerde osmose. SWI 99.166. Kiwa-VEWIN report.

Imamura Y., Chandra J., Mukherjee P. K., Lattif A. A., Szczotka-Flynn L. B., Pearlman E., Lass J. H., O'Donnell K. and Ghannoum M. A. (2008). Fusarium and Candida albicans biofilms on soft contact lenses: Model development, influence of lens type, and susceptibility to lens care solutions. *Antimicrob. Agents Chemother.*, **52**, 171–182.

Ivnitsky H., Katz I., Minz D., Shimoni E., Chen Y., Tarchitzky J., Semiat R. and Dosoretz C. G. (2005). Characterization of membrane biofouling in nanofiltration processes of wastewater treatment. *Desalination*, **185**, 255–268.

Jacobs J. F., Hasan M. N., Van Loosdrecht M. C. M. and Hagen W. R. (2007a). Laboratorium experimenten met thermostabiel ferritine voor fosfaatverwijdering. STOWA report. ISBN 978-90-5773-365-9 (2007).

Jacobs J. F., Hasan M. N. and Hagen W. R. (2007b). Thermostabiel ferritine voor een bionanotechnologische fosfaatverwijdering. *Afvalwaterwetenschap*, **6**, 143–149.

Jacobs J. F., Nahid Hasan M., Paik K. H., Hagen W. R. and Van Loosdrecht M. C. M. (2010). Development of a bionanotechnological phosphate removal system with thermostable ferritin. *Biotechnol. Bioeng.*, **105**(5), 918–923.

Jacobson J. D., Kennedy M. D., Amy G. and Schippers J. C. (2008). Phosphate limitation in reverse osmosis: an option to control biofouling. Proceedings EuroMed 2008 conference – Desalination for Clean Water and Energy Cooperation among Mediterranean Countries of Europe and the MENA Region, 9–13 November 2008, Dead Sea, Jordan.

Jacobson J. D., Kennedy M. D., Amy G. and Schippers J. C. (2009). Phosphate limitation in reverse osmosis: An option to control biofouling? *Desalination and Water Treatment*, **5**, 198–206.

Jarusutthirak C. and Amy G. (2006). Role of soluble microbial products (SMP) in membrane fouling and flux decline. *Environ. Sci. Technol.*, **40**, 969–974.

Joret J. C., Levi Y. and Volk C. (1991). Biodegradable dissolved organic carbon (BDOC) content of drinking water and potential regrowth of bacteria. *Wat. Sci. Technol.*, **24**, 95–101.

Kamp P. C. (1995). UF RO plant Heemskerk. Proceedings AWWA Membrane Technology Conference 1995. Reno, USA, 31–38.

Kamp P. C., Kruithof J. C. and Folmer H. C. (2000). UF/RO treatment plant Heemskerk: From challenge to full scale application. *Desalination*, **131**, 27–35.

Kang S.-T., Subramani A., Hoek E. M. V., Deshusses M. A. and Matsumoto M. R. (2004). Direct observation of biofouling in cross-flow microfiltration: Mechanisms of deposition and release. *J. Membr. Sci.*, **244**, 151–165.

Kapellos G. E., Alexiou T. S. and Payatakes A. C. (2007). Hierarchical simulator of biofilm growth and dynamics in granular porous materials. *Adv. Water Res.*, **30**, 1648–1667.

Kappelhof J. W. N. M., Vrouwenvelder H. S., Schaap M., Kruithof J. C., Van der Kooij D. and Schippers J. C. (2003). An in situ biofouling monitor for membrane systems. *WSTWS*, **3**, 205–210.

Karime M., Bouguecha S. and Hamrouni B. (2008). RO membrane autopsy of Zarzis brackish water desalination plant. *Desalination*, **220**, 258–266.

Karode S. K. and Kumar A. (2001). Flow visualization through spacer filled channels by computational fluid dynamics I. Pressure drop and shear rate calculations for flat sheet geometry. *J. Membr. Sci.*, **193**, 69–84.

Kasahara S., Maeda K. and Ishikawa M. (2004). Influence of phosphorus on biofilm accumulation in drinking water distribution systems. *Wat. Sci. Technol.: Wat. Suppl.*, **4**, 389–398.

Katz I. and Dosoretz C. G. (2008). Desalination of domestic wastewater effluents: Phosphate removal as pretreatment. *Desalination*, **222**, 230–242.

Keinänen M. M., Korhonen L. K., Lehtola M. J., Miettinen I. T., Martikainen P. J., Vartiainen T. and Suutari M. H. (2002). The microbial community structure of drinking water biofilms can be affected by phosphorus availability. *Appl. Environ. Microbiol.*, **68**, 434–439.

Kim A. S., Chen H. and Yuan R. (2006). EPS biofouling in membrane filtration: An analytic modeling study. *J. Colloid Interface Sci.*, **303**, 243–249.

Kissinger J. C. (1970). Sanitation studies of a reverse osmosis unit used for concentration of maple sap. *J. Milk Food Technol.*, **33**, 326–329.

Kors L. J., Moorman J. H. N., Wind A. P. M. and Van der Hoek J. P. (1998). Nitrification and low temperature in a raw water reservoir and rapid sand filters. *Wat. Sci. Technol.*, **37**, 169–176.

Koutsou C. P., Yiantsios S. G. and Karabelas A. J. (2004). Numerical simulation of the flow in a plane-channel containing a periodic array of cylindrical turbulence promoters. *J. Membr. Sci.*, **231**, 81–90.

Koutsou C. P., Yiantsios S. G. and Karabelas A. J. (2007). Direct numerical simulation of flow in spacer-filled channels: Effect of spacer geometrical characteristics. *J. Membr. Sci.*, **291**, 53–69.

Koutsou C. P., Yiantsios S. G. and Karabelas A. J. (2009). A numerical and experimental study of mass transfer in spacer-filled channels: Effects of spacer geometrical characteristics and Schmidt number. *J. Membr. Sci.*, **326**, 234–251.

Kruithof J. C., Schippers J. C., Kamp P. C., Folmer H. C. and Hofman J. A. M. H. (1998). Integrated multi-objective membrane systems for surface water treatment: Pretreatment of reverse osmosis by conventional treatment and ultrafiltration. *Desalination*, **117**, 37–48.

Kühl M., Rickelt L. F. and Thar R. (2007). Combined imaging of bacteria and oxygen in biofilms. *Appl. Environ. Microbiol.*, **73**, 6289–6295.

Kwok W. K., Picioreanu C., Ong S. L., Van Loosdrecht M. C. M., Ng W. J. and Heijnen J. J. (1998). Influence of biomass production and detachment forces on biofilm structures in a biofilm airlift suspension reactor. *Biotechnol. Bioeng.*, **58**, 400–407.

Lebon L., Leblond J., Hulin J. P., Martys N. S. and Schwartz L. M. (1996). Pulsed field gradient NMR measurements of probability distribution of displacement under flow in sphere packings. *Magn. Reson. Imaging*, **14**, 989–991.

LeChevalier M. W. (1990). Coliform regrowth in drinking water: A review. *J. Am. Water Works Assoc.*, **82**, 74–86.

LeChevallier M. W., Babcock T. M. and Lee R. G. (1987). Examination and characterization of distribution system biofilms. *Appl. Environ. Microbiol.*, **53**, 2714–2724.

Lehtola M. J., Miettinen I. T., Vartiainen T. and Martikainen P. J. (1999). A new sensitive bioassay for determination of microbially available phosphorus in water. *Appl. Environ. Microbiol.*, **65**, 2032–2034.

Lehtola M. J., Miettinen I. T. and Martikainen P. J. (2002a). Biofilm formation in drinking water affected by low concentrations of phosphorus. *Can. J. Microbiol.*, **48**, 494–499.

Lehtola M. J., Miettinen I. T., Vartiainen T. and Martikainen P. J. (2002b). Changes in content of microbially available phosphorus, assimilable organic carbon and microbial growth potential during drinking water treatment processes. *Water Res.*, **36**, 3681–3690.

Lehtola M. J., Juhna T., Miettinen I. T., Vartiainen T. and Martikainen P. J. (2002c). Formation of biofilms in drinking water distribution networks, a case study in two cities in Finland and Latvia. *J. Ind. Microbiol. Biotechnol.*, **31**, 489–494.

Lehtola M. J., Miettinen I. T., Vartiainen T., Rantakokko P., Hirvonen A. and Martikainen P. J. (2003). Impact of UV disinfection on microbially available phosphorus, organic carbon, and microbial growth in drinking water. *Water Res.*, **37**, 1064–1070.

Lens P., Vergeldt F., Lettinga G. and Van As H. (1999). ^{1}H NMR characterisation of the diffusional properties of methanogenic granular sludge. *Wat. Sci. Technol.*, **39**, 187–194.

Li F., Meindersma W., De Haan A. B. and Reith T. (2002). Optimization of commercial net spacers in spiral wound membrane modules. *J. Membr. Sci.*, **208**, 289–302.

Li Y.-L. and Tung K.-L. (2008a). CFD simulation of fluid flow through spacer-filled membrane module: Selecting suitable cell types for periodic boundary conditions. *Desalination*, **233**, 351–358.

Li Y.-L. and Tung K.-L. (2008b). The effect of curvature of a spacer-filled channel on fluid flow in spiral-wound membrane modules. *J. Membr. Sci.*, **319**, 286–297.

Li Y.-L., Tung K.-L., Lu M.-Y. and Huang S.-H. (2009). Mitigating the curvature effect of the spacer-filled channel in a spiral-wound membrane module. *J. Membr. Sci.*, **328**, 106–118.

Lisitsin D., Hasson D. and Semiat R. (2005). Critical flux detection in a silica scaling RO system. *Desalination*, **186**, 311–318.

Liu Y. and Tay J.-H. (2002). The essential role of hydrodynamic shear force in the formation of biofilm and granular sludge. *Water Res.*, **36**, 1653–1665.

Liu L.-F., Yu S.-C., Wu L.-G. and Gao C.-J. (2006). Study on a novel polyamide-urea reverse osmosis composite membrane (ICIC-MPD) II. Analysis of membrane antifouling performance. *J. Membr. Sci.*, **283**, 133–146.

Lloyd B. (1937). Bacteria in stored sea water. *Journal of the Royal Technical College, Glasgow*, **4**, 173–177.

Loeb S. and Sourirajan S. (1960). UCLA Dept. of Engineering Report 60-60.

Loeb S. and Sourirajan S. (1963). Saline Water Conversion-II. In: Advances in Chemistry Series 38. *Amer. Chem. Soc.* Washington, DC, 117–132.

Loeb S. and Sourirajan S. (1964). High flow porous membranes for separating water from saline solutions. Patent US3133132.

Lonsdale H. K. (1982). The growth of membrane technology. *J. Membr. Sci.*, **10**, 81–181.

Louie J. S., Pinnau I., Ciobanu I., Ishida K. P., Ng A. and Reinhard M. (2006). Effects of polyether-polyamide block copolymer coating on performance and fouling of reverse osmosis membranes. *J. Membr. Sci.*, **280**, 762–770.

Lowry O. H., Rosebrough N. J., Farr A. L. and Randall R. J. (1951). Protein measurement with the folin phenol reagent. *J. Biol. Chem.*, **193**, 265–275.

Magic-Knezev A. and Van der Kooij D. (2004). Optimisation and significance of ATP analysis for measuring active biomass in granular activated carbon filters used in water treatment. *Water Res.*, **38**, 3971–3979.

Maher W. and Woo L. (1998). Procedures for the storage and digestion of natural waters for the determination of filterable reactive phosphorus, total filterable phosphorus and total phosphorus. *Anal. Chem. Acta.*, **375**, 5–47.

Mallevialle J., Odendaal P. E. and Wiesner M. R. (1996). Water Treatment Membrane Processes. McGraw-Hill, New York. ISBN 0-07-001559-7.

Mänttäri M. and Nyström M. (2000). Critical flux in NF of high molar mass polysaccharides and effluents from the paper industry. *J. Membr. Sci.*, **170**, 257–273.

Mänttäri M., Nuortila-Jokinen J. and Nyström M. (1997). Influence of filtration conditions on the performance of NF membranes in the filtration of paper mill total effluent. *J. Membr. Sci.*, **137**, 187–199.

Manz B., Volke F., Goll D. and Horn H. (2003). Measuring local flow velocities and biofilm structure in biofilm systems with magnetic resonance imaging (MRI). *Biotechnol. Bioeng.*, **84**, 424–432.

Marconnet C., Coriton G., Djafer M. and Heim V. (2009). How to control biofouling on nanofiltration membranes: Case studies of different pretreatment processes. Proceedings 2009 AWWA membrane technology conference, Memphis, Tenessee.

Marszalek D. S., Gerchakov S. M. and Udey L. R. (1979). Influence of substrate composition on marine microfouling. *Appl. Environ. Microbiol.*, **38**, 987–995.

Martín-Rosales W., Pulido-Bosch A., Vallejos Á., Gisbert J., Andreu J. M. and Sánchez-Martos F. (2007). Hydrological implications of desertification in southeastern Spain. *Hydrol. Sci. J.*, **52**, 1146–1161.

Mathieu L., Paquin J. L., Block J. C., Randon G., Maillard J. and Reasoner D. (1992). Parameters governing bacterial growth in water distribution systems [Parametres gouvernant la proliferation bacterienne dans les reseaux de distribution]. *Revue des Sciences de l'Eau*, **5**, 91–112.

McDonough F. E. and Hargrove R. E. (1972). Sanitation of reverse osmosis/ultrafiltration equipment. *J. Milk Food Technol.*, **35**, 102–106.

McDonogh R. G., Schaule G. and Flemming H.-C. (1994). The permeability of biofouling layers on membranes. *J. Membr. Sci.*, **87**, 199–217.

Melo L. F. and Bott T. R. (1997). Biofouling in water systems. *Exp. Them. Fluid Sci.*, **14**, 375–381.

Metge D. W., Brooks M. H., Smith R. L. and Harvey R. W. (1993). Effect of treated sewage contamination upon bacterial energy charge, adenine nucleotides and DNA content in a sandy aquifer on Cape Cod. *Appl. Environ. Microbiol.*, **59**, 2304–2310.

Miettinen I. T., Vartiainen T. and Martikainen P. J. (1997). Phosphorus and bacterial growth in drinking water. *Appl. Environ. Microbiol.*, **63**, 3342–3245.

Miller J. E. (2003). Review of water resources and desalination technologies. Public report from Sandia National Laboratories. NM 87185-1349 – SAND 2003-0800: 3–54.

Mittelman M. C. (1991). Bacterial growth and biofouling control in purified water systems. In: *Biofouling and biocorrosion in industrial water systems,* Flemming H. C. and Geesey G. G. (eds.). Springer, Heidelberg, pp. 113–134.

Mohamed M. N. and Robarts R. D. (2003). Sestonic bacterial nutrient limitation in a northern temperate river and the impact of pulp-mill effluents. *Aquat. Microb. Ecol.*, **33**, 19–28.

Mohamed M. N., Lawrence J. R. and Robarts R. D. (1998). Phosphorus limitation of heterotrophic biofilms from the Fraser River, Britisch Columbia, and the Effect of Pulp Mill Effluent. *Microb. Ecol.*, **36**, 121–130.

Montgomery M. A. and Elimelech M. (2007). Water and sanitation in developing countries: Including health in the equation – Millions suffer from preventable illnesses and die every year. *Environ. Sci. Technol.*, **41**, 17–24.

Murga R., Forster T. S., Brown E., Pruckler J. M., Fields B. S. and Donlan R. M. (2001). Role of biofilms in the survival of Legionella pneumophila in a model potable-water system. *Microbiology*, **147**, 3121–3126.

Niquette P., Servais P. and Savoir R. (2000). Impacts of pipe materials on densities of fixed bacterial biomass in a drinking water distribution system. *Water Res.*, **34**, 1952–1956.

Nishijima W., Shoto E. and Okada M. (1997). Improvement of biodegradation of organic substance by addition of phosphorus in biological activated carbon. *Wat. Sci. Technol.*, **36**, 251–257.

Nollet J. A. (1748). Lecons de physique experimentale, Hippolyte-Louis Guerin and Louis-Francios Delatour, Paris.

Nollet J. A. (1752). Investigations on the Causes for the Ebullition of Liquids. In: Recherches sur les causes du Bouillonnement des Liquides, Histoire de l'Académie Royale des Sciences, Année MDCCXLVIII, Paris: 57–104.

OZ REEF – Natural Sea Water Composition. (2007). http://ozreef.org.llibrary/tables/natural_sea_water_composition.html.

Packer K. J. and Tessier J. J. (1996). The characterization of fluid transport in a porous solid by pulsed gradient stimulated echo NMR. *Mol. Phys.*, **87**, 267–272.

Paetsch B. and Koetter K. (1980). Reduction of algae growth in the reservoir Haltern as the consequence of phosphorus removal by chemical precipitation [Verminderung der algenentwicklung in der talsperre Haltern durch phosphat-fallung]. *GWF, Wasser – Abwasser*, **121**, 496–498.

Pang C. M. and Liu W.-T. (2007). Community Structure Analysis of Reverse Osmosis Membrane Biofilms and the Significance of Rhizobiales Bacteria in Biofouling. *Environ. Sci. Technol.*, **41**, 4728–4734.

Pang C. M., Hong P., Guo H. and Liu W.-T. (2005). Biofilm Formation Characteristics of Bacterial Isolates Retrieved from a Reverse Osmosis Membrane. *Environ. Sci. Technol.*, **39**, 7541–7550.

Pankratz T. (2000). Large scale desalination plants: a drought-proof water supply. *IDA News*, **9**, 3–4.

Patching J. W. and Fleming G. T. A. (2003). Industrial biofilms: Formation, problems and control. In: Lens P., Moran A. P., Mahony T., Stoodley P. and O'Flaherty V. (eds). *Biofilms in Medicine, Industry and Environmental Biotechnology*. IWA publishing, UK, pp. 568–572.

Paul D. H. (1991). Reverse osmosis: scaling, fouling & chemical attack. *Desalination and Water Reuse*, **1**, 8–11.

Paul D. H. (1996). Membranes: biofouling of reverse osmosis units. *Ultrapure water*, 64–67.

Paul D. H. (2002). Back to basics: chemical cleaners – generic versus proprietary product. *Ultrapure water*, 50–51.

Pellerin E., Michelitsch E., Darcovich K., Lin S. and Tam C. M. (1995). Turbulent transport in membrane modules by CFD simulation in two dimensions. *J. Membr. Sci.*, **100**, 139–153.

Pereira M. O., Kuehn M., Wuertz S., Neu T. and Melo L. F. (2002). Effect of flow regime on the architecture of a Pseudomonas fluorescens biofilm. *Biotechnol. Bioeng.*, **78**, 164–171.

Pettersson A. and Blomqvist P. (1992). Bioassay for phosphate demand in phytoplankton from acidified lakes: Lake Njupfatet, an example of phosphate deficiency induced by liming. *Hydrobiologia*, **246**, 99–110.

Peyton B. M. (1996). Effects of shear stress and substrate loading rate on Pseudomonas aeruginosa biofilm thickness and density. *Water Res.*, **30**, 29–36.

Picioreanu C., Kreft J.-U. and Van Loosdrecht M. C. M. (2004). Particle-based multidimensional multispecies model. *Appl. Environ. Microbiol.*, **70**, 3024–3040.

Picioreanu C., Van Loosdrecht M. C. M. and Heijnen J. J. (1998a). A new combined differential-discrete cellular automaton approach for biofilm modeling: Application for growth in gel beads. *Biotechnol. Bioeng.*, **57**, 718–731.

Picioreanu C., Van Loosdrecht M. C. M. and Heijnen J. J. (1998b). Mathematical modeling of biofilm structure with a hybrid differential-discrete cellular automaton approach. *Biotechnol. Bioeng.*, **58**, 101–116.

Picioreanu C., Van Loosdrecht M. C. M. and Heijnen J. J. (1999). Discrete-differential modelling of biofilm structure. *Wat. Sci. Technol.*, **39**, 115–122.

Picioreanu C., Van Loosdrecht M. C. M. and Heijnen J. J. (2000a). A theoretical study on the effect of surface roughness on mass transport and transformation in biofilms. *Biotechnol. Bioeng.*, **68**, 355–369.

Picioreanu C., Van Loosdrecht M. C. M. and Heijnen J. J. (2000b). Effect of diffusive and convective substrate transport on biofilm structure formation: A two-dimensional modeling study. *Biotechnol. Bioeng.*, **69**, 504–515.

Picioreanu C., Van Loosdrecht M. C. M. and Heijnen J. J. (2001). Two-dimensional model of biofilm detachment caused by internal stress from liquid flow. *Biotechnol. Bioeng.*, **72**, 205–218.

Picioreanu C., Vrouwenvelder J. S. and Van Loosdrecht M. C. M. (2009). Three-dimensional numerical modeling of biofouling and fluid dynamics in feed spacer channels of membrane devices. *J. Membr. Sci.*, **345**, 340–354.

Picioreanu C., van Loosdrecht M. C. M., Curtis T. P. and Scott K. (2010). Model based evaluation of the effect of pH and electrode geometry on microbial fuel cell performance. *Bioelectrochemistry*, **78**, 8–24.

Postel S. (1997). Last Oasis: Facing Water Scarcity, Norton, New York.

Prévost M., Rompré A., Coallier J., Servais P., Laurent P., Clément B. and Lafrance P. (1998). Suspended bacterial biomass and activity in full-scale drinking water distribution systems: Impact of water treatment. *Water Res.*, **32**, 1393–1406.

Purevdorj-Cage L. B. and Stoodley P. (2004). Biofilm structure, behavior, and hydrodynamics. In: Ghannoum M. and O'Toole G. A. (eds.) *Microbial Biofilms*. ASM press, Washington DC, US, pp. 160–173.

Ranade V. V. and Kumar A. (2006a). Comparison of flow structures in spacer-filled flat and annular channels. *Desalination*, **191**, 236–244.

Ranade V. V. and Kumar A. (2006b). Fluid dynamics of spacer filled rectangular and curvilinear channels. *J. Membr. Sci.*, **271**, 1–15.

Reasoner D. J. and Geldreich E. E. (1985). A new medium for the enumeration and subculture of bacteria from potable water. *Appl. Environ. Microbiol.*, **49**, 1–7.

Ridgway H. F. (1997). Membrane biofouling: An International Workshop. National Water Res. Institute occasional paper number NWRI-97-3. Sydney, Australia: 75–79.

Ridgway H. F. (1998). Microbial adhesion and biofouling of reverse osmosis membranes. In: Parekh B. S. (ed.), *Reverse Osmosis Technology*. Marcel Dekker, New York. pp. 429–481.

Ridgway H. F. (2003). Biological fouling of separation membranes used in water treatment applications, AWWA research foundation.

Ridgway H. F. and Flemming H. F. (1996). Membrane biofouling. In: Mallevialle J., Odendaal P. E. and Wiesner M. R. (eds), *Water Treatment Membrane Processes*. McGraw-Hill, New York, pp. 6.1–6.62.

Ridgway H. F., Argo D. G. and Olson B. H. (1981). Factors Influencing Biofouling of Reverse Osmosis Membranes at Water Factory 21; Chemical, microbiological, and ultrastructural characterization of the fouling layer. Vol III-B, final report prepared for U.S. Department of the interior, office of water research and technology, contract no. 14-34-0001-8520, Washington, D.C.

Ridgway H. F., Rigby M. G. and Argo D. G. (1984). Adhesion of a Mycobacterium sp. to cellulose diacetate membranes used in reverse osmosis. *Appl. Environ. Microbiol.*, **47**, 61–67.

Ridgway H. F., Rigby M. G. and Argo D. G. (1985). Bacterial adhesion and fouling of reverse osmosis membranes. *J. Am. Water Works Assoc.*, **77**, 97–106.

Ridgway H. F., Kelly A., Justice C. and Olson B. H. (1983). Microbial fouling of reverse-osmosis membranes used in advanced wastewater treatment technology: Chemical, bacteriological, and ultrastructural analyses. *Appl. Environ. Microbiol.*, **45**, 1066–1084.

Riviere J. W. M. (1989). Threats to the World's Water. *Sci. Am.*, **261**, 80–94.

Rogers A. H. (2008). Molecular Oral Microbiology. Caister Academic Press, ISBN 978-1-904455-24-0.

Santos J. L. C., Geraldes V., Velizarov S. and Crespo J. G. (2007). Investigation of flow patterns and mass transfer in membrane module channels filled with flow-aligned spacers using computational fluid dynamics (CFD). *J. Membr. Sci.*, **305**, 103–117.

Sathasivan A. and Ohgaki S. (1999). Application of new bacterial regrowth potential method for water distribution system – A clear evidence of phosphorus limitation. *Water Res.,* **33,** 137–144.

Sathasivan A., Ohgaki S. and Otaki M. (1997). Evaluation of water treatment processes in terms of bacterial regrowth control. Submitted for 19th biennial conference of IAWQ.

Sathasivan A., Ohgaki S. and Otaki M. (1998). Can phosphorus control be a feasible option to control regrowth in Tokyo drinking water distribution system? *Water Supply*, **16**, 249.

Scavia D. and Laird G. A. (1987). Bacterioplankton in Lake Michigan: Dynamics, controls, and significance to carbon flux. *Limnol. Oceanogr.*, **32**, 1017–1033.

Schaule G. (1992). Bakterielle Adhäsion und Aggregation auf inerten Oberflächen: Ihre Rolle bei der Bildung von Biofilmen auf Umkehrosmose-Membranen. PhD Thesis. ISBN 3-486-26357-9. Forschungs- und Entwicklungsinstitut für Industrie- und Siedlungswasserwirtschaft sowie Abfallwirtschaft e.V., Stuttgart, Germany.

Schaule G., Kern A. and Flemming H. C. (1993). RO treatment of dump trickling water: Membrane biofouling. *Desalination and Water Reuse*, **3**, 17–23.

Schneider R. P., Ferreira L. M., Binder P., Bejarano E. M., Góes K. P., Slongo E., Machado C. R. and Rosa G. M. Z. (2005). Dynamics of organic carbon and of bacterial populations in a conventional pretreatment train of a reverse osmosis unit experiencing severe biofouling. *J. Membr. Sci.*, **266**, 18–29.

Schock G. and Miquel A. (1987). Mass transfer and pressure loss in spiral wound modules. *Desalination*, **64**, 339–352.

Schwinge J., Neal P. R., Wiley D. E. and Fane A. G. (2002a). Estimation of foulant deposition across a leaf of a spiral-wound module. *Desalination*, **146**, 203–208.

Schwinge J., Wiley D. E. and Fletcher D. F. (2002b). A CFD study of unsteady flow in narrow spacer-filled channels for spiral-wound membrane modules. *Desalination*, **146**, 195–201.

Schwinge J., Wiley D. E. and Fletcher D. F. (2002c). Simulation of the flow around spacer filaments between narrow channel walls. 1. Hydrodynamics. *Ind. Eng. Chem. Res.*, **41**, 2977–2987.

Schwinge J., Wiley D. E. and Fletcher D. F. (2002d). Simulation of the flow around spacer filaments between channel walls. 2. Mass-transfer enhancement. *Ind. Eng. Chem. Res.*, **41**, 4879–4888.

Schwinge J., Neal P. R., Wiley D. E, Fletcher D. F. and Fane A. G. (2004). Spiral wound modules and spacers: Review and analysis. *J. Membr. Sci.*, **242**, 129–153.

Servais P., Laurent P. and Randon G. (1995). Comparison of the bacterial dynamics in various French distribution systems. *Aqua London*, **44**, 10–17.

Shakaib M., Hasani S. M. F. and Mahmood M. (2007). Study on the effects of spacer geometry in membrane feed channels using three-dimensional computational flow modeling. *J. Membr. Sci.*, **297**, 74–89.

Shakaib M., Hasani S. M. F. and Mahmood M. (2009). CFD modeling for flow and mass transfer in spacer-obstructed membrane feed channels. *J. Membr. Sci.*, **326**, 270–284.

Shrivastava A., Kumar S. and Cussler E. L. (2008). Predicting the effect of membrane spacers on mass transfer. *J. Membr. Sci.*, **323**, 247–256.

Seymour J. D., Gage J. P., Codd S. L. and Gerlach R. (2004). Anomalous fluid transport in porous media induced by biofilm growth. *Phys. Rev. Lett.*, **93**, 198103 1–4.

Seymour J. D., Gage J. P., Codd S. L. and Gerlach R. (2007). Magnetic resonance microscopy of biofouling induced scale dependent transport in porous media. *Adv. Water Res.*, **30**, 1408–1420.

Shannon M. A., Bohn P. W., Elimelech M., Georgiadis J. G., Marinãs B. J. and Mayes A. M. (2008). Science and technology for water purification in the coming decades. *Nature*, **452**, 301–310.

Speth T. F., Gusses A. M. and Scott Summers R. (2000). Evaluation of nanofiltration pretreatments for flux loss control. *Desalination*, **130**, 31–44.

Stejskal E. O. and Tanner J. E. (1965). Spin diffusion measurements: Spin echoes in the presence of a time-dependent field gradient. *J. Chem. Phys.*, **42**, 288–292.

Stikker A. (2002). Desal technology can help quench the world's thirst. *Water policy*, **4**, 47–55.

Stoodley P., Lewandowski Z., Boyle J. D. and Lappin-Scott H. M. (1998). Oscillation characteristics of biofilm streamers in turbulent flowing water as related to drag and pressure drop. *Biotechnol. Bioeng.*, **57**, 536–544.

Stoodley P., Dodds I., Boyle J. D. and Lappin-Scott H. M. (1999a). Influence of hydrodynamics and nutrients on biofilm structure. *J. Appl. Microbiol. Symp. Suppl.*, **85**, 19S–28S.

Stoodley P., Lewandowski Z., Boyle J. D. and Lappin-Scott H. M. (1999b). Structural deformation of bacterial biofilms caused by short-term fluctuations in fluid-shear: An in situ investigation of biofilm rheology. *Biotechnol. Bioeng.*, **65**, 83–92.

Stoodley P., Cargo R., Rupp C. J., Wilson S. and Klapper I. (2002). Biofilm material properties as related to shear-induced deformation and detachment phenomena. *J. Ind. Microbiol. Biotechnol.*, **29**, 361–367.

Stoodley P., Dodds I., De Beer D., Scott H. L. and Boyle J. D. (2005). Flowing biofilms as a transport mechanism for biomass through porous media under laminar and turbulent conditions in a laboratory reactor system. *Biofouling*, **21**, 161–168.

Subramani A., Kim S. and Hoek E. M. V. (2006). Pressure, flow, and concentration profiles in open and spacer-filled membrane channels. *J. Membr. Sci.*, **277**, 7–17.

Taherzadeh D., Picioreanu C., Küttler U., Simone A., Wall W. A. and Horn H. (2009). Computational study of the drag and oscillatory movement of biofilm streamers in fast flows. *Biotechnol. Bioeng.*. In press.

Tallarek U., Van Dusschoten D., Van As H., Bayer E. and Guiochon G. (1998). Study of transport phenomena in chromatographic columns by pulsed field gradient NMR. *J. Phys. Chem. B.*, **102**, 3486–3497.

Tallarek U., Vergeldt F. J. and Van As H. (1999). Stagnant mobile phase mass transfer in chromatographic media: intrapartcile diffusion and exchange kinetics. *J. Phys. Chem. B.*, **103**, 7654–7664.

Tasaka K., Katsura T., Iwahori H. and Kamiyama Y. (1994). Analysis of RO elements operated at more than 80 plants in Japan. *Desalination*, **96**, 259–272.

Tchobanoglous G., Burton F. L. and Stensel H. D. (2003). Wastewater Engineering: Treatment and Reuse, Metcalf and Eddy, 4th edn., McGraw-Hill, Inc., New York.

Tibbetts J. (2000). Water World 2000. *Environ. Health Perspect.*, **108**, A69–A73.

Toolan T., Wehr J. D. and Findlay S. (1991). Inorganic phosphorus stimulation of bacterioplankton production in a meso-eutrophic lake. *Appl. Environ. Microbiol.*, **57**, 2074–2078.

Torvinen E., Lehtola M. J., Martikainen P. J. and Miettinen I. T. (2007). Survival of Mycobacterium avium in drinking water biofilms as affected by water flow velocity, availability of phosphorus, and temperature. *Appl. Environ. Microbiol.*, **73**, 6201–6207.

Traube M. (1867). Experimente zur Theorie der Zellenbildung und Endosmose. Physiologie und Wissenschaftliche Medicin. In: Reischert and DuBois-Reynolds (eds.). Archiv fur Anatomie, Leipzig.

Trisep. www.trisep.com.

Tijhuis L., Hijman B., Van Loosdrecht M. C. M. and Heijnen J. J. (1996). Influence of detachment, substrate loading and reactor scale on the formation of biofilms in airlift reactors. *Appl. Microbiol. Biotechnol.*, **45**, 7–17.

Uchymiak M., Alex B., Christofides P., Daltrophe N., Weissman M., Gilron J., Rallo R. and Cohen Y. (2008). RO membrane desalting in a feed flow reversal mode. In: Proceedings 8th International Congress on Membranes and membrane processes (ICOM 2008) held in Honolulu, Hawaii, 12–18 July.

Ummenhofer C. C., England M. H., McIntosh P. C., Meyers G. A., Pook M. J., Risbey J. S., Gupta A. S. and Taschetto A. S. (2009). What causes southeast Australia's worst droughts? *Geophys. Res. Lett.*, **36**, L04706.

United Nations Environmental Programme (UNEP). (2008). Vital Water Graphics – An Overview of the State of the World's Fresh and Marine Waters. 2nd edn. UNEP, Nairobi, Kenya. ISBN: 92-807-2236-0. http://www.unep.org/dewa/vitalwater/index.html.

United Nations Millennium Declaration. (2000). http://www.un.org/millennium/declaration/ares552e.pdf.

University of California. (1965). High flow semi-permeable membranes for separating water from saline solutions. Patent GB1009172.

Van As H. and Lens P. (2001). Use of ^{1}H NMR to study transport processes in porous biosystems. *J. Ind. Microbiol. Biotechnol.*, **26**, 43–52.

Van der Aa L. T. J., Kors L. J., Wind A. P. M., Hofman J. A. M. H. and Rietveld L. C. (2002). Nitrification in rapid sand filter: Phosphate limitation at low temperatures. *Wat. Sci. Technol.: Wat. Suppl.*, **2**, 37–46.

Van der Hoek J. P., Hofman J. A. M. H., Bonné P. A. C., Nederlof M. M. and Vrouwenvelder H. S. (2000). RO treatment: Selection of a pretreatment scheme based on fouling characteristics and operating conditions based on environmental impact. *Desalination*, **127**, 89–101.

Van der Kooij D. (1992). Assimilable organic carbon as an indicator of bacterial regrowth. *J. Am. Wat. Works Assoc.*, **84**, 57–65.

Van der Kooij D., Visser A. and Hijnen W. A. M. (1982). Determination the concentration of easily assimilable organic carbon in drinking water. *J. Am. Wat. Works Assoc.*, **74**, 540–545.

Van der Kooij D., Vrouwenvelder H. S. and Veenendaal H. R. (1997). Bepaling en betekenis van de biofilmvormende eigenschappen van drinkwater. *H2O*, **30**, 767–771.

Van der Kooij D., Veenendaal H. R., Baars-Lorist C., Van der Klift D. W. and Drost Y. C. (1995). Biofilm formation on surfaces of glass and teflon exposed to treated water. *Water Res.*, **29**, 1655–1662.

Van der Kooij D., Hijnen W., Cornelissen E., Van Agtmaal S., Baas K. and Galjaard G. (2007). Elucidation of membrane biofouling processes using bioassays for assessing the microbial growth potential of feed water. Proceedings AWWA membrane technology conference, Tampa Bay, Florida.

Van der Meer W. G. J. (2003a). Mathematical modeling of NF and RO membrane filtration plants and modules. PhD Thesis, Delft University of Technology, The Netherlands.

Van der Meer W. G. J., Van Paassen J. A. M., Riemersma M. C. and Van Ekkendonk F. H. J. (2003b). Optiflux: From innovation to realization. *Desalination*, **157**, 159–165.

Van der Wende E., Characklis W. G. and Smith D. B. (1989). Biofilms and bacterial drinking water quality. *Water Res.*, **23**, 1313–1322.

Van Donk E., Mur L. R. and Ringelberg. (1990). A study of phosphate limitation in Lake Maarsseveen: phosphate uptake kinetics versus bioassays J. Environmental bioassay techniques and their application. *Proc. conference, Lancaster*, **1988**, 201–209.

Van Dijk J. C. and Van der Kooij D. (2004). Water quality 21 research programme of the water supplies in the Netherlands. *Wat. Sci. Technol.: Wat. Suppl.*, **4**, 181–188.

Van Gauwbergen D. and Baeyens J. (1997). Macroscopic fluid flow distribution distributions in spiral wound membrane elements. *Desalination*, **110**, 287–299.

Van Hoof S. C. J. M., Minnery J. G. and Mack B. (2002). Performing a membrane autopsy. *Desalination and Water Reuse*, **11**, 40–46.

Van Loosdrecht M. C. M., Lyklema J., Norde W. and Zehnder A. J. B. (1990). Influence of interfaces on microbial activity. *Microbiol. Rev.*, **54**, 75–87.

Van Loosdrecht M. C. M., Eikelboom D., Gjaltema A., Mulder A., Tijhuis L. and Heijnen J. J. (1995). Biofilm structures. *Wat. Sci. Technol.*, **32**, 35–43.

Van Loosdrecht M. C. M., Picioreanu C. and Heijnen J. J. (1997). A more unifying hypothesis for biofilm structures. *FEMS Microbiol. Ecol.*, **24**, 181–183.

Van Paassen J. A. M., Kruithof J. C., Bakker S. M. and Schoonenberg Kegel F. (1998). Integrated multi-objective membrane systems for surface water treatment: Pretreatement of nanofiltration by riverbank filtration and conventional ground water treatment. *Desalination*, **118**, 239–248.

Van Paassen J. A. M., Van der Meer W. G. J. and Post J. (2005). Optiflux: From innovation to realisation. *Desalination*, **178**, 325–331.

Verdouw J. and Folmer H. C. (1997). Effect of water temperature on the normalized flux for RO- membranes. AWWA Membrane technology conference proceedings. February 23–36, (1997), New Orleans, LA. ISBN 0-89867-891-X: 1001–1011.

Verliefde A. R. D. (2008). Rejection of organic micropollutants by high pressure membranes (NF/RO). PhD Thesis, Delft University of Technology, The Netherlands. Gildeprint. ISBN 978-90-8957-005-5.

Visvanathan C., Boonthanon N., Sathasivan A. and Jegatheesan V. (2002). Pretreatment of seawater for biodegradable organic content removal using membrane bioreactor. *Desalination*, **153**, 133–140.

Von Rège H. and Sand W. (1998). Evaluation of biocide efficacy by microcalorimetric determination of microbial activity in biofilms. *J. Microbiol. Methods*, **33**, 227–235.

Vrede K., Heldal M., Norland S. and Bratbak G. (2002). Elemental Composition (C, N, P) and Cell Volume of Exponentially Growing and Nutrient-Limited Bacterioplankton. *Appl. Environ. Microbiol.*, **68**(6), 2965–2971.

Vrouwenvelder H. S., Van Paassen J. A. M., Folmer H. C., Hofman J. A. M. H., Nederlof M. M. and Van der Kooij D. (1998). Biofouling of membranes for drinking water production. *Desalination*, **118**, 157–166.

Vrouwenvelder J. S., Manolarakis S. A., Veenendaal H. R. and Van Der Kooij D. (2000). Biofouling potential of chemicals used for scale control in RO and NF membranes. *Desalination*, **132**, 1–10.

Vrouwenvelder J. S. and Van der Kooij D. (2001). Diagnosis, prediction and prevention of biofouling of NF and RO membranes. *Desalination*, **139**, 65–71.

Vrouwenvelder J. S. and Van der Kooij D. (2003a). Diagnosis of fouling problems of NF and RO membrane installations by a quick scan. *Desalination*, **153**, 121–124.

Vrouwenvelder J. S., Kappelhof J. W. N. M., Heijman S. G. J., Schippers J. C. and Van der Kooij D. (2003b). Tools for fouling diagnosis of NF and RO membranes and assessment of the fouling potential of feed water. *Desalination*, **157**, 361–365.

Vrouwenvelder J. S., Lampe M. and Hijnen W. A. M. (2006a). Groeibevorderende eigenschappen van antiscalants toegepast op productiebedrijf Jan Lagrand. KWR 06.029. In Dutch. Kiwa Water Research, Nieuwegein, The Netherlands.

Vrouwenvelder J. S., Van Paassen J. A. M., Wessels L. P., Van Dam A. F. and Bakker S. M. (2006b). The membrane fouling simulator: A practical tool for fouling prediction and control. *J. Membr. Sci.*, **281**, 316–324.

Vrouwenvelder J. S., Bakker S. M., Cauchard M., Le Grand R., Apacandié M., Idrissi M., Lagrave S., Wessels L. P., Van Paassen J. A. M., Kruithof J. C. and Van Loosdrecht M. C. M. (2007a). The membrane fouling simulator: A suitable tool for prediction and characterisation of membrane fouling. *Wat. Sci. Technol.*, **55** (8–9), 197–205.

Vrouwenvelder J. S., Bakker S. M., Wessels L. P. and Van Paassen J. A. M. (2007b). The membrane fouling simulator as a new tool for biofouling control of spiral wound membranes. *Desalination*, **204**, 170–174.

Vrouwenvelder J. S., Hinrichs C., Sun A. R., Royer F., van Paassen J. A. M., Bakker S. M., Van der Meer W. G. J., Kruithof J. C. and Van Loosdrecht M. C. M. (2008a). Monitoring and control of biofouling in nanofiltration and reverse osmosis membranes. *Wat. Sci. Technol.: Wat. Suppl.*, **8**, 449–458.

Vrouwenvelder J. S., Manolarakis S. A., Van der Hoek J. P., Van Paassen J. A. M., Van der Meer W. G. J., Van Agtmaal J. M. C., Prummel H. D. M., Kruithof J. C. and Van Loosdrecht M. C. M. (2008b). Quantitative Biofouling Diagnosis in Full Scale Nanofiltration and Reverse Osmosis Installations. *Water Res.*, **42**, 4856–4868.

Vrouwenvelder J. S., Hinrichs C., Van der Meer W. G. J., Van Loosdrecht M. C. M. and Kruithof J. C. (2009a). Pressure drop increase by biofilm accumulation in spiral wound RO and NF membrane systems: Role of substrate concentration, flow velocity, substrate load and flow direction. *Biofouling*, **25**, 543–555.

Vrouwenvelder J. S., van Paassen J. A. M., van Agtmaal J. M. C., van Loosdrecht M. C. M. and Kruithof J. C. (2009b). A critical flux to avoid biofouling of spiral wound nanofiltration and reverse osmosis membranes: Fact or fiction? *J. Membr. Sci.*, **326**, 36–44.

Vrouwenvelder J. S., Graf von der Schulenburg D. A., Kruithof J. C., Johns M. L. and van Loosdrecht M. C. M. (2009c). Biofouling of spiral wound nanofiltration and reverse osmosis membranes: A feed spacer problem. *Water Res.*, **43**, 583–594.

Vrouwenvelder J. S., Picioreanu C., Kruithof J. C. and Van Loosdrecht M. C. M. (2010a). Biofouling in spiral wound membrane systems: Three-dimensional numerical model based evaluation of experimental data. *J. Membr. Sci.*, **346**, 71–85.

Vrouwenvelder J. S., Beyer F., Dahmani K., Galjaard G., Hasan M. N., Kruithof J. C. and Van Loosdrecht M. C. M. (2010c). Phosphate limitation to control biofouling. *Water Res.*, **44**(11), 3454–3466.

Vrouwenvelder J. S., Buiter J., Riviere M., Van der Meer W. G. J., Van Loosdrecht M. C. M. and Kruithof J. C. (2010b). Impact of flow regime on pressure drop increase and biomass accumulation and morphology in membrane systems. *Water Res.*, **44**(3), 689–702.

Vrouwenvelder, J. S., Kruithof, J. and Van Loosdrecht, M. C. M. (2010). Integrated approach for biofouling control. *Wat. Sci. Technol.* 62 (11), 2477–2490.

Vrouwenvelder J. S., Van Loosdrecht M. C. M. and Kruithof J. C. (2011). Early warning of biofouling in spiral wound nanofiltration and reverse osmosis membranes. *Desalination*, **265**, 206–212.

Vrouwenvelder J. S., Van Paassen J. A. M., Kruithof J. C. and Van Loosdrecht M. C. M. (2009d). Sensitive pressure drop measurements of individual lead membrane elements for accurate early biofouling detection. *J. Membr. Sci.*, **338**, 92–99.

Wangnick/GWI. (2005). Worldwide desalting plants inventory 2004. Global Water Intelligence. Oxford, England. Data Table 22.

Wanner O., Eberl H. J., Morgenroth E., Noguera D., Picioreanu C., Rittmann B. E. and Van Loosdrecht M. C. M. (2006). Mathematical modeling of biofilms. IWA Scientific and Technical Report No.18, IWA Publishing, London, UK.

Wardeh S. and Morvan H. P. (2008). CFD simulations of flow and concentration polarization in spacer-filled channels for application to water desalination. *Chem. Eng. Res. Des.*, **86**, 1107–1116.

Wäsche S., Horn H. and Hempel D. C. (2002). Influence of growth conditions on biofilm development and mass transfer at the bulk/biofilm interface. *Water Res.*, **36**, 4775–4784.

Wend C. F., Stewart P. S., Jones W. and Camper A. K. (2003). Pretreatment for membrane water treatment systems: A laboratory study. *Water Res.*, **37**, 3367–3378.

Wessels L. P., Rietman B. M. and Jong R. C. M. (2001). Method and device for the purification of surface water. Dutch patent NL1019130. International patent document WO 03031342.

Whittaker C., Ridgway H. F. and Olson B. H. (1984). Evaluation of cleaning strategies for removal of biofilms from reverse-osmosis membranes. *Appl. Environ. Microbiol.*, **48**, 395–403.

Wijeyekoon S., Mino T., Satoh H. and Matsuo T. (2004). Effects of substrate loading rate on biofilm structure. *Water Res.*, **38**, 2479–2488.

Wilf M. and Alt A. (2000). Application of low fouling RO membrane elements for reclamation of municipal wastewater. *Desalination*, **132**, 11–19.

Wills A., Bott T. R. and Gibbard I. J. (2000). The control of biofilms in tubes using wire-wound inserts. *Can. J. Chem. Eng.*, **78**, 61–64.

Wingender J., Neu T. R. and Flemming H.-C. (1999). Microbial extracellular polymeric substances: Characterization, structure, and function. Springer-Verlag. Berlin Heidelberg. ISBN 3-540-65720-7.

Winters H. (1997). Twenty years experience in seawater reverse osmosis and how chemicals in pretreatment affect fouling of membranes. *Desalination*, **110**, 93–96.

Winters H. (2001). Identification of critical flux and cross flow conditions for the control of bacterial and organic fouling of seawater reverse osmosis membranes, The Middle East Desalination Research Center. MEDRC series of R&D reports. Project 97-AS-004b.

Winters H. and Isquith I. R. (1979). In plant microfouling in desalination. *Desalination*, **30**, 387–399.

World Health Organization. (2009). http://www.who.int/topics/drinking_water/en/.

WWDR 2. (2006). Water, a shared responsibility. The United Nations World Water Development Report 2 http://www.unesco.org/water/wwap.

Xavier J. B., Picioreanu C. and Van Loosdrecht M. C. M. (2005). A framework for multidimensional modelling of activity and structure of multispecies biofilms. *Environ. Microbiol.*, **7**, 1085–1103.

Xu P., Drewes J. E., Kim T.-U., Bellona C. and Amy G. (2006). Effect of membrane fouling on transport of organic contaminants in NF/RO membrane applications. *J. Membr. Sci.*, **279**, 165–175.

Yao S., Costello M., Fane A. G. and Pope J. M. (1995). Non-invasive observation of flow profiles and polarisation layers in hollow fibre membrane filtration modules using NMR micro-imaging. *J. Membr. Sci.*, **99**, 207–216.

Zobell C. E. (1943). The effect of solid surfaces upon bacterial activity. *J. Bacteriol.*, **46**, 39–56.

Zobell C. E. and Anderson D. Q. (1936). Observations on the multiplication of bacteria in different volumes of stored seawater and the influence of oxygen tension and solid surfaces. *Biol. Bull*, **71**, 324–342.

Zondervan E. and Roffel B. (2007). Evaluation of different cleaning agents used for cleaning ultra filtration membranes fouled by surface water. *J. Membr. Sci.*, **304**, 40–49.

Zohary T., Herut B., Krom M. D., Mantoura R. F. C., Pitta P., Psarra S., Rassoulzadegan F., Stambler N., Tanaka T., Frede Thingstad T., Malcolm S. and Woodward E. (2005). P-limited bacteria but N and P co-limited phytoplankton in the Eastern Mediterranean – A microcosm experiment. *Deep-Sea Research Part II: Topical Studies in Oceanography*, **52**, 3011–3023.

Nomenclature

This section provides a list of the most commonly used symbols in this book.

Abbreviations

2-d	two-dimensional	(−)
3-d	three-dimensional	(−)
ABC	approach for biofouling control	(−)
ADK	waterwork research facility in Andijk	(−)
AS	anti-scalant	(−)
ATP	adenosinetriphosphate	(pg cm^{-2} or pg mL^{-1})
AOC	assimilable organic carbon; µg acetate-C equivalents/L	(µg C L^{-1})
BFR	biofilm formation rate in glass ring biofilm monitor	(pg ATP cm^{-2} day^{-1})
CFD	computational fluid dynamics	(−)
CFU	colony forming unit	(CFU)
DOC	dissolved organic carbon	(mg C L^{-1})
EPS	extracellular polysaccharides or exopolymeric substances	(−)
FCP	feed channel pressure drop	(kPa)
ΔFCP	feed channel pressure drop increase	(kPa or %)
HMK	waterwork Jan Lagrand in Heemskerk	(−)
HPC	heterotrophic plate count	(CFU cm^{-2} or CFU mL^{-1})
ICP-MS	inductively coupled plasma mass spectrometry	(−)

MFI	modified fouling index	($s L^{-2}$)
MFS	membrane fouling simulator	(−)
mil	unit of length equal to one thousandth (10^{-3}) of an inch	(m)
MRI	magnetic resonance imaging	(−)
MTC	normalized flux (mass transfer coefficient)	($m s^{-1} Pa^{-1}$)
NF	nanofiltration	(−)
NMR	nuclear magnetic resonance	(−)
NPD	normalized pressure drop over the feed spacer channel	(kPa)
ΔNPD	relative increase of the normalized pressure drop × 100	(%)
P	phosphate	(−)
PT1	standard pressure transmitter	(−)
PT2	high precision pressure transmitter	(−)
PT3, PT4	sensitive pressure difference transmitter	(−)
RO	reverse osmosis	(−)
S	substrate (dosed to feed water)	(−)
SD	standard deviation	(−)
SEM	scanning electron microscope	(−)
SEM-EDX	SEM energy-dispersive X-ray spectroscopy analysis	(−)
S-MFS	Spacer MFS (plastic) used for e.g. NMR studies	(−)
TDC	total direct cell count	(cells cm^{-2})
TMP	trans-membrane pressure drop	(kPa)
TOC	total organic carbon	($mg C L^{-1}$)

List of symbols

a	coefficient	(−)
A_{eff}	effective surface area	(m^2)
b	coefficient	(−)
c	coefficient	(−)
C_{bulk}	concentration nutrient in bulk liquid	($kg m^{-3}$)
$C_{membrane}$	concentration nutrient at biofilm surface	($kg m^{-3}$)

D	diffusion coefficient	($m^2 s^{-1}$)
d_h	hydraulic diameter (of channel)	(m)
h	feed channel height	(m)
J_{dif}	diffusive flux	($kg m^{-2} s^{-1}$)
J_{con}	convective flux	($kg m^{-2} s^{-1}$)
J_v	permeation rate	($m s^{-1}$)
k	mass transport coefficient (of acetate)	($m s^{-1}$)
L	length (of membrane or MFS)	(m)
ΔP	pressure drop	($kg m^{-1} s^{-2}$)
Q	volumetric feed flow rate	($m^3 s^{-1}$)
Re	Reynolds number	(−)
S	specific surface of feed spacer	($m^2 m^{-3}$)
Sc	Schmidt number	(−)
Sh	Sherwood number	(−)
v	linear flow velocity	($m s^{-1}$)
w	membrane width (element or MFS)	(m)

Greek

λ	friction coefficient	(−)
ρ	specific density	($kg m^{-3}$)
ε	porosity of feed spacer	(−)
η	dynamic viscosity	($kg m^{-1} s^{-1}$)
υ	kinematic viscosity	($m^2 s^{-1}$)
τ	shear stress	(Pa)

Permissions

All chapters in this book were first published by IWA Publishing; hereby published with permission under the Creative Commons Attribution License or equivalent. Every chapter published in this book has been scrutinized by our experts. Their significance has been extensively debated. The topics covered herein carry significant information for a comprehensive understanding. They may even be implemented as practical applications or may be referred to as a beginning point for further studies.

The contributors of this book come from diverse backgrounds, making this book a truly international effort. We would like to thank all the contributing authors for lending their expertise to make the book truly unique. They have played a crucial role in the development of this book. Without their invaluable contributions this book wouldn't have been possible. They have made vital efforts to compile up to date information on the varied aspects of this subject to make this book a valuable addition to the collection of many professionals and students.

This book was conceptualized with the vision of imparting up-to-date and integrated information in this field. To ensure the same, a matchless editorial board was set up. Every individual on the board went through rigorous rounds of assessment to prove their worth. After which they invested a large part of their time researching and compiling the most relevant data for our readers.

The editorial board has been involved in producing this book since its inception. They have spent rigorous hours researching and exploring the diverse topics which have resulted in the successful publishing of this book. They have passed on their knowledge of decades through this book. To expedite this challenging task, the publisher supported the team at every step. A small team of assistant editors was also appointed to further simplify the editing procedure and attain best results for the readers.

Apart from the editorial board, the designing team has also invested a significant amount of their time in understanding the subject and creating the most relevant covers. They scrutinized every image to scout for the most suitable representation of the subject and create an appropriate cover for the book.

The publishing team has been an ardent support to the editorial, designing and production team. Their endless efforts to recruit the best for this project, has resulted in the accomplishment of this book. They are a veteran in the field of academics and their pool of knowledge is as vast as their experience in printing. Their expertise and guidance has proved useful at every step. Their uncompromising quality standards have made this book an exceptional effort. Their encouragement from time to time has been an inspiration for everyone.

The publisher and the editorial board hope that this book will prove to be a valuable piece of knowledge for students, practitioners and scholars across the globe.

Index

A

Adenosinetriphosphate, 22, 25, 41, 51, 75, 145, 167, 189, 315

Ammonia, 15, 253, 279

Anaerobic Groundwater, 23, 212, 236

Antiscalants, 18, 66, 254-255, 261, 264-265, 269, 271-272, 289, 294, 312

Assimilable Organic Carbon, 22, 26, 301, 303, 311, 315

B

Biocide Dosage, 59, 159, 228, 274

Biodegradable Compounds, 30, 33-34, 37, 50, 52, 70, 82, 84, 142, 144, 151, 157, 171, 210, 212, 219, 226, 235, 240, 258, 288-289

Biodegradable Substrate, 175, 263

Biofouling, 1, 4, 6, 8, 12-18, 20-22, 24, 26, 30, 32-35, 209-213, 248-255, 261-264, 269-272, 274-290, 292-294, 296-303, 305-309, 311-313, 315

Biofouling Detection, 17, 69, 84-85, 156, 278, 286, 313

Biofouling Process, 68, 83, 108, 156, 165, 173, 206, 226, 228, 275, 294

Biomass, 12-13, 16-18, 21-22, 24-35, 37-41, 47, 50-51, 226-232, 234, 236, 239-242, 244, 246-254, 261-269, 271-272, 277-289, 306-307, 309, 313

Biomass Growth, 50, 106, 111, 115, 123, 127, 135, 188, 190-191, 194-195, 200, 204, 206-207, 236, 246, 250, 261, 263, 272, 279-280, 289, 294

Biomass Parameters, 16, 21-22, 26-27, 29-31, 34-35, 37, 39, 41, 51, 72, 145-146, 162-163, 212, 223, 249, 263, 266

C

Capillary Membranes, 288, 293-294

Cell Counts, 22, 25, 51, 75, 146, 223

Channel Formation, 105, 194, 196, 203-204

Channel Pressure, 5, 7-8, 11, 18, 24, 26-27, 39, 44-45, 198, 204-207, 211, 213, 223, 227-229, 231-232, 238, 251, 253, 261-263, 265, 271-272, 275, 293, 315

Chemical Dosage, 14, 60, 65-66, 261-262, 279, 285, 287

Chloramination, 15, 253, 279

Chlorine Damages, 14, 253, 279

Cleaning Strategies, 35, 45, 275, 280, 282, 286, 313

Computational Domain, 106-107, 109, 112, 114, 116, 130, 139, 184-185, 194

Concentration Polarization, 10, 105, 129, 134, 138-139, 178-179, 191, 205-206, 225, 249-250, 270-271, 275, 290, 295, 313

D

Desalination, 1-4, 6, 72, 88, 254, 270, 296-300, 302-314

Disinfection, 2, 71, 252, 254, 304

F

Feed Pressure, 5-6, 12, 24, 49, 58, 65, 71-72, 109, 144-145, 167, 178, 181

Feed Side, 9-10, 23-25, 27, 32, 40, 46-47, 57, 75, 80-81, 83-84, 142, 151-152, 154, 157, 177, 179, 194, 218, 220, 225-226, 262, 269-272, 283, 286

Feed Spacers, 9, 61, 89, 139, 143, 157, 162, 177, 179-180, 182-184, 187, 189-190, 203, 205-207, 235, 251, 257, 283, 287-288, 292-293

Flow Cell, 89, 92-93, 96

Fouling Simulator, 17, 41, 44-45, 47-49, 57, 59, 187, 210-211, 233, 257, 272, 278, 282, 284-285, 287, 292, 294, 312, 316

G

Gradient Increments, 92-93, 166

Growth Rate, 111, 190-191, 276, 289, 291

H

High Pressure Membrane, 44, 69, 85, 141, 181, 209, 270, 274, 285

Human Carcinogen, 15, 253, 279

Hydrodynamics, 16-17, 38, 47, 67, 103, 105, 108, 182-183, 187, 206, 210, 223, 227-228, 230, 248-249, 280, 288, 295, 300, 307-309

L

Lead Membrane, 27-28, 50, 56-57, 71, 80-82, 84-86, 117, 152-153, 170, 179, 186, 194, 218, 228-229, 251, 277-278, 282-283, 286-287, 313

Liquid Velocity, 108-109, 116, 122, 139

Luminometer, 25, 51, 75, 145-146, 167, 189, 212

M

Mass Transfer, 38, 105, 107, 111-112, 123, 134-135, 147, 149, 161, 182, 205, 277, 290, 298, 303, 308-310, 313, 316

Mass Transport, 9, 92, 103, 107, 112-113, 122-123, 130, 134, 137, 147-148, 186, 205-206, 297, 307, 317

Mean Velocity, 95, 101-102

Membrane Elements, 4, 9-12, 16-18, 21-28, 30-31, 34-35, 155, 157-158, 162-164, 167, 169-170, 257, 261-262, 270, 272, 274, 277, 297, 311, 313

Membrane Fouling, 4, 11-12, 17, 20, 24, 41, 44-45, 47-49, 179-180, 183, 187, 207, 209-211, 227, 277-278, 282, 284-285, 287, 292, 294, 302, 312, 314, 316

Membrane Installation, 8, 21-22, 26, 30, 37, 40-41, 47, 80-82, 84-86, 139, 144, 177-178, 218, 228-229, 254, 257, 262, 271

Membrane Length, 46, 49, 186, 238

Membrane Modules, 57, 72, 78-79, 81-85, 87, 101, 167, 169-170, 177-179, 182-184, 186-188, 190, 194, 196, 204-205, 207, 269, 282-289, 293, 304, 306, 308

Membrane Oxidation, 15, 253, 279

Membrane Performance, 5, 7, 16, 18, 20, 38-39, 45, 61, 66, 80, 95, 104-105, 138, 155, 168, 177-179, 206-207, 223, 248-249, 251, 275

Membrane Rejection, 5, 15, 253, 279

Membrane Resistance, 15, 253, 279

Membrane Surface, 8-9, 81, 101-103, 117, 121, 134-135, 146-148, 155, 170, 185, 191, 194, 200, 205-206

Microbial Growth, 115, 186, 252-253, 270, 280, 289-290, 303-304, 311

Molecular Weight, 6, 39, 71, 144

Monochloramines, 15, 253

Morphology, 17-18, 104, 172, 223, 227, 231-232, 238, 244, 247-249, 251, 280, 289, 291, 293, 313

O

Organic Fouling, 12, 20, 44, 59, 104, 181, 209, 314

Osmotic Pressure, 7-8, 58, 61, 64, 105, 139, 177, 206, 225, 229, 297, 301

P

Passive Tracer, 89, 92, 103

Peristaltic Pump, 50, 143, 164, 188, 212, 236, 261

Pesticides, 2, 7, 254

Phase Shift Velocity, 92-93, 166, 189

Phosphate Concentration, 263, 268-271, 294

Plant Performance, 8, 39, 223, 229, 271, 293

Polyamide, 7-8, 22, 143, 161-162, 187, 297, 299, 304

Pressure Drop, 5, 7-8, 11, 13, 16-18, 20-24, 26-28, 30, 163, 166-167, 170-171, 173, 253, 257, 259, 287-288, 293-294, 302, 309, 312-313, 315-317

Pressure Transmitter, 50, 65, 72-73, 76, 78, 80-81, 85, 145, 166-167, 189, 212, 236, 259, 278, 316

Pressure Vessel, 10, 22, 24, 35, 56, 78, 80, 86-87, 117, 152, 162, 164, 177, 186, 213, 218, 220-221, 225-226, 228-230, 251, 269, 282, 285-287

Pulse Gradient, 92-93, 166

R

Relative Friction Factor, 238-244, 246

Reverse Osmosis, 4-9, 11-12, 20-21, 41, 44, 67-69, 83, 149, 156-160, 162, 178-182, 190, 207, 209, 230-232, 251, 258, 271, 274, 276, 294-308, 312-314, 316

S

Saline Concentration, 6-7, 52

Saline Solute, 6-7

Salt Rejection, 4, 7, 105

Sensitive Pressure Drop, 16-17, 42, 58, 85, 227, 313

Sodium Acetate, 143, 164, 212

Spacer Geometry, 10, 18, 89-90, 97, 107-108, 112-113, 118-124, 138-139, 179, 183-185, 191, 207, 251, 300, 309

Spacer Junctions, 96, 99-101, 171-173, 194

Spacer Sheets, 67, 146, 163, 169, 171, 211, 235, 257, 261

Spatial Orientation, 143, 162, 187, 235, 257

Spiral-wound Membrane, 46-47, 51-52, 57, 61, 68, 125-126, 130, 143, 160-162, 164, 169, 178-179, 187, 196, 235, 257, 275, 280, 282, 304, 308

Stock Solution, 92, 143, 164, 261

Substrate Load, 17-18, 104, 209-211, 213, 219-222, 230, 233-234, 238-240, 244, 247-248, 277, 312

Substrate Removal, 228, 263, 271, 278, 280, 292

V

Valve System, 65, 261

Velocity Direction, 116, 118-120

Velocity Distribution, 17-18, 101-102, 116-117, 139, 161, 171, 173-174, 180, 187, 189, 193, 204, 207, 277

Visual Inspection, 79, 157

Volume Reduction, 100-101, 173

W

Water Quality, 2, 16, 20, 22-23, 32, 37, 39, 44, 69, 74-75, 77, 85, 88, 163, 181, 184, 191, 207, 226, 254, 260, 262, 268, 299, 311

www.ingramcontent.com/pod-product-compliance
Lightning Source LLC
LaVergne TN
LVHW051049220926
848485LV00001BA/51